# ReAL-LiFe SeA MONSTeRS

Judith
Jango-Cohen

Illustrated by
Ryan Durney

On My Own

SCIENCE

Millbrook Press/Minneapolis

Millbrook Press, Inc.
A division of Lerner Publishing Group, Inc.
241 First Avenue North
Minneapolis, MN 55401 U.S.A.

Website address: www.lernerbooks.com


Library of Congress Cataloging-in-Publication Data

Jango-Cohen, Judith.
        Real-life sea monsters / by Judith Jango-Cohen ; illustrations by Ryan Durney.
            p.    cm. —   (On my own science)
        Includes bibliographical references.
        ISBN-13: 978–0–8225–6747–9 (lib. bdg. : alk. paper)
        ISBN-10: 0–8225–6747–4 (lib. bdg. : alk. paper)
        1.  Sea monsters—Juvenile literature.  2.  Dangerous marine animals—Juvenile
    literature.  I. Durney, Ryan, ill.  II. Title.  III. Series.
        QL89.J37  2008
        591.77—dc22                                                    2006009817

Manufactured in the United States of America
1  2  3  4  5  6  –  JR  –  13  12  11  10  09  08

*To Eliot—who helps me slay my monsters*


*Thanks to Professor John Olney, ichthyologist at
the Virginia Institute of Marine Science in Gloucester Point,
and to Carol Hinz, editor, for her enthusiastic partnership*
—J.J.-C.



*To my cousin, Danny Bardell, who was more like a brother
to me. I will always remember the fun we had at the ocean,
jumping the waves and running from jellyfish.*
—R.D.

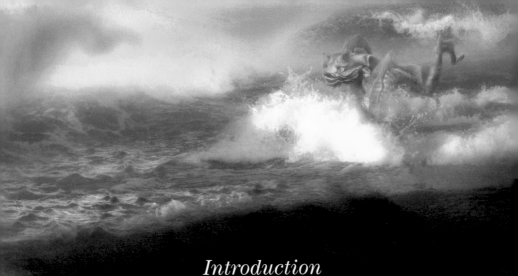

## Introduction

Once upon a time,
sailors whispered strange stories.
They told tales about monsters
with terrible tentacles that tackled ships.
They warned of creatures that whipped
waves into storms.
They spoke of sea serpents with
sharp fangs and bloodred eyes.
Do sea monsters like this truly exist?
Or do they live only in the world of stories?
Read on and learn the truth behind the tales.

## The Mighty Kraken

Curious sailors stare into the black,
bubbling sea.
What kind of creature is hiding beneath?
Suddenly, a beast rises from the waves.
The sailors freeze in horror.
It is the mighty Kraken.
The Kraken's glowing eyes glare at
the sailors.
Its 10 slimy arms squirm like
angry snakes.
The monster lunges and lashes out
its arms.
It snatches the ship.
Then the mighty Kraken slowly sinks
beneath the sea.

No one has proven that this ship-sinking
sea creature exists.
But scientists do know of a huge animal
that looks like the Kraken.
This animal sometimes washes up onshore.

The largest body ever measured was
57 feet long.

That's about the length of 10 adults
lying head to toe.

This amazing creature is the giant squid.

The giant squid has eight arms
and two longer tentacles.
Inside the ring of arms and tentacles
is a sharp, beaked mouth.
The giant squid catches food by
flinging out its tentacles.
Suckers with jagged edges line the
tentacle tips and arms.
The suckers work like
suction cups to grip
fish and crabs.

Giant squid have a funnel
near their head.
The squid sucks water into
the funnel and then shoots it out.
The jets of water push the squid
through the ocean.

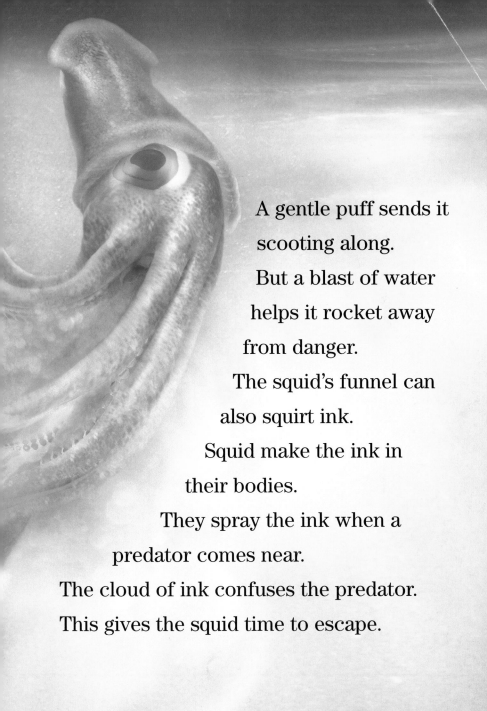

A gentle puff sends it
scooting along.
But a blast of water
helps it rocket away
from danger.
The squid's funnel can
also squirt ink.
Squid make the ink in
their bodies.
They spray the ink when a
predator comes near.
The cloud of ink confuses the predator.
This gives the squid time to escape.

An adult giant squid's only predator is the sperm whale.

Sperm whales can grow up to 60 feet long.

In 1965, a crew of whalers watched the two giants battle.

The whale seized the squid in its gigantic jaw.

But the squid smothered the whale's head with its arms and tentacles.

Both animals sank below the sea.

Later, the whalers found the dead whale with a tentacle around its throat.

The squid's head was inside the whale's stomach.

Scientists believe that giant squid live
in most of the world's oceans.
But scientists almost never see
living giant squid.
Because of this, many mysteries remain.
For example, we do not know if giant squid
live in groups or alone.
We do not know how big they can get or
how fast they can swim.

One day, we may find better ways to explore the deep sea.
Until then, the mighty Kraken will keep its secrets.

## Bewitching Mermaids

A sailing ship glides through the waves
where mermaids sing.
The mermaids send spells swirling into
the sky and sea.
Clouds as black as smoke appear.
Waves spring up and lick at the ship
like flames.
The water claws at the sailors and
swallows them.
Later, the mermaids will gather the
dead sailors' bones.
They will carve them into haunted harps.

Long ago, people believed in
mermaids, creatures that are
half human and half fish.
Stories warned of wicked mermaids.
One was Jenny Green Teeth,
who drowned children.

Sailors thought that evil mermaids stirred up storms.

Many sailors, including Christopher Columbus, claimed that they saw mermaids.

Columbus did not think the mermaids were pretty.

He said, "They have a face like a man."

What did Columbus see?

He might have seen an animal that modern sailors sometimes mistake for a person swimming in the water.

As sailors get closer, they realize that the creature is a manatee.

Like a mermaid, the manatee has a long, fish-shaped tail.

But manatees are not fish.

They are mammals.

All mammals must breathe oxygen
from the air.

That is why people sometimes spot
manatees at the water's surface.

Usually, manatees poke out just
the tops of their snouts to breathe.

The average manatee is about 10 feet long.
It weighs more than 1,000 pounds.
Manatees must eat about 100 pounds of
food each day.
They nibble on the stems, roots, and leaves
of water plants.

They also gulp down acorns
that have fallen into the water.
A manatee uses its two flexible flippers
to push plants into its mouth.
Its upper lip is divided into two parts
to help it grab and grip its food.

Manatees spend much of their time alone. But sometimes they form groups. Manatees squeak and chirp as a way of talking to one another.

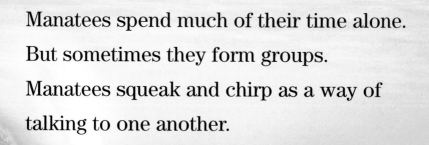

Mother manatees and their calves
nuzzle noses.
They hold one another with their flippers.

Manatees only live in warm, shallow water near land.

These areas are also popular with people.

This causes serious problems for manatees.

Motorboats can ram into manatees or cut them with their propellers.

Manatees also get killed by eating trash, such as plastic bags.

But there is hope for the manatee.

Marine parks are being set aside.

In the parks, manatees can live in peace.

Scientists are studying new ways to
protect manatees.

Many people are working together to help
these modern-day mermaids survive.

## Sinister Sea Serpents

On a beach, a crew of sailors
is making a fire.
They do not notice a sea serpent lurking
near the shore.
The serpent raises its head, topped with a
blazing red mane.
Then it slithers from the water.
The serpent springs upon a sailor and
coils around him.
It twists tighter and tighter.
The gasping sailor stares into the serpent's
red eyes.
Then the monster digs its fangs
into the sailor's neck.

People have told stories of sea serpents for more than 2,000 years.

Even in modern times, reports about sea serpents continue.

Late one night in 1963, a woman was walking her dog along a California beach. Suddenly, she began to scream. Neighbors came running. On the beach was the dead body of an 18-foot "sea serpent."

Police called in scientists to examine the beast.
They identified the creature as an oarfish.

Oarfish look a lot like sea serpents.
They have blue heads and silvery,
snakelike bodies.
Some of them are 50 feet long.
Oarfish have a bright red fin that runs from
their heads to their tails.
Oarfish ripple this fin to swim.
Unlike most fish, the oarfish
swims upright in the water.

The oarfish has long spines on its head.
They look like a red crown.
The oarfish also has one long, red spine on
each side of its body.
The tips of these spines look like paddles.
Oarfish rotate them, similar to the way a
person uses oars to row a boat.
But some scientists think that these "oars"
are not used for swimming.
They believe that the oarfish uses them to
sense chemicals in the water.
This may help it find food such as
tiny fish, shrimp, and jellyfish.
To eat these creatures, the
oarfish sticks out its upper
jaw and sucks them in.

We do not know a lot about oarfish because humans rarely see them.
Oarfish sometimes wash up onshore after a storm.
People also find oarfish with damaged tails near the sea's surface.
Scientists believe that boats may cause these injuries.

Normally, oarfish live between 60
and 3,000 feet underwater.
That is why people don't see healthy
oarfish very often.
Those who do meet these silvery,
shimmering creatures are always impressed.
Some people describe the oarfish as the
most beautiful creature they have ever seen.

## Mysteries Remain

Have scientists solved all the sea
monster mysteries?
No, not all of them.
We still have stories of creatures
that cannot be explained.
People have reported seeing
beasts with alligator heads,
flippers, and bristly manes.
Could there be real-life sea monsters
that scientists have not yet found,
described, and named?
Some people believe it is possible.

Humans have explored just a tiny
part of Earth's oceans.
If all the world's oceans were
contained in 100 equal-sized tanks,
we would have 95 left to explore.

Every year, scientists find more than
100 new species of ocean life.
Some species are quite large.
In 1976, scientists discovered the
14-foot-long megamouth shark.
Before then, no one had any idea that
such a shark existed.

What else may be living in the dark, deep seas?
About 250 years ago, a scientist said that the ocean hides many things. That also means that there is much for us to find.

# Glossary

**calves (CAVZ):** baby manatees

**fin (FIN):** a thin body part that helps fish move, balance, and steer

**flippers (FLIP-purz):** wide, flat body parts that help underwater animals swim and eat

**mammals (MAH-mulz):** warm-blooded animals that have backbones, fur or hair on their bodies, give birth to live young, and make milk to feed their young

**marine (muh-REEN):** something relating to the sea

**oxygen (OX-i-gin):** a chemical that mammals need to breathe in the air

**predator (PREH-duh-tur):** an animal that hunts and eats other animals

**snouts (SNOWTZ):** noses

**species (SPEE-sees):** a type of animal

**tentacles (TEN-tih-kulz):** long, armlike body parts that animals use for feeling or grasping food and other objects

**whalers (WALE-urz):** people who work on a ship hunting whales

# Fun Facts about Sea Monsters

A squid called the colossal squid has swiveling hooks on its suckers. Scientists believe it may be even larger than the giant squid.

Scientists know more about many dinosaurs than they do about the giant squid.

In September 2005, Japanese researchers became the first scientists to photograph a living giant squid. Their robotic camera took the pictures almost 3,000 feet below the surface of the North Pacific Ocean.

Female manatees will adopt a baby manatee that has lost its mother. They will stay close to it, protect it, and feed it their milk.

Scientists believe that ancient relatives of manatees were land animals closely related to elephants.

Some kinds of manatees have elephant-like "toenails" on the tips of their flippers.

Part of the scientific name for the oarfish—*Regalecus*—comes from a Latin word meaning "royal." Japanese fishermen call it king of the palace under the sea.

A relative of the oarfish scares off predators by producing an electric shock. Scientists do not know if the oarfish can do the same.

# Selected Bibliography

Cherry, John, ed. *Mythical Beasts*. San Francisco: Pomegranate Artbooks / British Museum Press, 1995.

Coleman, Loren, and Patrick Huyghe. *The Field Guide to Lake Monsters, Sea Serpents, and Other Mystery Denizens of the Deep*. New York: Jeremy P. Tarcher / Putnam, 2003.

Ellis, Richard. *Monsters of the Sea*. New York: Knopf, 1994.

Ellis, Richard. *The Search for the Giant Squid: The Biology and Mythology of the World's Most Elusive Sea Creature*. New York: Penguin, 1999.

Heuvelmans, Bernard. *In the Wake of the Sea-Serpents*. Translated by Richard Garnett. New York: Hill and Wang, 1968.

Powell, James. *Manatees*. Stillwater, MN: Voyageur Press, 2002.

# Further Reading and Websites

Climo, Shirley. *A Treasury of Mermaids: Mermaid Tales from around the World*. New York: HarperCollins Publishers, 1997.

Dussling, Jennifer. *Giant Squid: Mystery of the Deep*. New York: Grosset & Dunlap, 1999.

Edwards, Katie. *Myths and Monsters: Secrets Revealed*. Watertown, MA: Charlesbridge, 2004.

Innes, Brian. *Water Monsters*. Austin, TX: Steck-Vaughn Company, 1999.

Markle, Sandra. *Outside and Inside Giant Squid*. New York: Walker, 2003.

Walker, Sally M. *Manatees*. Minneapolis: Carolrhoda Books, Inc., 1999.


Creature Feature: West Indian Manatees
http://www.nationalgeographic.com/kids/creature_feature/0307/manatees.html

In Search of Giant Squid
http://seawifs.gsfc.nasa.gov/squid.html

Monsters of the Deep
http://www.seasky.org/monsters/sea7a.html

Oarfish
http://www.amonline.net.au/FISHES/fishfacts/fish/rglesne.htm

# Fodor's 98

# Cape Cod, Martha's Vineyard, Nantucket

**The complete guide, thoroughly up-to-date**

Packed with details that will make your trip

**The must-see sights, off and on the beaten path**

What to see, what to skip

**Mix-and-match vacation itineraries**

City strolls, countryside adventures

**Smart lodging and dining options**

Essential local do's and taboos

**Transportation tips, distances, and directions**

Key contacts, savvy travel tips

**When to go, what to pack**

Clear, accurate, easy-to-use maps

**Books to read, videos to watch, background essays**

Fodor's Travel Publications, Inc.
New York • Toronto • London • Sydney • Auckland
www.fodors.com/

# Fodor's Cape Cod, Martha's Vineyard, Nantucket

**EDITOR:** Matthew Lore

**Editorial Contributors:** Robert Andrews, Dorothy Antczak, David Brown, Alan W. Petrucelli, Seth Rolbein, Heidi Sarna, Helayne Schiff, M. T. Schwartzman (Gold Guide editor), Dinah A. Spritzer

**Editorial Production:** Stacey Kulig

**Maps:** David Lindroth, *cartographer*; Steven Amsterdam, *map editor*

**Design:** Fabrizio La Rocca, *creative director*; Guido Caroti, *associate art director*; Jolie Novak, *photo editor*

**Production/Manufacturing:** Rebecca Zeiler

**Cover Photograph:** Jake Rajs/Tony Stone Images

## Copyright

ISBN 0-679-03454-4

## Special Sales

Fodor's Travel Publications are available at special discounts for bulk purchases for sales promotions or premiums. Special editions, including personalized covers, excerpts of existing guides, and corporate imprints, can be created in large quantities for special needs. For more information, contact your local bookseller or write to Special Markets, Fodor's Travel Publications, 201 East 50th Street, New York, NY 10022. Inquiries from Canada should be directed to your local Canadian bookseller or sent to Random House of Canada, Ltd., Marketing Department, 1265 Aerowood Drive, Mississauga, Ontario L4W 1B9. Inquiries from the United Kingdom should be sent to Fodor's Travel Publications, 20 Vauxhall Bridge Road, London SW1V 2SA, England.

PRINTED IN THE UNITED STATES OF AMERICA

10 9 8 7 6 5 4 3 2 1

# CONTENTS

## Maps

# ON THE ROAD WITH FODOR'S

**W**E'RE ALWAYS THRILLED to get letters from readers, especially one like this:

*It took us an hour to decide what book to buy and we now know we picked the best one. Your book was wonderful, easy to follow, very accurate, and good on pointing out eating places, informal as well as formal. When we saw other people using your book, we would look at each other and smile.*

Our editors and writers are deeply committed to making every Fodor's guide "the best one"—not only accurate but always charming, brimming with sound recommendations and solid ideas, right on the mark in describing restaurants and hotels, and full of fascinating facts that make you view what you've traveled to see in a rich new light.

## About Our Writers

Our success in achieving our goals—and in helping to make your trip the best of all possible vacations—is a credit to the hard work of our extraordinary writers and the editors.

Eastham resident **Alan W. Petrucelli** updated the Upper and Mid Cape sections of our Cape Cod chapter and our Martha's Vineyard chapter. He has written for the *New York Times, USA Weekend,* the *Cape Cod Times,* and many other publications.

We're pleased to have as our new dining critic for 1998 writer and documentary filmmaker **Seth Rolbein,** who has lived on the Cape for most of the past 20 years. His subjects and formats vary widely, but his books include a novel, *Sting of the Bee,* published by St. Martin's Press. He has also been published by many regional and national publications, including the *Boston Globe Sunday* magazine and *Yankee* magazine.

**Dorothy Antczak** is a freelance writer and editor who lives and works in Provincetown. She updated our coverage of the Lower Cape and Nantucket.

For their help in putting together this guide, the authors and editor wish especially to thank Cape Air; Gina Barboza, Steamship Authority; Kevin M. Boyle, *Entertainment Report;* Karl and Lynn Buder, Thorncroft Inn; Nancy Wurlitzer Craig and Carleton Davis, the Cape Playhouse; Vincent Longo and Jan Preus, Cape Cod Melody Tent; Elaine Perry, Cape Cod Chamber of Commerce; and Lisa Schwartz and Randi Vega, Martha's Vineyard Chamber of Commerce.

## New This Year

For 1998 we've added three special "Close-Ups" to the guide. The first is a handy week-at-a-glance guide to some of the best regular weekly events that take place during the busy high-season months of July and August. The second new close-up is on the Cape Cod Baseball League, and the third describes the pleasures of the Cape's many seasons. Our new dining reviewer, Seth Rolbein, has given us almost two dozen new restaurants on both the Cape and the islands, and we've also added a wealth of lodging establishments, giving you many additional B&Bs, hotels, and campgrounds from which to choose.

And this year, Fodor's joins Rand McNally, the world's largest commercial mapmaker, to bring you a detailed color map of the Cape and the islands. Just detach it along the perforation and drop it in your tote bag.

We're also proud to announce that the American Society of Travel Agents has endorsed Fodor's as its guidebook of choice. ASTA is the world's largest and most influential travel trade association, operating in more than 170 countries, with 27,000 members pledged to adhere to a strict code of ethics reflecting the Society's motto, "Integrity in Travel." ASTA shares Fodor's devotion to providing smart, honest travel information and advice to travelers, and we've long recommended that our readers consult ASTA member agents for the experience and professionalism they bring to the table.

On the Web, check out Fodor's site (www.fodors.com/) for information on major destinations around the world and travel-savvy interactive features. The Web site also lists the 85-plus radio stations nationwide that carry the Fodor's Travel Show, a live call-in program that airs every weekend. Tune in to hear guests discuss their wonderful adventures—or call in to get answers for your most pressing travel questions.

## How to Use This Book

### Organization

Up front is the **Gold Guide,** an easy-to-use section divided alphabetically by topic. Under each listing you'll find tips and information that will help you accomplish what you need to in the Cape and the islands. You'll also find addresses and telephone numbers of organizations and companies that offer destination-related services and detailed information and publications.

The first chapter in the guide, **Destination: Cape Cod, Martha's Vineyard, Nantucket** helps get you in the mood for your trip. New and Noteworthy cues you in on trends and happenings, What's Where gets you oriented, Pleasures and Pastimes describes the activities and sights that really make the Cape and the islands unique, Fodor's Choice showcases our top picks, and Festivals and Seasonal Events alerts you to special events you'll want to seek out.

Chapters in *Cape Cod, Martha's Vineyard & Nantucket '98* are arranged geographically based on how you will approach each area. Each regional chapter is divided by geographical area. Within each area, towns are covered in logical geographical order, and attractive stretches of road and minor points of interest between them are indicated by the designation *En Route*. Throughout, Off the Beaten Path sights appear after the places from which they are most easily accessible. And within town sections, all restaurants and lodgings are grouped together.

To help you decide what to visit in the time you have, all chapters begin with recommended itineraries. You can mix and match those from several chapters to create a complete vacation. The A to Z section that ends each chapter covers getting there, getting around, and helpful contacts and resources.

At the end of the book you'll find Portraits, including a wonderful essay about the history of the Cape and Islands, followed by suggestions for pretrip reading, both fiction and nonfiction.

### Icons and Symbols

★    Our special recommendations
✕    Restaurant
🏠    Lodging establishment
✕🏠   Lodging establishment whose restaurant warrants a detour
⚠    Campgrounds
☺    Good for kids (rubber duckie)
☞    Sends you to another section of the guide for more information
✉    Address
☎    Telephone number
FAX   Fax number
☯    Opening and closing times
🎟    Admission prices (those we give apply to adults; substantially reduced fees are almost always available for children, students, and senior citizens)

Numbers in white and black circles (e.g., ② and ❷ ) that appear on the maps, in the margins, and in town sections all correspond to one another.

### Dining and Lodging

The restaurants and lodgings we list are the cream of the crop in each price range. Price categories are as follows:

**For restaurants:**

| CATEGORY | CAPE COD AND MARTHA'S VINEYARD | NANTUCKET |
|---|---|---|
| $$$$ | over $40 | over $45 |
| $$$ | $20–$40 | $35–$45 |
| $$ | $15–$25 | $25–$35 |
| $ | under $15 | under $25 |

*per person for a three-course meal, excluding drinks, service, and 5% sales tax.*

**For hotels:**

| CATEGORY | COST* |
|---|---|
| $$$$ | over $200 |
| $$$ | $150–$200 |
| $$ | $100–$150 |
| $ | under $100 |

*All prices are for a standard double room in high season and do not include tax or gratuities. Some inns add a 15% service charge. The state tax on lodging is 5.7%; Nantucket and Martha's Vineyard (Down-Island only) impose an additional 4% tax.*

### Hotel Facilities

We always list the facilities that are available—but we don't specify whether they cost extra: When pricing accommodations, always ask what's included. In addition, assume that all rooms have private baths unless otherwise noted.

Assume that hotels operate on the **European Plan** (EP, with no meals) unless we note that they use the **Continental Plan** (CP, with a Continental breakfast daily), the **Breakfast Plan** (BP, with a full breakfast daily), **Modified American Plan** (MAP, with breakfast and dinner daily), or the **Full American Plan** (FAP, with all meals).

Finally, because many lodging establishments, especially B&Bs, do not allow pets, it's always a good idea to phone in advance if you plan to take your pet on the road with you. Also, because many B&Bs do not allow children under a certain age—breakable bric-a-brac and precipitous stairways can create hazards for youngsters—always call ahead to find out a B&B's policy.

### Restaurant Reservations and Dress Codes

Reservations are always a good idea, especially on the Cape and islands in summer. We note only when they're essential or when they are not accepted. Book as far ahead as you can, and reconfirm when you get to town. Unless otherwise noted, the restaurants listed are open daily for lunch and dinner. We mention dress only when men are required to wear a jacket or a jacket and tie. Look for an overview of local habits in the Pleasures and Pastimes section that follows each chapter introduction.

### Credit Cards

The following abbreviations are used: **AE,** American Express; **D,** Discover; **DC,** Diners Club; **MC,** MasterCard; and **V,** Visa.

## Please Write to Us

You can use this book in the confidence that all prices and opening times are based on information supplied to us at press time; Fodor's cannot accept responsibility for any errors. Time inevitably brings changes, so always confirm information when it matters—especially if you're making a detour to visit a specific place. In addition, when making reservations be sure to mention if you have a disability or are traveling with children, if you prefer a private bath or a certain type of bed, or if you have specific dietary needs or other concerns.

Were the restaurants we recommended as described? Did our hotel picks exceed your expectations? Did you find a museum we recommended a waste of time? If you have complaints, we'll look into them and revise our entries when the facts warrant it. If you've discovered a special place that we haven't included, we'll pass the information along to our correspondents and have them check it out. So send us your feedback, positive *and* negative: E-mail us at editors@fodors.com (specifying the name of the book on the subject line) or write the Cape Cod editor at Fodor's, 201 East 50th Street, New York, NY 10022. Have a wonderful trip!

Karen Cure
*Editorial Director*

# Cape Cod, Martha's Vineyard, and Nantucket

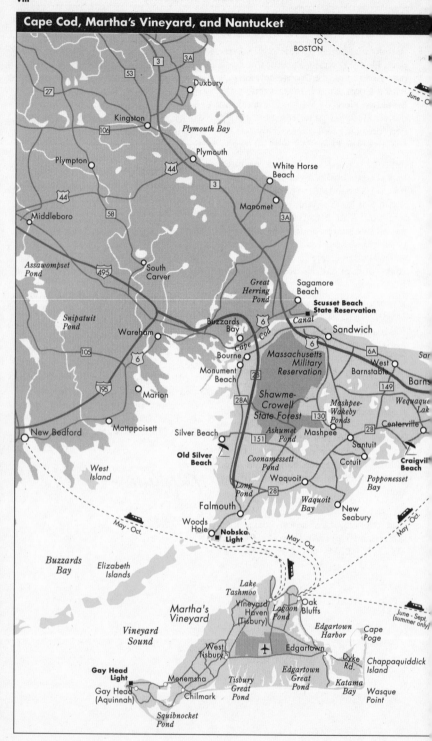

TO
BOSTON

June - O

3
3A
53
27
Duxbury
Kingston
106
Plymouth Bay
Plympton
Plymouth
44
White Horse
Beach
44
3
Middleboro
58
Manomet
3A
Assawompset
Pond
495
South
Carver
Great
Herring
Pond
Sagamore
Beach
Scusset Beach
State Reservation
Snipatuit
Pond
Wareham
Buzzards
Bay
6
Canal
Sandwich
Cape
Cod
Bourne
6
6A
Sar
105
Cape
Massachusetts
Military
Reservation
West
Barnstable
Barns
195
Monument
Beach
23
149
Marion
28A
Shawme-
Crowell
State Forest
Mashpee-
Wakeby
Ponds
130
Wequaque
Lak
New Bedford
Mattapoisett
Silver Beach
151
Ashumet
Pond
Mashpee
28
Centerville
Old Silver
Beach
Coonamessett
Pond
Santuit
Cotuit
Craigvil
Beach
West
Island
Long
Pond
Waquoit
28
Waquoit
Bay
Popponesset
Bay
Falmouth
New
Seabury
Woods
Hole
Nobska
Light
May - Oct.
May - Oct.
May - Oct.
Buzzards
Bay
Elizabeth
Islands
Lake
Tashmoo
Vineyard
Haven
(Tisbury)
Lagoon
Pond
Oak
Bluffs
June - Sept.
(summer only)
Martha's
Vineyard
Edgartown
Harbor
Cape
Poge
Vineyard
Sound
West
Tisbury
Edgartown
Dyke
Rd.
Chappaquiddick
Island
Gay Head
Light
Gay Head
(Aquinnah)
Menemsha
Chilmark
Tisbury
Great
Pond
Edgartown
Great
Pond
Katama
Bay
Wasque
Point
Squibnocket
Pond

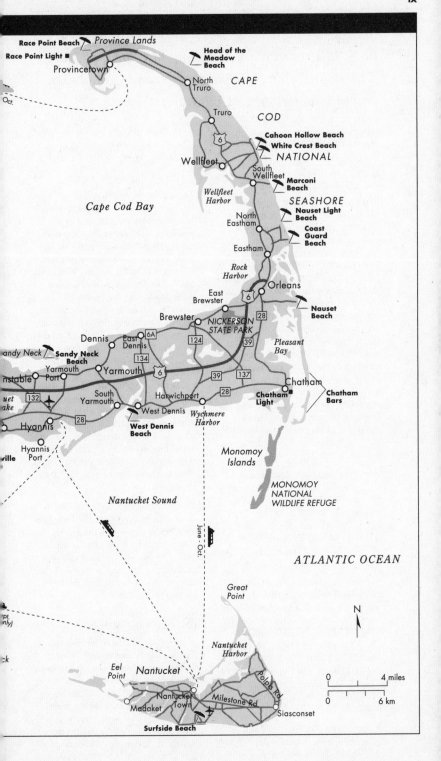

Race Point Beach
Race Point Light ■
Province Lands
Provincetown
Oct.
Head of the Meadow Beach
North Truro
CAPE
Truro
COD
6
Cahoon Hollow Beach
White Crest Beach
Wellfleet
NATIONAL
South Wellfleet
Marconi Beach
Wellfleet Harbor
Cape Cod Bay
SEASHORE
North Eastham
Nauset Light Beach
Coast Guard Beach
Eastham
Rock Harbor
Orleans
East Brewster
6
Brewster
NICKERSON STATE PARK
28
Nauset Beach
Dennis
East Dennis
6A
124
39
Pleasant Bay
andy Neck
Sandy Neck Beach
134
Yarmouth Port
Yarmouth
6
39
137
Chatham
nstable
132
South Yarmouth
Harwichport
28
Chatham Light ■
Chatham Bars
uet
ake
Hyannis
28
West Dennis
West Dennis Beach
Wychmere Harbor
ville
Hyannis Port
Monomoy Islands
MONOMOY NATIONAL WILDLIFE REFUGE
Nantucket Sound
June - Oct.
ATLANTIC OCEAN
pt.
nly)
Great Point
ck
Eel Point
Nantucket
Nantucket Harbor
Polpis Rd.
0        4 miles
N
Madaket
Nantucket Town
Milestone Rd.
Siasconset
0        6 km
Surfside Beach

# SMART TRAVEL TIPS A TO Z

*Basic Information on Traveling in Cape Cod, Savvy Tips to Make Your Trip a Breeze, and Companies and Organizations to Contact*

## A

### AIR TRAVEL

#### MAJOR AIRLINE OR LOW-COST CARRIER?

Most people choose a flight based on price. Yet there are other issues to consider. Major airlines offer the greatest number of departures; smaller airlines—including regional, low-cost and no-frill airlines—usually have a more limited number of flights daily. Major airlines have frequent-flyer partners, which allow you to credit mileage earned on one airline to your account with another. Low-cost airlines offer a definite price advantage and fewer restrictions, such as advance-purchase requirements. Safety-wise, low-cost carriers as a group have a good history, but **check the safety record before booking** any low-cost carrier; call the Federal Aviation Administration's Consumer Hotline (☞ Airline Complaints, *below*).

➤ MAJOR AIRLINES: **American** (☎ 800/433–7300) to Logan. **Continental** (☎ 800/525–0280) to Logan, Hyannis. **Delta** (☎ 800/221–1212) to Logan. **Northwest** (☎ 800/225–2525) to Logan. **TWA** (☎ 800/221–2000) to Logan. **US Airways** (☎ 800/428–4322) to Logan, Hyannis.

➤ SMALLER AIRLINES: **Business Express** (☎ 508/862–0556 or 800/345–4400). **Cape Air/Nantucket Airlines** (☎ 508/771–6944, 508/228–6252, 508/228–6234, 800/352–0714, or 800/635–8787 in MA). **Colgan Air** (☎ 508/325–5100 or 800/272–5488). **Continental Express** (☎ 800/525–0280). **Island Airlines** (☎ 508/228–7575, 800/698–1109 in MA, or 800/248–7779).

➤ CHARTERS: **Air New England** (☎ 508/693–8899). **Coastal Air** (☎ 508/228–3350, 203/448–1001 or 800/262–7858). **Ocean Wings** (☎ 508/325–5548 or 800/253–5039). **Westchester Air** (☎ 914/761–3000 or 800/759–2929).

#### GET THE LOWEST FARE

The least-expensive airfares to Cape Cod are priced for round-trip travel. Major airlines usually require that you **book in advance and buy the ticket within 24 hours,** and you may have to **stay over a Saturday night.** It's smart to **call a number of airlines, and when you are quoted a good price, book it on the spot**—the same fare may not be available on the same flight the next day. Airlines generally allow you to change your return date for a fee $25–$50. If you don't use your ticket you can apply the cost toward the purchase of a new ticket, again for a small charge. However, most low-fare tickets are nonrefundable. To get the lowest airfare, **check different routings.** If your destination or home city has more than one gateway, compare prices to and from different airports. Also price off-peak flights, which may be significantly less expensive.

To save money on flights from the United Kingdom and back, **look into an APEX or Super-PEX ticket.** APEX tickets must be booked in advance and have certain restrictions. Super-PEX tickets can be purchased at the airport on the day of departure—subject to availability.

#### DON'T STOP UNLESS YOU MUST

When you book, **look for nonstop flights** and **remember that "direct" flights stop at least once.** Try to **avoid connecting flights,** which require a change of plane. Two airlines may jointly operate a connecting flight, so ask if your airline operates every segment—you may find that your preferred carrier flies you only part of the way.

## USE AN AGENT

Travel agents, especially those who specialize in finding the lowest fares (☞ Discounts & Deals, *below*), can be especially helpful when booking a plane ticket. When you're quoted a price, **ask your agent if the price is likely to get any lower.** Good agents know the seasonal fluctuations of airfares and can usually anticipate a sale or fare war. However, waiting can be risky: The fare could go *up* as seats become scarce, and you may wait so long that your preferred flight sells out. A wait-and-see strategy works best if your plans are flexible, but if you must arrive and depart on certain dates, don't delay.

## AVOID GETTING BUMPED

Airlines routinely overbook planes, knowing that not everyone with a ticket will show up, but sometimes everyone does. When that happens, airlines ask for volunteers to give up their seats. In return these volunteers usually get a certificate for a free flight and are rebooked on the next flight out. If there are not enough volunteers the airline must choose who will be denied boarding. The first to get bumped are passengers who checked in late and those flying on discounted tickets, **so get to the gate and check in as early as possible,** especially during peak periods.

Always **bring a photo ID to the airport.** You may be asked to show it before you are allowed to check in.

## ENJOY THE FLIGHT

For better service, **fly smaller or regional carriers,** which often have higher passenger-satisfaction ratings. Sometimes you'll find leather seats, more legroom, and better food.

For more legroom, **request an emergency-aisle seat;** don't however, sit in the row in front of the emergency aisle or in front of a bulkhead, where seats may not recline.

If you don't like airline food, **ask for special meals when booking.** These can be vegetarian, low-cholesterol, or kosher, for example.

## COMPLAIN IF NECESSARY

If your baggage goes astray or your flight goes awry, complain right away. Most carriers require that you file a claim immediately.

➤ AIRLINE COMPLAINTS: U.S. Department of Transportation **Aviation Consumer Protection Division** (✉ C-75, Room 4107, Washington, DC 20590, ☎ 202/366–2220). **Federal Aviation Administration (FAA) Consumer Hotline** (☎ 800/322–7873).

AIRPORTS

The major gateway to Cape Cod is Boston's **Logan International Airport.** Smaller airports include **Barnstable Municipal Airport** in Hyannis, **Martha's Vineyard Airport,** and **Nantucket Memorial Airport.**

Flying time is one hour from New York, 2½ hours from Chicago, six hours from Los Angeles, and 3½ hours from Dallas.

➤ AIRPORT INFORMATION: Boston: **Logan International Airport** (☎ 617/561–1806 or ☎ 800/235–6426). Hyannis: **Barnstable Municipal Airport** (☎ 508/775–2020). Martha's Vineyard: **Martha's Vineyard Airport** (☎ 508/693–7022). Nantucket: **Nantucket Memorial Airport** (☎ 508/325–5300). Provincetown: **Provincetown Municipal Airport** (☎ 508/487–0241)

## B

## BUS TRAVEL

Bus service is available to Cape Cod, with stops at many Cape towns and some connecting service to the islands by ferry. **Greyhound** serves Boston from all over the country; from there you can connect to a local carrier, such as **Bonanza Bus Lines,** which offers direct service to Falmouth/Woods Hole from Boston and New York City. **Plymouth & Brockton Street Railway** travels to Provincetown from Boston and Logan Airport, with stops en route.

➤ BUS COMPANIES: Greyhound (☎ 800/231–2222). Bonanza Bus Lines (☎ 508/548–7588 or 800/556–3815). Plymouth & Brockton Street Railway (☎ 508/775–5524 or 508/746–0378).

## ON THE CAPE & ISLANDS

*See* the A to Z section of each chapter for local bus travel information.

THE GOLD GUIDE / SMART TRAVEL TIPS

## BUSINESS HOURS

Shop hours are generally 9 or 10 to 5, though in high season many tourist-oriented stores stay open until 10 PM or later. Except in the main tourist areas, shops are often closed on Sunday.

## C

## CAMERAS, CAMCORDERS, & COMPUTERS

Always **keep your film, tape, or computer disks out of the sun.** Carry an extra supply of batteries, and **be prepared to turn on your camera, camcorder, or laptop** to prove to security personnel that the device is real. Always **ask for hand inspection of film,** which becomes clouded after successive exposure to airport x-ray machines, and **keep videotapes and computer disks away from metal detectors.**

➤ PHOTO HELP: Kodak Information Center (☎ 800/242–2424). *Kodak Guide to Shooting Great Travel Pictures,* available in bookstores or from Fodor's Travel Publications (☎ 800/533–6478; $16.50 plus $4 shipping).

## CAR RENTAL

Rates in Boston begin at $31 a day and $149 a week for an economy car with air conditioning, an automatic transmission, and unlimited mileage. This does not include tax on car rentals, which is 5%.

➤ MAJOR AGENCIES: **Alamo** (☎ 800/327–9633, 0800/272–2000 in the U.K.). **Avis** (☎ 508/775–2888 or 800/331–1212, 800/879–2847 in Canada). **Budget** (☎ 508/771–2744 or 800/527–0700, 0800/181181 in the U.K.). **Dollar** (☎ 800/800–4000; 0990/565656 in the U.K., where it is known as Eurodollar). **Hertz** (☎ 508/775–5825 or 800/654–3131, 800/263–0600 in Canada, 0345/555888 in the U.K.). **National InterRent** (☎ 508/771–4353 or 800/227–7368; 0345/222525 in the U.K., where it is known as Europcar InterRent).

➤ LOCAL AGENCIES: Nantucket: **Nantucket Windmill** (☎ 508/228–1227 or 800/228–1227). Martha's Vineyard: **Adventure Rentals** (☎ 508/693–1959). **Vineyard Classic and Specialty Cars**(☎ 508/693–5551).

## CUT COSTS

To get the best deal, **book through a travel agent who is willing to shop around.** When pricing cars, **ask about the location of the rental lot.** Some off-airport locations offer lower rates, and their lots are only minutes from the terminal via complimentary shuttle. You also may want to **price local car-rental companies,** whose rates may be lower still, although their service and maintenance may not be as good as those of a name-brand agency. Remember to ask about required deposits, cancellation penalties, and drop-off charges if you're planning to pick up the car in one city and leave it in another.

Also **ask your travel agent about a company's customer-service record.** How has it responded to late plane arrivals and vehicle mishaps? Are there often lines at the rental counter, and, if you're traveling during a holiday period, does a confirmed reservation guarantee you a car?

Be sure to **look into wholesalers,** companies that do not own fleets but rent in bulk from those that do and often offer better rates than traditional car-rental operations. Prices are best during off-peak periods.

➤ RENTAL WHOLESALERS: The **Kemwel Group** (☎ 914/835–5555 or 800/678–0678, FAX 914/835–5126).

## NEED INSURANCE?

When driving a rented car you are generally responsible for any damage to or loss of the vehicle. You also are liable for any property damage or personal injury that you may cause while driving. Before you rent, **see what coverage you already have** under the terms of your personal auto-insurance policy and credit cards.

For about $14 a day, rental companies sell protection, known as a collision- or loss-damage waiver (CDW or LDW) that eliminates your liability for damage to the car; it's always optional and should never be automatically added to your bill.

In Massachusetts, the car-rental company must pay for damage to third parties up to a preset legal limit. Once that limit is reached your personal auto or other liability insurance kicks in. However, **make sure you have enough coverage to pay for the car.** If you do not have auto insurance or an umbrella policy that covers damage to third parties, purchasing CDW or LDW is highly recommended.

## BEWARE SURCHARGES

Before you pick up a car in one city and leave it in another, **ask about drop-off charges or one-way service fees,** which can be substantial. Note, too, that some rental agencies charge extra if you return the car before the time specified on your contract. To avoid a hefty refueling fee, **fill the tank just before you turn in the car,** but be aware that gas stations near the rental outlet may overcharge.

## MEET THE REQUIREMENTS

In the United States you must be 21 to rent a car, and rates may be higher if you're under 25. You'll pay extra for child seats (about $3 per day), which are compulsory for children under five, and for additional drivers (about $2 per day). Residents of the U.K. will need a reservation voucher, a passport, a U.K. driver's license, and a travel policy that covers each driver, in order to pick up a car.

## CHILDREN & TRAVEL

## CHILDREN IN CAPE COD

Cape Cod is very family oriented and provides every imaginable diversion for kids, including lodgings and restaurants that cater to them and that are affordable for families on a budget. Cottages and condominiums are popular with families, offering privacy, room, kitchens, and sometimes laundry facilities. Often cottage or condo communities have play yards and pools, sometimes even full children's programs.

Be sure to plan ahead and **involve your youngsters** as you outline your trip. When packing, include things to keep them busy en route. On sightseeing days try to schedule activities of special interest to your children. If you are renting a car don't forget to **arrange for a car seat** when you reserve.

Be sure to **look for the ♺ icon,** which identifies child-friendly historical, cultural, and natural sights. You should also **consult the Children's Activities heading in the A to Z sections of the Cape Cod**

and **Martha's Vineyard chapters.** For other kid-friendly events, **check the Festivals and Seasonal Events guide in Chapter 1 as well as the Week-at-a-Glance Box in Chapter 2.**

➤ ADDITIONAL LOCAL INFORMATION: The Bristol County Convention & Visitors Bureau (⊠ Box 976, New Bedford, MA 02741, ☎ 508/997–1250 or 800/288–6263) has a calendar of family-oriented events along south-coastal New England in its "Southern Coastal New England Americana Trail Guide." *Just for Kids: The New England Guide and Activity Book for Young Travelers,* by Ed and Roon Frost (⊠ Glove Compartment Books, Box 1602, Portsmouth, NH 03802), is available by mail for $7.95 plus $3 shipping and handling.

## HOTELS

Most hotels in Cape Cod allow children under a certain age to stay in their parents' room at no extra charge, but others charge them as extra adults; be sure to **ask about the cutoff age for children's discounts.**

Children under the age noted in parenthesis can share their parents' room for free at the following hotels: the **Admiralty Resort** (12) and **Quality Inn** (18) in Falmouth, the **Quality Inn Cape Cod** (18) and **Hyannis Inn Motel** (11) in Hyannis, the **Four Points by Sheraton** (17) in Eastham, and the **Best Western Chateau Motor Inn** (18) and the **Masthead** (12) in Provincetown. In season, the oceanfront **New Seabury Resort** offers summer day programs for children 4–14 at a local camp (sports, nature studies, arts and crafts classes, games, overnights), plus tennis and golf clinics, miniature golf, and activities like puppet shows and bands. West Dennis's **Lighthouse Inn** offers an extensive summer program of supervised activities daily 9:30–3 and children's dinners and entertainment 5:30–8:30, including Saturday-night pizza-and-movie parties; it also has a private beach, a game room, a playground, and miniature golf. **Chatham Bars Inn, Ocean Edge** in Brewster, and **Tara Hyannis Hotel** offer summer programs for kids, with arts and crafts, outdoor activities, and games. **Sea Crest** in North Falmouth offers a children's day camp. In July and August, the **Admiralty Resort** in

Falmouth has a supervised playroom open daily 1–5, where kids can do all kinds of fun activities. For more information on these properties, *see* Dining and Lodging *in* Chapter 2.

On Martha's Vineyard, the **Mattake-sett** condominium community has a full children's program and plenty of amenities for kids. At the **Colonial Inn** in Edgartown, children under 16 stay free, and the **Beach Plum Inn** in Menemsha, where you're likely to see even the owners' son running around, is good for families. For more information on these properties, *see* Dining and Lodging *in* Chapter 3.

On Nantucket, at **Beachside Resort,** children under 16 stay free. At the **Harbor House,** children under 13 dine free with their parents between 5 and 6:30 PM. For more information on these properties, *see* Dining and Lodging *in* Chapter 4.

### FLYING

As a general rule, infants under two not occupying a seat fly free. If your children are two or older **ask about children's airfares.**

In general the adult baggage allowance applies to children paying half or more of the adult fare.

According to the FAA it's a good idea to use safety seats aloft for children weighing less than 40 pounds. Airlines, however, can set their own policies: U.S. carriers allow FAA-approved models but usually require that you buy a ticket, even if your child would otherwise ride free, since the seats must be strapped into regular seats. Airline rules vary regarding their use, so it's important to **check your airline's policy about using safety seats during takeoff and landing.** Safety seats cannot obstruct any of the other passengers in the row, so get an appropriate seat assignment as early as possible.

When making your reservation, **request children's meals or a free-standing bassinet** if you need them; the latter are available only to those seated at the bulkhead, where there's enough legroom. Remember, however, that bulkhead seats may not have their own overhead bins, and there's no storage space in front of you—a major inconvenience.

### GROUP TRAVEL

If you're planning to take your kids on a tour, look for companies that specialize in family travel.

➤ FAMILY-FRIENDLY TOUR OPERATORS: **Families Welcome!** (⊠ 92 N. Main St., Ashland, OR 97520, ☎ 541/482–6121 or 800/326–0724, FAX 541/482–0660).

CONSUMER PROTECTION

Whenever possible, **pay with a major credit card** so you can cancel payment if there's a problem, provided that you can provide documentation. This is a good practice whether you're buying travel arrangements before your trip or shopping at your destination.

If you're doing business with a particular company for the first time, **contact your local Better Business Bureau and the attorney general's offices** in your state and the company's home state, as well. Have any complaints been filed?

Finally, if you're buying a package or tour, always **consider travel insurance** that includes default coverage (☞ Insurance, *below*).

➤ LOCAL BBBs: **Council of Better Business Bureaus** (⊠ 4200 Wilson Blvd., Suite 800, Arlington, VA 22203, ☎ 703/276–0100, FAX 703/525–8277).

### CUSTOMS & DUTIES

### ENTERING THE U.S.

Visitors age 21 and over may import the following into the United States: 200 cigarettes or 50 cigars or 2 kilograms of tobacco, 1 liter of alcohol, and gifts worth $100. Prohibited items include meat products, seeds, plants, and fruits.

➤ INFORMATION: **U.S. Customs Service** (Inquiries, ⊠ Box 7407, Washington, DC 20044, ☎ 202/927–6724; complaints, Office of Regulations and Rulings, 1301 Constitution Ave. NW, Washington, DC 20229; registration of equipment, ⊠ Resource Management, 1301 Constitution Ave. NW, Washington DC, 20229, ☎ 202/927–0540).

### ENTERING CANADA

If you've been out of Canada for at least seven days you may bring in

THE GOLD GUIDE / SMART TRAVEL TIPS

C$500 worth of goods duty-free. If you've been away for fewer than seven days but more than 48 hours, the duty-free allowance drops to C$200; if your trip lasts 24–48 hours, the allowance is C$50. You may not pool allowances with family members. Goods claimed under the C$500 exemption may follow you by mail; those claimed under the lesser exemptions must accompany you.

Alcohol and tobacco products may be included in the seven-day and 48-hour exemptions but not in the 24-hour exemption. If you meet the age requirements of the province or territory through which you reenter Canada you may bring in, duty-free, 1.14 liters (40 imperial ounces) of wine or liquor or 24 12-ounce cans or bottles of beer or ale. If you are 16 or older you may bring in, duty-free, 200 cigarettes and 50 cigars; these items must accompany you.

You may send an unlimited number of gifts worth up to C$60 each duty-free to Canada. Label the package UNSOLICITED GIFT—VALUE UNDER $60. Alcohol and tobacco are excluded.

➤ INFORMATION: **Revenue Canada** (✉ 2265 St. Laurent Blvd. S, Ottawa, Ontario K1G 4K3, ☎ 613/993–0534, 800/461–9999 in Canada).

### ENTERING THE U.K.

From countries outside the EU, including the United States, you may import, duty-free, 200 cigarettes or 50 cigars; 1 liter of spirits or 2 liters of fortified or sparkling wine or liqueurs; 2 liters of still table wine; 60 milliliters of perfume; 250 milliliters of toilet water; plus £136 worth of other goods, including gifts and souvenirs.

➤ INFORMATION: **HM Customs and Excise** (✉ Dorset House, Stamford St., London SE1 9NG, ☎ 0171/202–4227).

## D

### DISABILITIES & ACCESSIBILITY

### ACCESS IN CAPE COD

➤ LOCAL RESOURCES: **Cape Organization for Rights of the Disabled** (CORD; ☎ 508/775–8300) will supply information on accessibility of restaurants, hotels, beaches, and other tourist facilities on Cape Cod. **Sight Loss Services** (☎ 508/394–3904 or 800/427–6842 in MA) provides accessibility and other information and referrals for people with vision impairments. The **Cape Cod National Seashore** (✉ South Wellfleet 02663, ☎ 508/349–3785) has facilities, services, and programs accessible to visitors with disabilities. Write for information on what is available, or ask at the Seashore visitor centers (Eastham, ☎ 508/255–3421; Provincetown, ☎ 508/487–1256).

### TIPS & HINTS

When discussing accessibility with an operator or reservationist, **ask hard questions.** Are there any stairs, inside or out? Are there grab bars next to the toilet and in the shower/tub? How wide is the doorway to the room? To the bathroom? For the most extensive facilities meeting the latest legal specifications, **opt for newer accommodations,** which are more likely to have been designed with access in mind. Older buildings or ships may offer more limited facilities. Be sure to **discuss your needs before booking.**

➤ COMPLAINTS: **Disability Rights Section** (✉ U.S. Department of Justice, Box 66738, Washington, DC 20035–6738, ☎ 202/514–0301 or 800/514–0301, FAX 202/307–1198, TTY 202/514–0383 or 800/514–0383) for general complaints. **Aviation Consumer Protection Division** (☞ Air Travel, above) for airline-related problems. **Civil Rights Office** (✉ U.S. Department of Transportation, Departmental Office of Civil Rights, S-30, 400 7th St. SW, Room 10215, Washington, DC 20590, ☎ 202/366–4648) for problems with surface transportation.

### TRAVEL AGENCIES & TOUR OPERATORS

The Americans with Disabilities Act requires that travel firms serve the needs of all travelers. That said, you should note that some agencies and operators specialize in making travel arrangements for individuals and groups with disabilities.

➤ TRAVELERS WITH MOBILITY PROBLEMS: **Access Adventures** (✉ 206 Chestnut Ridge Rd., Rochester, NY 14624, ☎ 716/889–9096), run by a former physical-rehabilitation counselor. **Hinsdale Travel Service** (✉ 201

E. Ogden Ave., Suite 100, Hinsdale, IL 60521, ☎ 630/325–1335), a travel agency that benefits from the advice of wheelchair traveler Janice Perkins. **Wheelchair Journeys** (✉ 16979 Redmond Way, Redmond, WA 98052, ☎ 425/885–2210 or 800/313–4751), for general travel arrangements.

➤ Travelers with Developmental Disabilities: **Sprout** (✉ 893 Amsterdam Ave., New York, NY 10025, ☎ 212/222–9575 or 888/222–9575, FAX 212/222–9768).

### DISCOUNTS & DEALS

Be a smart shopper and **compare all your options before making a choice.** A plane ticket bought with a promotional coupon may not be cheaper than the least expensive fare from a discount ticket agency. For high-price travel purchases, such as packages or tours, keep in mind that what you get is just as important as what you save. Just because something is cheap doesn't mean it's a bargain.

### LOOK IN YOUR WALLET

When you use your credit card to make travel purchases you may get free travel-accident insurance, collision-damage insurance, and medical or legal assistance, depending on the card and the bank that issued it. American Express, MasterCard, and Visa provide one or more of these services, so **get a copy of your credit card's travel-benefits policy.** If you are a member of the American Automobile Association (AAA) or an oil-company-sponsored road-assistance plan, always **ask hotel or car-rental reservationists about auto-club discounts.** Some clubs offer additional discounts on tours, cruises, or admission to attractions. And don't forget that auto-club membership entitles you to free maps and trip-planning services.

### DIAL FOR DOLLARS

To save money, **look into "1-800" discount reservations services,** which use their buying power to get a better price on hotels, airline tickets, even car rentals. When booking a room, always **call the hotel's local toll-free number** (if one is available) rather than the central reservations number—you'll often get a better price.

Always ask about special packages or corporate rates.

➤ Airline Tickets: ☎ 800/FLY–4–LESS. ☎ 800/FLY–ASAP.

➤ Hotel Rooms: **Central Reservation Service (CRS)** (☎ 800/548–3311). **RMC Travel** (☎ 800/245–5738).

### SAVE ON COMBOS

Packages and guided tours can both save you money, but don't confuse the two. When you buy a package your travel remains independent, just as though you had planned and booked the trip yourself. Fly/drive packages, which combine airfare and car rental, are often a good deal.

### JOIN A CLUB?

Many companies sell discounts in the form of travel clubs and coupon books, but these cost money. You must use participating advertisers to get a deal, and only after you recoup the initial membership cost or book price do you begin to save. If you plan to use the club or coupons frequently you may save considerably. Before signing up, find out what discounts you get for free.

➤ Discount Clubs: **Entertainment Travel Editions** (✉ 2125 Butterfield Rd., Troy, MI 48084, ☎ 800/445–4137; $23–$48, depending on destination). **Great American Traveler** (✉ Box 27965, Salt Lake City, UT 84127, ☎ 800/548–2812; $49.95 per year). **Moment's Notice Discount Travel Club** (✉ 7301 New Utrecht Ave., Brooklyn, NY 11204, ☎ 718/234–6295; $25 per year, single or family). **Privilege Card International** (✉ 237 E. Front St., Youngstown, OH 44503, ☎ 330/746–5211 or 800/236–9732; $74.95 per year). **Sears's Mature Outlook** (✉ Box 9390, Des Moines, IA 50306, ☎ 800/336–6330; $14.95 per year). **Travelers Advantage** (✉ CUC Travel Service, 3033 S. Parker Rd., Suite 1000, Aurora, CO 80014, ☎ 800/548–1116 or 800/648–4037; $49 per year, single or family). **Worldwide Discount Travel Club** (✉ 1674 Meridian Ave., Miami Beach, FL 33139, ☎ 305/534–2082; $50 per year family, $40 single).

### DRIVING

Driving on the Cape and islands has sets of unique and sometimes frustrating aspects addressed in individual chapters. ☞ Cape Cod A to Z *in* Chapter 2 for information on Cape roads.

To get to Martha's Vineyard with your car, you'll have to take a ferry from Woods Hole, and to reach Nantucket, take the ferry from Hyannis (☞ Martha's Vineyard A to Z *in* Chapter 3 and Nantucket A to Z *in* Chapter 4).

## F

### FERRY TRAVEL

Ferries to Provincetown leave from Boston and Plymouth in season.

Ferries leave for Martha's Vineyard from Woods Hole year-round and from Falmouth, Hyannis, and New Bedford in season.

Hyannis ferries serve Nantucket year-round. In season, a passenger ferry connects the islands, and a cruise out of Hyannis makes a one-day round-trip with stops at both islands (☞ A to Z sections *in* individual chapters).

## G

### GAY & LESBIAN TRAVEL

Provincetown, at the tip of the Cape, is one of the East Coast's leading lesbian and gay seaside destinations and also has a large year-round lesbian and gay populace. Dozens of P-town establishments, from B&Bs to bars, cater specifically to lesbian and gay visitors, and most of the town's restaurants are gay-friendly; several are gay-owned and -operated. Hyannis has the one gay bar on the Cape that's not in P-town (*see* Nightlife in Hyannis in Chapter 2).

➤ BOOKS: *Fodor's Gay Guide to the USA,* available in bookstores or from Fodor's Travel Publications (☎ 800/533–6478; $19.50 plus $4 shipping), has a chapter on Provincetown.

➤ TOUR OPERATORS: New England Vacation Tours (✉ Box 560, W. Dover, VT 05356) offers vacation packages to Cape Cod and the islands.

➤ GAY- AND LESBIAN-FRIENDLY TRAVEL AGENCIES: Advance Damron (✉ 1 Greenway Plaza, Suite 800, Houston, TX 77046, ☎ 713/850–1140 or 800/695–0880, FAX 713/888–1010). Club

Travel (✉ 8739 Santa Monica Blvd., West Hollywood, CA 90069, ☎ 310/358–2200 or 800/429–8747, FAX 310/358–2222). Islanders/Kennedy Travel (✉ 183 W. 10th St., New York, NY 10014, ☎ 212/242–3222 or 800/988–1181, FAX 212/929–8530). Now Voyager (✉ 4406 18th St., San Francisco, CA 94114, ☎ 415/626–1169 or 800/255–6951, FAX 415/626–8626). Yellowbrick Road (✉ 1500 W. Balmoral Ave., Chicago, IL 60640, ☎ 773/561–1800 or 800/642–2488, FAX 773/561–4497). Skylink Women's Travel (✉ 3577 Moorland Ave., Santa Rosa, CA 95407, ☎ 707/585–8355 or 800/225–5759, FAX 707/584–5637), serving lesbian travelers.

## H

### HEALTH

#### STAYING WELL

A common problem on the East Coast is Lyme disease (named after Lyme, Connecticut, where it was first diagnosed). This bacterial infection is transmitted by deer ticks and can be very serious, leading to chronic arthritis and worse if left untreated. Pregnant women are advised to **avoid areas of possible Lyme tick infestation;** if contracted during early pregnancy, the disease can harm a fetus.

Deer ticks are most prevalent April–October but can be found year-round. They are about the size of a pinhead. Wear light-colored clothing, which makes it easier to spot any ticks that might have attached themselves to you. Anyone planning to explore wooded areas or places with tall grasses (including dunes) should **wear long pants, socks drawn up over pant cuffs, and a long-sleeve shirt with a close-fitting collar;** boots are also recommended. The National Centers for Disease Control recommends that DEET repellent be applied to skin (not face!) and that permethrin be applied to clothing directly before entering infested areas; **use repellents very carefully** and conservatively with small children. Ticks also attach themselves to pets.

Recent research suggests that if ticks are removed within 12 hours of attachment to the body, the infectious bacteria is not likely to enter the bloodstream. That makes evenings a good opportunity to check yourself for ticks—look at the warm spots and hairlines on the body that attract ticks.

To remove a tick, apply tweezers to where it is attached to the skin and pull on the mouth parts. Try not to squeeze the body of the tick, which can send the body fluids containing the bacteria into the bloodstream. (Heating the tip of the tweezers before grasping the tick will cause the bug to release its bite, allowing for removal of the entire tick, including the sometimes-embedded head.) Disinfect the bite with alcohol and save the tick in a closed jar in case symptoms of the disease develop.

The first symptom of Lyme disease may be a ringlike rash or flulike symptoms, such as general feelings of malaise, fever, chills, and joint or facial pains. If diagnosed early, it can be treated with antibiotics. If you suspect your symptoms may be due to a tick bite, inform your doctor and ask to be tested.

**Poison ivy** is a pervasive vinelike plant, recognizable by its leaf pattern: three shiny green leaves together. In spring, new poison ivy leaves are red, likewise they can take on a reddish tint as fall approaches. The oil from these leaves produces an itchy skin rash that spreads with scratching. If you think you may have touched some leaves, wash as soon as you can with soap and cool water.

➤ Lyme Disease Info: **Centers for Disease Control** (☎ 404/332–4555) or the **Massachusetts Department of Public Health** (✉ Southeast Office, 109 Rhode Island Rd., Lakeville, MA 02347, ☎ 508/947–1231). Brochures on the disease are available at many tourist information areas.

## I

### INSURANCE

**Travel insurance is the best way to protect yourself against financial loss.** The most useful policies are trip-cancellation-and-interruption, default, medical, and comprehensive insurance.

Without insurance you will lose all or most of your money if you cancel your trip, regardless of the reason. It's essential that you **buy trip-cancellation-and-interruption insurance,** particularly if your airline ticket, cruise, or package tour is nonrefundable and cannot be changed. When considering how much coverage you need, look for a policy that will cover the cost of your trip plus the nondiscounted price of a one-way airline ticket, should you need to return home early. Also **consider default or bankruptcy insurance,** which protects you against a supplier's failure to deliver.

Citizens of the United Kingdom can buy an annual travel-insurance policy valid for most vacations during the year in which it's purchased. If you are pregnant or have a preexisting medical condition, make sure you're covered. According to the Association of British Insurers, a trade association representing 450 insurance companies, it's wise to buy extra medical coverage when you visit the United States.

If you have purchased an expensive vacation, comprehensive insurance is a must. **Look for comprehensive policies that include trip-delay insurance,** which will protect you in the event that weather problems cause you to miss your flight, tour, or cruise. A few insurers sell waivers for preexisting medical conditions. Companies that offer both features include Access America, Carefree Travel, Travel Insured International, and Travel Guard (☞ *below*).

Always **buy travel insurance directly from the insurance company;** if you buy it from a travel agency or tour operator that goes out of business you probably will not be covered for the agency or operator's default, a major risk. Before you make any purchase, **review your existing health and home-owner's policies** to find out whether they cover expenses incurred while traveling.

➤ Travel Insurers: In the U.S., **Access America** (✉ 6600 W. Broad St., Richmond, VA 23230, ☎ 804/285–3300 or 800/284–8300), **Carefree Travel Insurance** (✉ Box 9366, 100 Garden City Plaza, Garden City, NY 11530, ☎ 516/294–0220 or 800/323–3149), **Near Travel Services** (✉ Box 1339, Calumet City, IL 60409, ☎ 708/868–6700 or 800/654–6700), **Travel Guard International** (✉

1145 Clark St., Stevens Point, WI 54481, ☎ 715/345–0505 or 800/826–1300), **Travel Insured International** (✉ Box 280568, East Hartford, CT 06128–0568, ☎ 860/528–7663 or 800/243–3174), **Travelex Insurance Services** (✉ 11717 Burt St., Suite 202, Omaha, NE 68154-1500, ☎ 402/445–8637 or 800/228–9792, FAX 800/867–9531), **Wallach & Company** (✉ 107 W. Federal St., Box 480, Middleburg, VA 20118, ☎ 540/687–3166 or 800/237–6615). In Canada, **Mutual of Omaha** (✉ Travel Division, 500 University Ave., Toronto, Ontario M5G 1V8, ☎ 416/598–4083, 800/268–8825 in Canada). In the U.K., **Association of British Insurers** (✉ 51 Gresham St., London EC2V 7HQ, ☎ 0171/600–3333).

## L

### LODGING

The Cape has a wide range of lodging options, from campsites to B&Bs to luxurious self-contained resorts offering all kinds of sporting facilities, restaurants, entertainment, services (including business services and children's programs), and all the assistance you'll ever need in making vacation arrangements. Single-night lodgings for those just passing through can be found at countless tacky but cheap and conveniently located little roadside motels, as well as at others that are spotless and cheery yet still inexpensive, or at chain hotels at all price levels; these places often have a pool, TVs, or other amenities to keep children entertained in the evening. Families may want to **consider condominiums, cottages, and efficiencies,** which offer more space, living areas, kitchens, and sometimes laundry facilities, children's play areas, or children's programs.

### APARTMENT, HOUSE, & VILLA RENTALS

If you want a home base that's roomy enough for a family and comes with cooking facilities, **consider a furnished rental.** In fact, many visitors to the Cape and the islands rent a house for a week or longer rather than stay at a B&B or hotel. These can save you money, however some rentals are luxury properties, economical only when your party is large. Many local Cape and islands agencies specialize in rentals. If you do decide to rent, **be sure to book a property well in advance of your trip,** as many properties are rented out to the same families or groups year after year. *See* the A to Z sections at the ends of chapters 2–4 for information on renting houses on the Cape and islands and for the names of agencies.

Home-exchange directories list rentals (often second homes owned by prospective house swappers), and some services search for a house or apartment for you (even a castle if that's your fancy) and handle the paperwork. Some send an illustrated catalog; others send photographs only of specific properties, sometimes at a charge. Up-front registration fees may apply.

➤ RENTAL AGENTS: **Property Rentals International** (✉ 1008 Mansfield Crossing Rd., Richmond, VA 23236, ☎ 804/378–6054 or 800/220–3332, FAX 804/379–2073). **Rent-a-Home International** (✉ 7200 34th Ave. NW, Seattle, WA 98117, ☎ 206/789–9377 or 800/488–7368, FAX 206/789–9379). **Hideaways International** (✉ 767 Islington St., Portsmouth, NH 03801, ☎ 603/430–4433 or 800/843–4433, FAX 603/430–4444) is a travel club whose members arrange rentals among themselves; yearly membership is $99.

### B&BS

Bed-and-breakfasts are popular on Cape Cod, Martha's Vineyard, and Nantucket. Many are housed in interesting old sea captains' homes and other 17th-, 18th-, and 19th-century buildings. In most cases, B&Bs are not appropriate for families. Noise travels easily, rooms are often small, and the furnishings are often fragile. Usually a B&B will not offer a phone or TV in guest rooms; also, more and more B&Bs do not allow smoking. Numerous B&B reservation agencies serve the Cape and the Islands; some leading agencies are listed under Contacts and Resources in the A to Z sections of chapters 2, 3, and 4. For additional information, get a free B&B guide from the Massachusetts Office of Travel & Tourism (☞ Visitor Information *in* Important Contacts A to Z)

## CAMPING

Camping is not allowed on Nantucket, but there are many private and state-park camping areas on Cape Cod and a few on Martha's Vineyard. Write to the Massachusetts Office of Travel & Tourism and the Cape Cod Chamber of Commerce (☞ Visitor Information *in* Important Contacts A to Z). For additional camping information, look for the ⚠ icon in Dining and Lodging sections in individual chapters and consult the Contacts and Resources section of the Cape Cod A to Z in Chapter 2.

## HOME EXCHANGES

If you would like to exchange your home for someone else's, **join a home-exchange organization,** which will send you its updated listings of available exchanges for a year and will include your own listing in at least one of them. Making the arrangements is up to you.

➤ EXCHANGE CLUBS: **HomeLink International** (✉ Box 650, Key West, FL 33041, ☎ 305/294–7766 or 800/638–3841, ℻ 305/294–1148) charges $83 per year.

## HOSTELING

There are very good hostels on the Cape and islands. In season, when all other rates are jacked up beyond belief, hostels are often the only budget accommodation option, but you must plan ahead to reserve space. You may luck out on last-minute cancellations, but it would be unwise to rely on them. See Dining and Lodging sections of individual chapters for specific information.

## M
### MONEY

## ATMS

Before leaving home, make sure that your credit cards have been programmed for ATM use.

➤ ATM LOCATIONS: **Cirrus** (☎ 800/424–7787). **Plus** (☎ 800/843–7587).

➤ ON MARTHA'S VINEYARD: ATMs are at the following locations: **Edgartown National Bank** (✉ 2 South Water St. and 251 Upper Main St.). **Park Avenue Mall** (✉ Oak Bluffs; ☎ 508/627–3343). By the **Compass Bank** (✉ Opposite steamship offices

in Vineyard Haven and Oak Bluffs; ✉ 19 Lower Main St., Edgartown; ✉ Up-Island Cronig's Market, State Rd., West Tisbury; ☎ 508/693–9400).

➤ ON NANTUCKET: ATMs are at the following locations: **Nantucket Memorial Airport** (☎ 508/325–5300). **Nantucket Bank** (✉ 2 Orange St. or 104 Pleasant St.). **Pacific National Bank** (✉ 61 Main St.). **A&P** (✉ Straight Wharf). **Finast** (✉ Lower Pleasant St.).

## N
### NATIONAL PARKS

You may be able to **save money on park entrance fees** by getting a discount pass. The Golden Eagle Pass ($50) gets you and your companions free admission to all parks for one year. (Camping and parking are extra.) Both the Golden Age Passport, for U.S. citizens or permanent residents age 62 and older, and the Golden Access Passport, for travelers with disabilities, entitle holders to free entry to all national parks plus 50% off fees for the use of many park facilities and services. Both passports are free; you must show proof of age and U.S. citizenship or permanent residency (such as a U.S. passport, driver's license, or birth certificate) or proof of disability. All three passes are available at all national park entrances. Golden Eagle and Golden Access passes are also available by mail.

➤ PASSES BY MAIL: **National Park Service** (✉ Department of the Interior, Washington, DC 20240).

## P
### PACKING FOR CAPE COD

Only a few restaurants on Cape Cod, Martha's Vineyard, and Nantucket require formal dress, as do some dinner cruises. The area prides itself on informality. Do **pack a sweater or jacket, even in summer, for nights can be cool.** For suggested clothing regarding deer ticks and Lyme disease (☞ Health, *above*).

Bring an extra pair of eyeglasses or contact lenses in your carry-on luggage, and if you have a health problem, **pack enough medication** to last the entire trip. It's important that you **don't put prescription drugs or valu-**

ables in luggage to be checked: it might go astray.

Perhaps most important of all, don't forget your swimsuit!

### LUGGAGE

In general you are entitled to check two bags on flights within the United States. A third piece may be brought on board, but it must fit easily under the seat in front of you or in the overhead compartment.

Airline liability for baggage is limited to $1,250 per person on flights within the United States. On international flights it amounts to $9.07 per pound or $20 per kilogram for checked baggage (roughly $640 per 70-pound bag) and $400 per passenger for unchecked baggage. Insurance for losses exceeding these amounts can be bought from the airline at check-in for about $10 per $1,000 of coverage; note that this coverage excludes a rather extensive list of items, which is shown on your airline ticket.

Before departure, **itemize your bags' contents** and their worth, and label the bags with your name, address, and phone number. (If you use your home address, cover it so that potential thieves can't see it readily.) Inside each bag, **pack a copy of your itinerary.** At check-in, **make sure that each bag is correctly tagged** with the destination airport's three-letter code. If your bags arrive damaged or fail to arrive at all, file a written report with the airline before leaving the airport.

### PASSPORTS & VISAS

### CANADIANS

A passport is not required to enter the United States.

### U.K. CITIZENS

British citizens need a valid passport to enter the United States. If you are staying for fewer than 90 days on vacation, with a return or onward ticket, you probably will not need a visa. However, you will need to fill out the Visa Waiver Form, 1-94W, supplied by the airline.

➤ INFORMATION: **London Passport Office** (☎ 0990/21010) for fees and documentation requirements and to request an emergency passport. **U.S. Embassy Visa Information Line** (☎ 01891/200–290) for U.S. visa information; calls cost 49p per minute or 39p per minute cheap rate. **U.S. Embassy Visa Branch** (☒ 5 Upper Grosvenor St., London W1A 2JB) for U.S. visa information; send a self-addressed, stamped envelope. Write the **U.S. Consulate General** (☒ Queen's House, Queen St., Belfast BTI 6EO) if you live in Northern Ireland.

## S

### SENIOR-CITIZEN TRAVEL

To qualify for age-related discounts, **mention your senior-citizen status up front** when booking hotel reservations (not when checking out) and before you're seated in restaurants (not when paying the bill). Note that discounts may be limited to certain menus, days, or hours. When renting a car, **ask about promotional car-rental discounts,** which can be cheaper than senior-citizen rates.

➤ LOCAL RESOURCES: **Elder Services of Cape Cod and the Islands** (☒ 68 Rte. 134, South Dennis 02660, ☎ 508/ 394–4630 or 800/244–4630 in MA; on Martha's Vineyard, ☒ Linton La., Box 2337, Oak Bluffs 02557, ☎ 508/ 693–4393; on Nantucket, ☒ 144 Orange St., Nantucket 02554, ☎ 508/228–4647) offers information and referrals.

➤ EDUCATIONAL TRAVEL PROGRAMS: **Elderhostel** (☒ 75 Federal St., 3rd floor, Boston, MA 02110, ☎ 617/ 426–8056).

### SHOPPING

On the islands the majority of shops close down in winter. Because of the Cape's large year-round population, its shops tend to remain open, though most Provincetown and Wellfleet shops and galleries do close. Throughout the Cape and islands, shop owners respond to the flow of tourists as well as to their own inclinations. Especially off-season, it's best to **phone a shop before going out of your way to visit it.**

➤ LOCAL RESOURCES: For a directory of area antiques dealers and auctions, contact the **Cape Cod Antique Dealers Association** (☒ Send business-size SASE to Box 196, Harwich 02645). On the Cape, Provincetown and

Wellfleet are the main centers for art. Both the **Provincetown Gallery Guild** (✉ Box 242, Provincetown 02657) and the **Wellfleet Art Galleries Association** (✉ Box 916, Wellfleet 02667) issue pamphlets on local galleries. For a listing of crafts shops on the Cape, write to the **Society of Cape Cod Craftsmen** (✉ Box 1709, Wellfleet 02667-1709), the **Artisans' Guild of Cape Cod** (✉ Send business-size SASE to 46 Debs Hill Rd., Yarmouth 02675), or **Cape Cod Potters** (✉ Box 76, Chatham 02633).

## STUDENTS

There are very good hostels on the Cape and islands. ☞ Lodging sections of individual chapters.

➤ STUDENT IDs AND SERVICES: **Council on International Educational Exchange** (✉ CIEE, 205 E. 42nd St., 14th floor, New York, NY 10017, ☎ 212/822–2600 or 888/268–6245, ℻ 212/822–2699), for mail orders only, in the United States. **Travel Cuts** (✉ 187 College St., Toronto, Ontario M5T 1P7, ☎ 416/979–2406 or 800/667–2887) in Canada.

➤ HOSTELING: **Hostelling International—American Youth Hostels** (✉ 733 15th St. NW, Suite 840, Washington, DC 20005, ☎ 202/783–6161, ℻ 202/783–6171). **Hostelling International—Canada** (✉ 400-205 Catherine St., Ottawa, Ontario K2P 1C3, ☎ 613/237–7884, ℻ 613/237–7868). **Youth Hostel Association of England and Wales** (✉ Trevelyan House, 8 St. Stephen's Hill, St. Albans, Hertfordshire AL1 2DY, ☎ 01727/855215 or 01727/845047, ℻ 01727/844126). Membership in the U.S., $25; in Canada, C$26.75; in the U.K., £9.30).

## T

## TAXES

Massachusetts state sales tax is 5%.

## TELEPHONES

### CALLING HOME

AT&T, MCI, and Sprint long-distance services make calling home relatively convenient and let you avoid hotel surcharges. Typically you dial an 800 number in the United States.

➤ TO OBTAIN ACCESS CODES: **AT&T USADirect** (☎ 800/874–4000). **MCI**

Call USA (☎ 800/444–4444). **Sprint Express** (☎ 800/793–1153).

## TOUR OPERATORS

Buying a prepackaged tour or independent vacation can make your trip to Cape Cod less expensive and more hassle-free. Because everything is prearranged you'll spend less time planning.

Operators that handle several hundred thousand travelers per year can use their purchasing power to give you a good price. Their high volume may also indicate financial stability. But some small companies provide more personalized service; because they tend to specialize, they may also be more knowledgeable about a given area.

### A GOOD DEAL?

The more your package or tour includes, the better you can predict the ultimate cost of your vacation. Make sure you know exactly what is covered, and **beware of hidden costs.** Are taxes, tips, and service charges included? Transfers and baggage handling? Entertainment and excursions? These can add up.

If the package or tour you are considering is priced lower than in your wildest dreams, **be skeptical.** Also, **make sure your travel agent knows the accommodations** and other services. Ask about the hotel's location, room size, beds, and whether it has a pool, room service, or programs for children, if you care about these. Has your agent been there in person or sent others you can contact?

### BUYER BEWARE

Each year consumers are stranded or lose their money when tour operators—even very large ones with excellent reputations—go out of business. So **check out the operator.** Find out how long the company has been in business, and ask several agents about its reputation. **Don't book unless the firm has a consumer-protection program.**

Members of the National Tour Association and United States Tour Operators Association are required to set aside funds to cover your payments and travel arrangements in case the company defaults. Nonmembers may carry insurance instead. Look for the details, and for the name of an underwriter with a solid reputation, in the operator's brochure. Note: When it comes to tour operators, **don't trust escrow accounts.** Although the Department of Transportation watches over charter-flight operators, no regulatory body prevents tour operators from raiding the till. You may want to protect yourself by buying travel insurance that includes a tour-operator default provision. For more information, *see* Consumer Protection, *above.*

It's also a good idea to choose a company that participates in the American Society of Travel Agents Tour Operator Program (TOP). This gives you a forum if there are any disputes between you and your tour operators; ASTA will act as mediator.

➤ TOUR-OPERATOR RECOMMENDATIONS: **National Tour Association** (✉ NTA, 546 E. Main St., Lexington, KY 40508, ☎ 606/226–4444 or 800/755–8687). **United States Tour Operators Association** (✉ USTOA, 342 Madison Ave., Suite 1522, New York, NY 10173, ☎ 212/599–6599, 𝖥𝖠𝖷 212/599–6744). **American Society of Travel Agents** (☞ Travel Agencies, *below*).

## USING AN AGENT

Travel agents are excellent resources. In fact, large operators accept bookings made only through travel agents. But it's a good idea to **collect brochures from several agencies,** because some agent's suggestions may be influenced by relationships with tour and package firms that reward them for volume sales. If you have a special interest, **find an agent with expertise in that area**; ASTA (☞ Travel Agencies, *below*) has a database of specialists worldwide. Do some homework on your own, too: Local tourism boards can provide information about lesser-known and small-niche operators, some of which may sell only direct.

## SINGLE TRAVELERS

Prices for packages and tours are usually quoted per person, based on two sharing a room. If traveling solo, you may be required to pay the full double-occupancy rate. Some operators eliminate this surcharge if you agree to be matched with a roommate

of the same sex, even if one is not found by departure time.

## GROUP TOURS

Among companies that sell tours to Cape Cod, the following are nationally known, have a proven reputation, and offer plenty of options. The classifications used below represent different price categories, and you'll probably encounter these terms when talking to a travel agent or tour operator. The key difference is usually in accommodations, which run from budget to better, and better-yet to best.

➤ DELUXE: **Globus** (✉ 5301 S. Federal Circle, Littleton, CO 80123-2980, ☎ 303/797–2800 or 800/221–0090, 𝖥𝖠𝖷 303/347–2080). **Maupintour** (✉ 1515 St. Andrews Dr., Lawrence, KS 66047, ☎ 913/843–1211 or 800/255–4266, 𝖥𝖠𝖷 913/843–8351). **Tauck Tours** (✉ Box 5027, 276 Post Rd. W, Westport, CT 06881–5027, ☎ 203/226–6911 or 800/468–2825, 𝖥𝖠𝖷 203/221–6828).

➤ FIRST-CLASS: **Collette Tours** (✉ 162 Middle St., Pawtucket, RI 02860, ☎ 401/728–3805 or 800/832–4656, 𝖥𝖠𝖷 401/728–1380). **Mayflower Tours** (✉ Box 490, 1225 Warren Ave., Downers Grove, IL 60515, ☎ 630/960–3793 or 800/323–7604, 𝖥𝖠𝖷 630/960–3575).

## PACKAGES

Like group tours, independent vacation packages are available from major tour operators and airlines. The companies listed below offer vacation packages in a broad price range.

## THEME TRIPS

➤ FALL FOLIAGE: To catch New England's autumn color, contact any of the tour operators listed under Group Tours, above.

➤ LIGHTHOUSES: **U.S. Lighthouse Society** (✉ 244 Kearny St., San Francisco, CA 94108, ☎ 415/362–7255, 𝖥𝖠𝖷 415/362–7464).

➤ WHALE-WATCHING: **Oceanic Society Expeditions** (✉ Fort Mason Center, Bldg. E, San Francisco, CA 04123, ☎ 415/441–1106 or 800/326–7491).

➤ YACHT CHARTERS: **Huntley Yacht Vacations** (✉ 210 Preston Rd., Wernersville, PA 19565, ☎ 610/678–2628 or 800/322–9224, FAX 610/670–1767). **Lynn Jachney Charters** (✉ Box 302, Marblehead, MA 01945, 617/639–0787 or 800/223–2050, FAX 617/639–0216). **Nicholson Yacht Charters** (✉ 78 Bolton St., Cambridge, MA 02140–3321, ☎ 617/661–0555 or 800/662–6066, FAX 617/661–0554). **Ocean Voyages** (✉ 1709 Bridgeway, Sausalito, CA 94965, ☎ 415/332–4681 or 800/299–4444, FAX 415/332–7460). **Russell Yacht Charters** (✉ 404 Hulls Hwy., #175, Southport, CT 06490, ☎ 203/255–2783 or 800/635–8895).

## TRAIN TRAVEL

Because of continuing financial difficulties, Amtrak service to the Cape was suspended in 1997; it is not currently scheduled to resume in 1998, but call for an update.

➤ TRAIN TRAVEL: **Amtrak** (☎ 800/872–7245).

## TRAVEL AGENCIES

A good travel agent puts your needs first. Look for an agency that has been in business at least five years, and emphasizes customer service, and has someone on staff who specializes in your destination. In addition, **make sure the agency belongs to the American Society of Travel Agents** (ASTA). If your travel agency is also acting as your tour operator, *see* Tour Operators, *above*.

➤ LOCAL AGENT REFERRALS: **American Society of Travel Agents** (✉ ASTA, ☎ 800/965–2782, 24-hour hot line, FAX 703/684–8319). **Alliance of Canadian Travel Associations** (✉ Suite 201, 1729 Bank St., Ottawa, Ontario K1V 7Z5, ☎ 613/521–0474, FAX 613/521–0805). **Association of British Travel Agents** (✉ 55–57 Newman St., London W1P 4AH, ☎ 0171/637–2444, FAX 0171/637–0713).

## TRAVEL GEAR

Travel catalogs specialize in useful items, such as compact alarm clocks and travel irons, that can **save space when packing.**

➤ MAIL-ORDER CATALOGS: **Magellan's** (☎ 800/962–4943, FAX 805/568–5406). **Orvis Travel** (☎ 800/541–3541, FAX 540/343–7053).

**TravelSmith** (☎ 800/950–1600, FAX 800/950–1656).

## U

## U.S. GOVERNMENT

The U.S. government can be an excellent source of inexpensive travel information. When planning your trip, find out what government materials are available.

➤ PAMPHLETS: **Consumer Information Center** (✉ Consumer Information Catalogue, Pueblo, CO 81009, ☎ 719/948–3334) for a free catalog that includes travel titles.

## VISITOR INFORMATION

Before you go, contact the state's office of tourism and the area's chambers of commerce for general information, seasonal events, and brochures. For specific information on Cape Cod's state forests and parks, the area's farmers' markets and fairs, or wildlife, contact the special-interest government offices below. When you arrive, you can also pay a visit to the chambers of commerce for additional information.

➤ CHAMBERS OF COMMERCE: **Cape Cod Chamber of Commerce** (✉ Junction of Rtes. 6 and 132, Hyannis 02601, ☎ 508/362–3225, FAX 508/362–3698). **Martha's Vineyard Chamber of Commerce** (✉ Box 1698, Vineyard Haven 02568, ☎ 508/693–0085, FAX 508/693–7589). **Nantucket Chamber of Commerce** (✉ 48 Main St., Nantucket 02554, ☎ 508/228–1700, FAX 508/325–4925). For a list of local Cape Cod chambers of commerce, which can also provide information, *see* Essential Information in Chapter 2.

➤ STATE: **Massachusetts Office of Travel & Tourism** (✉ 100 Cambridge St., 13th floor, Boston 02202, ☎ 617/727–3201 or 800/227—6277, FAX 617/727–6525); 800/447–6277 for brochures.

➤ SPECIAL INTERESTS: The **Department of Environmental Management** (✉ Division of Forests and Parks, 100 Cambridge St., Room 1905, Boston 02202, ☎ 617/727–3000, FAX 617/727–9402). The **Department of Food and Agriculture** (✉ 100 Cambridge St., Suite 2103, Boston 02202, ☎ 617/727–3000, FAX 617/727–7235).

The **Division of Fisheries and Wildlife** (✉ Field Headquarters, Route 135, Westborough 01581, ☎ 508/792–7270, FAX 508/792–7275).

# W

Memorial Day through Labor Day (in some cases, Columbus Day) is high season on Cape Cod, Martha's Vineyard, and Nantucket. This is summer with a capital *S*, a time for barbecues, beach bumming, water sports, and swimming. During summer everything is open for business on the Cape, but you can also expect high-season evils: high prices, crowds, and traffic.

The Cape and the islands, however, are increasingly a year-round destination. See our Shoulder Season Close-Up in Chapter 2 for some of the highlights of the off-season.

### CLIMATE

Although there are plenty of idyllic beach days to go around on the Cape, rain or fog is not an uncommon part of even an August vacation here.

Visitors who do not learn to appreciate the beauty of the land and sea in mist and rain may find themselves mighty cranky.

Temperatures in winter and summer are milder on the Cape and islands than on the mainland, due in part to the warming influence of the Gulf Stream and the moderating ocean breezes. As a rule (and there have been dramatically anomalous years— 1994 in particular), the Cape and islands get much less snow than the mainland, and what falls generally does not last. Still, winter can bring bone-chilling dampness, especially on the windswept islands.

The following are average daily maximum and minimum temperatures for Hyannis; it's likely to be two or three degrees cooler on the coast and on the islands.

➤ FORECASTS: For local Cape weather, coastal marine forecasts, and today's tides, call the weather line of **WQRC** in Hyannis (☎ 508/771-5522). **Weather Channel Connection** (☎ 900/932–8437), 95¢ per minute from a Touch-Tone phone.

## Climate

| Jan. | 40F | +4C | May | 62F | 17C | Sept. | 70F | 21C |
|------|-----|-----|------|-----|-----|-------|-----|-----|
|      | 25  | −4  |      | 48  | +9  |       | 56  | 13  |
| Feb. | 41F | +5C | June | 71F | 22C | Oct.  | 59F | 15C |
|      | 26  | −3  |      | 56  | 13  |       | 47  | 8   |
| Mar. | 42F | +6C | July | 78F | 26C | Nov.  | 49F | 9C  |
|      | 28  | −2  |      | 63  | 17  |       | 37  | 3   |
| Apr. | 53F | 12C | Aug. | 76F | 24C | Dec.  | 40F | 4C  |
|      | 40  | +4  |      | 61  | 16  |       | 26  | −3  |

# 1 Destination: Cape Cod, Martha's Vineyard, Nantucket

# THE SEASONS OF AMERICA PAST AND PRESENT

THE WORLD TO-DAY is sick to its thin blood for lack of elemental things," wrote Henry Beston in his 1928 Cape Cod classic, *The Outermost House*, "for fire before the hands, for water welling from the earth, for air, for the dear earth itself underfoot." It is this that the Cape and its neighboring islands most have to offer an increasingly complex and artificial world: the chance to reconnect with elemental things. Walking along the shore poking at the washed-up sea life or watching birds fish in the surf, listening to the rhythm of the waves, experiencing the tranquillity of night on the beach or the power of a storm on water—all this is somehow life-affirming and satisfyingly real.

Cape Cod—the craggy arm of peninsula 60 mi southeast of Boston—and the islands of Martha's Vineyard and Nantucket share their geologic origins as debris deposited by a retreating glacier in the last ice age. They also share a moderate coastal climate and a diversity of terrain that foster an equally diverse assortment of plant and animal life, some of which exist nowhere else in northern climes.

Barrier beaches (sandbars that protect an inner harbor from the battering of the ocean), such as Monomoy on the Cape and Coatue on Nantucket, are breeding and resting grounds for a stunning variety of shore and seabirds, and the marshes and ponds are rich in waterfowl. Stellwagen Bank, just north of Provincetown, is a prime feeding ground for whales and dolphins, and shallow sandbars are favorite playgrounds for harbor and gray seals.

Among the flotsam and jetsam along the shores, beachcombers find horseshoe crabs, starfish, sea urchins, sponges, jellyfish, and a plethora of shells: white quahogs, elegant scallops, blue mussels, long straight razor clams, spiraling periwinkles, pointy turret shells, smooth round moon snails, conical whelks, rough-ridged oysters.

Much of the land, including a third of Nantucket's acreage and a quarter of the Vineyard's, is protected from development. Nature preserves encompassing pine forests, marshes, swamps, cranberry bogs, and many other varieties of terrain are laced with well-marked walking and bicycling trails. On Nantucket, acres of moorland are spread with a rough tapestry of gnarled scrub oaks, low-lying blueberry bushes, fragrant bayberry, bearberry, and heather (the last originally brought to Nantucket from Scotland by accident in a shipment of pine trees).

Thanks to the establishment of the Cape Cod National Seashore in 1961, you can walk for almost 30 mi along the Atlantic beach and rarely see a trace of human habitation—besides a few old shacks in the dunes of Provincetown, or the lighthouses that stand watch over the Cape's dangerous shoals. Across dunes anchored by poverty grass sprawl beach plums, pink salt-spray roses, and purple beach peas.

Through the creation of many National Historic Districts—in which change is kept to a minimum to preserve the historical integrity of the area—similar protection has been extended to the Cape and islands' oldest and loveliest man-made landscapes. One of the most important, as well as most visually harmonious, is along the Old King's Highway, where some of the Cape's first towns were incorporated in 1639. Lining this tree-shaded country road are simple saltboxes from the earliest days, fancier houses built later by prosperous sea captains, and the traditional Cape cottages, shingles weathered to a silvery gray, with soft pink roses spilling across them. Here, too, are the Cape's windmills, as well as the white-steepled churches, taverns, and village greens that savor of old New England.

Nearly the entire island of Nantucket is part of its historic district. A rigid enforcement of district guidelines has created a town architecturally almost frozen in time, and one of the world's great treasures. Among the neat clapboard and weathered-shingle houses that line its cobblestone streets and narrow lanes are former warehouses, factories, and mansions that date from the golden age of whaling. Provincetown, too, has been designated a historic district, preserving for posterity its cheerful mix of tiny waterfront shops (former fish shacks) and captains' mansions, with everything from a 1746 Cape house to a mansarded French Second Empire building.

The Cape and islands also preserve their past in a wealth of small museums—nearly every town has one—that document local history, often back to Indian days. (In 1620, when the Pilgrims first anchored at Provincetown, exploring the Cape before heading on to Plymouth, an estimated 30,000 Wampanoags lived on Cape Cod.) Often set in houses that are themselves historic, these museums provide a visual history of the lives of the English settlers and their descendants, including their economic pursuits: from farming to the harvesting of salt, salt hay, and cranberries (still an important local crop) to fishing and whaling to tourism, which began as far back as the late-19th century.

The importance of whaling to the area—Nantucket was the world's premier whaling port in the early to mid-19th century—is reflected in the historical museums. The travels of the area's whaling and packet-schooner seamen and captains are illustrated with such items as antique nautical equipment, harpoons, charts, maps, journals, scrimshaw created during the often years-long whaling voyages, and gifts brought back from exotic ports for wives who had waited so patiently (those who *had* waited, that is—some women chose to go along with their husbands for the ride).

Remote Nantucket has just one town, with a few outlying villages, one of them full of rose-covered cottages that once housed an actors' colony. Large tracts of undeveloped moorland and nature preserves give the island an open, breezy feel. It has long been a summer bastion of the wealthy, who are likely to be seen dressed down to the hilt and tooling around on beat-up bicycles.

Martha's Vineyard, on the other hand, is known as the celebrity island for its famous summer (and year-round) residents in the arts and entertainment, many of whom participate in the annual Possible Dreams Auction and other high-visibility charitable events. More than twice the size of Nantucket, the Vineyard offers more variety than its smaller neighbor to

19 mi to the east: Its six towns range from a young and rowdy seaside hamlet of Victorian cottages to a rural New England village to an elegant, well-manicured town of sea captains' homes and flower gardens. The landscape, too, is more varied, including a 5,146-acre pine forest, rolling farmland enclosed by drystone walls, and dramatic clay cliffs.

Both islands are ringed with beautiful sandy beaches, some backed with high or low dunes, others bordering moorland or marshes. You can visit them in a day trip from a Cape Cod base, but three days is a practical minimum for getting a real sense of island life.

ALL THREE AREAS are noted for their shopping (for crafts, art, and antiques especially); for theater, both small community groups and professional summer stock companies; and for plenty of recreational offerings that take advantage of the marine environment, including water sports, fishing charters, and even Jeep safaris to isolated beaches for surf casting. (Piping plovers nest on beaches, and access to beaches is sometimes severely limited, so check about restrictions before renting that Jeep!) All three are also family oriented—especially the Cape, which has endless amusements to offer children beyond the ever-beckoning beach. There are also such typical New England entertainments as chowder suppers and clambakes.

The "season" used to be strictly from Memorial Day to Labor Day, but the boundaries have blurred; many places now open in April or earlier and close as late as November, and a core remain open year-round. Unfortunately, most of the historic sites and museums, largely staffed by volunteers, still adhere to the traditional dates and so are inaccessible in the off-season.

In fact, all seasons invite a different kind of visit. In summer, you have your choice of plunking down somewhere near a beach and never moving, filling your schedule with museums and activities, or combining the two in whatever mix suits you. In fall, the water may be warm enough for swimming as late as October, crowds are gone, and prices are lower. Turning foliage, though nothing like the dramatic displays found elsewhere in New England, is still an enjoyable addition to a fall visit; it usually reaches its peak around the end of October (the peak can be as much as a few weeks earlier or later than this, though, depending on the weather in the preceding months). Moors turn purple and gold and rust; burning bush along roadsides flames a brilliant red. Cranberries ripen to a bright burgundy color and are harvested by a method fascinating to watch. Trees around freshwater marshes, ponds, and swamps tend to color earlier and brighter; the red maple swamps, Beech Forest in Provincetown, and Route 6A from Sandwich to Orleans are particularly colorful spots.

Fall and winter are oyster and scallop season, and restaurants that remain open serve a wide selection of dishes made with these freshly harvested delicacies. Winter is a quiet time, when many tourist-oriented activities and facilities shut down, but prices are low and you can walk the beaches in often total solitude. For a quiet or romantic weekend getaway, country inns offer cozy rooms with canopy beds, where you can curl up before the fireplace after returning from a leisurely, candlelit dinner.

As for spring, it gets a wee bit wet, and on Nantucket expect a good dose of fog. Still, when daffodils come bursting up from roadsides, especially on Nantucket, everything begins to turn green. By April, seasonal shops and restaurants begin to open, and locals prepare for yet another summer.

# NEW AND NOTEWORTHY

## Cape Cod

**Hyannis**'s Main Street is gradually becoming a more attractive place for an afternoon stroll. A design improvement plan has been responsible for rose plantings, a gazebo beside the post office, and signs that are more complementary to the surroundings. "Walkway to the Sea," a flower-lined promenade designed by Ben Thompson, the architect responsible for Boston's Faneuil Hall Marketplace, should

be finished by the end of the 1998 season, and will connect Main Street with the harbor. More shops showcasing American artisans are opening, and there are now several coffee shops, as well as a Ben & Jerry's, along the way for refreshment.

In **Mashpee**, the **Boch Center for the Performing Arts** is slated to complete a new complex of buildings by 1999. The new facilities will include a 2,000-seat theater, studios, and classroom space for educational programs that will focus on playwriting, acting, and other workshops. In 1996 the Boch Center announced that the first anchor tenant was none other than the prestigious Boston Ballet Company. Until the complex is completed, performances are being held under a tent at Mashpee Commons or in the auditoriums of local high schools.

In 1997 renovations were completed on the former **Provincetown Marina building** on MacMillan Wharf, and it opened as the Whydah Museum, the new home for artifacts and treasure recovered from the *Whydah,* a pirate ship that sank off the coast of Wellfleet more than 265 years ago. The new museum is both entertaining and educational, with a display featuring the conservation and restoration processes at work on actual artifacts, and a history component that gives the untold story of the pirating life as it was in the 18th century. Eventually the museum hopes to collect and house all of the recovered pieces, some of which are presently on loan to other museums across the country.

In conjunction with the annual Blessing of the Fleet in June, 1997, **Provincetown** celebrated the first annual **Portuguese Week,** a series of events designed to honor the history of the Portuguese in Provincetown and their role in the fishing industry. There were ethnic foods and crafts, dances, concerts, fireworks, children's games, and a poetry reading (in both Portuguese and English) of Pesoa's work.

## Martha's Vineyard

The Vineyard has become one of the nation's leading summer hot spots for the elite from the worlds of entertainment, media, and politics. Sharon Stone, Spike Lee, Walter Cronkite, Ted Danson and Mary Steenburgen, Beverly Sills, Billy Joel, Art Buchwald, Patricia Neal, and Diane Sawyer and Mike Nichols spend their summers here, rubbing creative elbows with such year-round residents as Carly Simon and James Taylor. President Clinton, First Lady Hillary Rodham Clinton, and daughter Chelsea returned to the island in August, 1997, for the fourth time since Clinton took office.

The art scene on the Vineyard has really blossomed, with more than 35 galleries displaying the work of internationally recognized artists and photographers. Weekly openings are eagerly awaited social events, where guests can nibble, sip, and meet-and-greet the men and women behind the works. The Granary Gallery in Chilmark still carries the late photographer Alfred Eisenstaedt's work. And don't miss the breathtaking photography of Peter Simon (Carly's brother) that can be found at his gallery located at the Feast of Chilmark restaurant.

The Trustees of Reservations report an increased rate of erosion on south-facing beaches—up to 75 ft in some spots. Although 60 to 75 acres of beach were created at **Wasque** over the past 20 years, the current cycle of erosion reclaims an average of 17 ft annually. In **Oak Bluffs,** the town, the state, and the Army Corps of Engineers are continuing the battle to save Beach Road, a narrow, view-rich strip between the ocean and Sengekontacket Pond.

## Nantucket

The second annual **Nantucket Film Festival,** a weeklong series of screenings, readings, panel discussions, and Q&A sessions, in June 1997, was another overwhelming success. From 10 AM until midnight, an array of films, from 20-minute shorts to full-length features, were shown at two venues in town. Actors and actresses involved in some of the productions flooded the town and added celebrity status to the festivities.

Restoration of the **African Meeting House** in Nantucket Town began in 1997; the property should be open to the public in 1998. Plans are to incorporate a museum and space for lectures, concerts, and readings that will help to portray the experiences of the black community on Nantucket.

The first segment of the new **Polpis bike path,** a 2-mi stretch between Nantucket

Town and the Life Saving Museum, was completed for the 1997 season; the remainder, a winding path along Polpis road to Siasconset, is scheduled to be completed by the 1998 season.

Visitors will be pleased to know that the first **Lyme disease vaccine** for human beings is being tested on Nantucket. Developed at the Harvard School of Public Health and 95% effective on mice, the vaccine should be available, if all goes well, before the end of the decade.

# WHAT'S WHERE

## Cape Cod

Continually shaped by ocean currents, this windswept land of sandy beaches and dunes has an amazing natural beauty. Everyone comes for the seaside, yet crimson cranberry bogs, forests of birch and beech, and grassy meadows and marshlands gracing the interior are equally splendid. Local history is fascinating, whale-watching provides an exhilarating experience of the natural world, cycling trails lace the landscape, shops purvey everything from the useful to the kitsch and tacky, and you can dine on simple, as-fresh-as-it-comes seafood, delicious, thoughtfully prepared cuisine, or most anything in between.

## Martha's Vineyard

With all of the star appeal that's come to it in recent years, you might think that the Vineyard has changed. Yes, more and more visitors come to the island every year, but Vineyarders have risen to the occasion. As a result, some restaurants have acquired a pleasant, easygoing sophistication, while others hold on to their local traditions. The bustle and crush of Vineyard Haven, Oak Bluffs, and Edgartown's summertime crowds continue to belie the quieter feeling that off-season visitors have come to love. In season, you can always step back into rural time up-island—at the fantastic West Tisbury Farmers' Market or on a conservation area's delightful walking trails. And the fresh fish loaded onto the docks in Menemsha is as delicious as it's ever been.

Herman Melville may never have set foot on Nantucket, but he was right about its exuberance—in summer, the place brims over with activity. To the eye, Nantucket Town's museum-quality houses and the outlying beaches and rolling moors make the island an aesthetic world unto itself. So hop a ferry to the Gray Lady of the Sea to measure your gait on its historic cobblestone streets, stand knee-deep in surf casting for stripers, pick up some old whaling lore, or set yourself up on the sand as the sun arcs its way across the sky.

# PLEASURES AND PASTIMES

## Beaches

Cape Cod, Martha's Vineyard, and Nantucket are known for long, dune-backed sand beaches, with both surf and calm water. Swimming season is approximately mid-June–September (sometimes into October). The Cape Cod National Seashore on the ocean side of The Lower Cape has the Cape's best beaches, with high dunes, wide strands of sand, and no development on the shores.

Beaches on the Vineyard and Nantucket are no less spectacular. The botanically curious will find plants that do not appear on the Cape, and it's even more possible on the islands to feel the delight of having an isolated beach all to yourself. There may, in fact, be something about island beaches that heightens their appeal—knowing that you must cross the water in front of you in order to return to your other life gives them a more precious quality.

## Conservation Areas

With all of the bustle that can seem unavoidable on a Cape Cod, Martha's Vineyard, or Nantucket vacation, a sure bet for escape is any one of the numerous Audubon, Land Bank, Sherrif's Meadow, Nature Conservancy, or Trustees of the Reservations refuges. The efforts of these groups have enhanced the lives of hundreds of animal and plant species and created opportunities for all of us to share their world. All ages will delight in seeing an osprey nest or the slow-motion stalking of a great blue heron, the head of a river

otter coursing through the water, great shorebird colonies, beach plum bushes in bloom, a meadow in late summer, or that stray blueberry bush with fruit on it. There isn't a better way to experience the vitality of the region than to visit one of its conservation areas.

Flora and fauna of local interest to look out for: beach plum bushes that bloom in late May and bear fruit in the fall; spring-blooming, white-flowered shadbush and its red-purple June berries; shade- and moisture-loving cinnamon ferns; June-blooming pasture rose and the lovely rosa rugosa that bloom throughout the summer; the brilliant orange butterfly weed, so named for the affinity that monarch butterflies have for the nectar of its summer flowers; the fragrant midsummer blooms of the sweet pepper bush; blueberry and huckleberry bushes; the low-slung, waxy-leaved, dark green teaberry with its early fall red berries; fall-flowering yellow seaside goldenrod and purple sea lavender; caribou moss, also near the sea; the plentiful beach grass (which you should avoid treading upon in order to keep it plentiful); oval-leafed bayberry bushes, whose scent is a wintertime household delight; and tupelo, sassafras, pygmy beech, cottonwood, Norway spruce, red cedar, pitch pine, tamarack, and numerous other trees. Unfortunately, you need to watch out for poison ivy, that invasive spoiler of human comfort (its berries are a great boon to birds, however).

As for wildlife, there are hawks and harriers, osprey, pheasant, quail, numerous ducks and geese, terns, bobwhites, meadowlarks, catbirds, towhees, swallows, orioles, goldfinches, yellowthroats, a great variety of warblers, and more spring and fall migrants than we can mention; a kingdom of mollusks and sea creatures—horseshoe, hermit, and blue crabs, oysters, scallops, quahogs, and so on; and rabbits, otter, muskrats, mice, and deer. From beach to marsh to meadow to salt-sprayed sand plains, the variety of habitats is tremendous.

Note: Along with the poison ivy alert, check yourself for deer ticks after a day's walk or hike in the outdoors (☞ Health *in* the Gold Guide).

## Restaurants

What's new in Cape and Island cuisine? Influenced by arrivals from Boston and New York, a generation of young chefs—many of whom own their restaurants—is serving eclectic, inventive, cosmopolitan, upscale menus that begin with seafood but may end up with everything from lime and ginger to cilantro and balsamic vinegar. First-rate, sophisticated dining is now available in most towns, but particularly in Provincetown and on the islands. The old days of nothing but pot roast, baked cod, or chicken potpie are long gone.

Although the fish that gave the Cape its name is much harder to find now than in Pilgrim days, seafood remains the area's major culinary attraction. Flounder and hake are more often available than haddock and cod, but for freshness and preparation, no one does it better than local chefs. Bass has made a good comeback; natives much prefer it to salmon or even swordfish steaks. The very best New England clam chowder—made from large native clams called quahogs (pronounced *ko*-hawgs), potatoes, and cream—doesn't hide the taste with a flour thickener. Roadside stands still serve tasty fried clams or fish and chips, either of which, eaten at a picnic table overlooking the bay, may be more memorable than the fanciest meal in town.

For generations, Cotuit and Wellfleet have vied to produce the world's best-tasting oyster. That neither has emerged victorious over the other is to everyone's benefit; try a half-dozen of both and judge for yourself. Other specialties include scallops (at their very best at the end of the summer season), mussels, and of course the nation's favorite crustacean, the large-clawed North Atlantic lobster.

## Shopping

Art galleries and crafts shops abound on Cape Cod, Martha's Vineyard, and Nantucket, a reflection of the long attraction the area has held for artists and craftspeople. The region is also a popular antiquing spot—which means both the genuine article and kitschy wanna-bes. Coastal environments and a seafaring past account for the proliferation of sea-related crafts (as well as marine-antiques dealers) on the Cape and the islands. A craft form that originated on the years-long voyages to whaling grounds is scrimshaw, the etching of finely detailed designs of sailing ships and sea creatures onto a hard surface. In the beginning, the bones or teeth of whales were used; today's ecologically minded (and legally constrained) scrimshanders use a synthetic substitute like Corian, a DuPont countertop material.

Another whalers' pastime was the sailor's valentine: a glass-enclosed wood box, often in an octagonal shape (derived from the shape of the compass boxes that were originally used), containing an intricate mosaic of tiny seashells. The shells were collected on stopovers in the West Indies and elsewhere; sorted by color, size, and shape; and then glued into elaborate patterns during the long hours aboard ship. Exquisite examples can be found in Nantucket museums.

The Nantucket lightship basket was developed in the mid-19th century on a lightship off the island's coast. In good weather there was little to do, and so (the story goes) crew members began weaving intricately patterned baskets of cane, a trade some continued onshore and passed on. Later a woven lid and decoration were added, and the utilitarian baskets were on their way to becoming the handbags that today cost hundreds of dollars. Although Nantucket is still the locus of the craft, with a dozen active basket makers, antique and new baskets can be found on the Cape and the Vineyard as well.

Cranberry glass, a light ruby glass made by fusing gold with glass or crystal, is sold in gift shops all over the Cape. It was once made by the Sandwich Glass company (among others) but now retains only the association with Cape Cod, as it is made elsewhere in the United States and in Europe.

For some great vacation reading, there are also a number of highly browseable new and used-book stores on the Cape and islands.

## Sports and the Outdoors

The Cape and the islands are top spots for swimming, surfing, windsurfing, sailing, and virtually all other water sports. Shipwrecks make for interesting dive sites, but don't expect a tropical underwater landscape. Golfers have a number of excellent courses to choose from, including championship layouts, and most remain open nearly year-round. Bicycling is a joy on the mostly level roads, along paved and scenic bike paths and through the many nature preserves. Bird-watchers have an endless variety of habitats to choose from, often in a single nature preserve. Fishing, especially for bluefish and striped bass, brings many people to the area year after year, and there are some major derbies around. Remember, you will need a license (available at most tackle shops for a nominal fee) for freshwater fishing.

Spectators can choose from a plethora of bike and running races, golf competitions, horse shows, sailboat races, and the well-attended Cape Cod Baseball League games, breeding ground of champions (☞ "The Cape Cod Baseball League" Close-Up box *in* Chapter 2).

## Theater

In 1916 a young aspiring writer arrived in Provincetown to try his hand at writing plays. In July of that same year, Eugene O'Neill's first play, *Bound East for Cardiff*, made its debut in a waterfront fish house to tremendous success. More than 80 years later, theater continues to thrive on the Cape. The country's most renowned summer theater, the **Cape Playhouse** in Dennis—where Bette Davis got her start first as an usher, then as an actress (in 1928, in *Mr. Pim Passes By*)—tops the list for the best and brightest stage fare. The **Wellfleet Harbor Actors Theater** has staged many important New England and world premieres. The **Cape Cod Theatre Project** out of Woods Hole showcases readings of new works by leading playwrights. Then there's the newest kid on the creative block: **Provincetown Repertory Theatre.** Founded in 1995 by Ken Hoyt, PRT brought legendary director Jose Quintero out of retirement in the summer of 1996 to direct two one-act O'Neill plays. (The event was so anticipated that Kirstie Alley and Jason Robards traveled to the tip of the Cape for opening night.) In 1997 three-time Pulitzer Prize–winning playwright Edward Albee appeared at a benefit reading to help raise money for PRT's new theater on the grounds of the Pilgrim Monument and Provincetown Museum. Community theater abounds as well, most notably at the **Academy Playhouse** in Orleans, **Cape Repertory** in Brewster, and the **Harwich Junior Theatre.**

# FODOR'S CHOICE

## Old New England

⭐ **The town of Sandwich,** with its abundance of old houses, its cemetery, museums, salt marsh, and its lovely setting on Shawme Pond, is unsurpassed for regional charm.

⭐ **Bright crimson cranberries** floating on flooded bogs just before harvest on Cape Cod and Nantucket are a perfect reminder of the handwork that was one of the joys of the seasonal round-up.

⭐ **Hallet's century-old drugstore** in Yarmouth Port—how sweet life can be, especially when you're sipping an ice cream soda while swiveling on a stool at a marble bar!

⭐ **The Wednesday night community sing-along** (in July and August only) at the Oak Bluffs Camp Ground on Martha's Vineyard is great old-time fun, as much for the melody as for the sense of community you can't help but feel.

⭐ **The brick lighthouse** overlooking the dramatic cliffs at Gay Head on Martha's Vineyard quietly speaks, like a time-tested sentinel, of the elemental forces of wind and sea.

⭐ More than the past works of any other town in America, **the cobblestone streets and historic architecture of Nantucket** continue to glow with the dignity and charm of early American life.

## Great Views

⭐ **Sunset over Cape Cod Bay** on any bay beach from Eastham to Provincetown is an unforgettable delight at any time of year.

⭐ **Stargazing** from Nantucket's Loines Observatory is so astonishing because the island's isolation from mainland lights makes the sky virtually blaze with starlight.

⭐ **From the Provincetown Monument,** the panoramic view of the stretch of shoreline along the lower Cape.

## Natural Phenomena

⭐ If you crave a little rejuvenation, or just want to commune with nature, spend an afternoon observing life on the salt marsh at **Bass Hole Boardwalk** in Yarmouth Port or taking a walk onto the tidal flats when the water is out.

From Chatham Light, looking out at the **"Chatham Break"** in the sandbar is a reminder of the power of the sea and a fascinating display of the process of geological change.

**Seeing harbor seals off Race Point** in Provincetown in winter is one of the pleasures of the seaside Cape at a time when you feel like you have the place to yourself.

**Whales** breaching alongside your whale-watch boat will fill you with a sense of wonder unlike any you've ever felt—the creatures are simply marvelous. It's also quite a treat to see porpoises jumping in and out of the boat's bow waves or in its wake.

**The moors of Nantucket** in fall, next to cranberry bogs mid-harvest, turn colors that you can almost hear, if you listen closely enough.

## Conservation Areas

**Monomoy Wildlife Refuge,** two islands off Chatham, is a bird and birders' haven, particularly during spring and fall migrations.

**Massachusetts Audubon Wellfleet Bay Sanctuary** in South Wellfleet, with its numerous adult and kids' programs and its beautiful salt-marsh setting, is a favorite migration stop for Cape vacationers year-round.

**Felix Neck** on Martha's Vineyard is another Massachusetts Audubon Society jewel, its ponds, marshes, fields and woods teeming with plant and animal life.

**Wasque Reservation** on the southeastern tip of the Vineyard consists of miles of beaches. Fishing and birding are tops, and there isn't a better place for a long walk to take in the beauty of sand and sea.

At the **Coatue–Coskata–Great Point** reserves in Nantucket, you may see a marsh hawk or an oystercatcher floating in the air or coming and going from stands of cedar and oak, or you might just land a striped bass or bluefish if you brought along a surf-casting pole. Or you could just put your feet in the sand and walk out to the Great Point Light.

**Eel Point,** on the other end of Nantucket, blossoms at various times of year with wild roses, bayberry, heather, and goldenrod.

It is another great surf-casting and birding area, with small, sandbar islands close to shore serving as perches for shorebirds.

## Beaches

**Nauset Light, Coast Guard, and Race Point** beaches stretch across the length of the Cape Cod National Seashore. They are the truly classic Cape beaches.

**Lucy Vincent Beach** in Chilmark on Martha's Vineyard is quite simply the best beach on the island. It *is* restricted to resident use in summer, at which point you may want to venture out to **Gay Head** to take in the sights on the beach under the phenomenal cliffs and lighthouse.

**Eel Point** has one of Nantucket's great beaches, with shallow water and sandbars on the Sound side that are fun to explore.

## Shopping

**Sandwich Auction House** has been holding weekly auctions for 24 years, with periodic theme sales of toys, silver, paperweights, and collectibles.

**The weekly flea market** at the Wellfleet Drive-In Theater is one big browse, whether or not you take anything home with you.

**Herridge Books** in Wellfleet reflects the town's longtime appeal to vacationing writers, with an outstanding collection of interesting reading. A book lover could easily take a day or more to do justice to what's on the shelves in this store.

The **West Tisbury Farmers' Market** on Martha's Vineyard is the largest in Massachusetts, with about 50 vendors of local produce, homemade jam and preserves.

**Bunch of Grapes** in Vineyard Haven on the Vineyard contains an amazing array of books, books, and more books, including island-related titles. Don't be surprised to find yourself browsing the shelves next to the Hollywood celebs and assorted literati who shop here. It's a browser's dream come true.

**Rafael Osona auction house** in Nantucket Town is for buyers and spectators alike. Whether or not you buy, the auction can be great entertainment.

**Lightship baskets and scrimshaw** on Nantucket continue a crafts tradition unique to the area.

**Farm stands** everywhere provide one of the best ways to get close to the land and the rhythm of rural life.

## Dining

⋆ **Chillingsworth, Brewster, Cape Cod.** One of the Cape's finest restaurants, Chillingsworth continues to uphold the standards that set it apart. For a special night out, don't think twice; just make a reservation well ahead of time. $$$$

⋆ **Front Street, Provincetown, Cape Cod.** In a town that consistently continues to raise the level of the Cape's cuisine, this small, unassuming venue serves nothing short of superb, original food, offered by a marvelous staff. You couldn't ask for much more at Cape's end. $$$

⋆ **Sagamore Inn, Sandwich, Cape Cod.** Rustic, roomy, family run, not pretentious or expensive—this is the kind of place you might stop in for lunch on a rainy day, and then dinner the following night. $

⋆ **Mojo's, Provincetown, Cape Cod.** If every hole-in-the-wall was as inventive as this one, and cared as much about its menu, the world would be a cheaper, better place. $

⋆ **L'étoile, Edgartown, Martha's Vineyard.** Set in the incomparable Charlotte Inn, this longtime favorite is one of the finest restaurants outside of major cities on the east coast. $$$$

⋆ **Savoir Fare, Edgartown, Martha's Vineyard.** In its short rise to the top on the Vineyard, the menu devised by the uncle-nephew team here has been constantly inventive and delicious. $$$–$$$$

⋆ **The Sweet Life, Oak Bluffs, Martha's Vineyard.** Relaxed, gracious, young but not trendy, intimate but not cloying, the Sweet Life serves terrific new American food. $$$

⋆ **Larson's, Menemsha, Martha's Vineyard.** There is no seating here, but you may find that raw oysters, steamers, lobster, crab cakes, and chowder have never tasted better than they do on the dock outside of Larson's market. $

⋆ **American Seasons, Nantucket.** Fare from all four corners of the country, wine from the best of the American vineyards, together served in a romantic, artistic setting. $$$

## Lodging

⋆ **Captain's House Inn, Chatham, Cape Cod.** Friendly and professional service, colonial atmosphere, and wonderful old buildings make this one of the Cape's most pleasant B&Bs. $$$–$$$$

⋆ **Augustus Snow House, Harwich Port, Cape Cod.** This elegant Victorian inn takes you back to another era, with a tearoom where you can savor afternoon tea on weekends. $$–$$$

⋆ **Heaven on High, Barnstable, Cape Cod.** Perfectly named, perfectly decorated, and perfectly posh (without being pretentious), Heaven on High is such a blessed haven you might never want to leave to explore the town's nearby sights and sounds. $$

⋆ **Charlotte Inn, Edgartown, Martha's Vineyard.** Words are hard to come by when a dream is as real as the Charlotte Inn, an Edwardian period piece where friendliness and relaxed elegance speak volumes. Go ahead and treat yourself. $$$$

⋆ **The Inn at Blueberry Hill, Chilmark, Martha's Vineyard.** Rural, Up-Island escapes are a favorite reason to go to the Vineyard, and this laid-back yet stylish retreat puts you smack in the middle of a nature preserve threaded with walking trails. $$$$

⋆ **Summer House, Nantucket.** These rose-covered cottages in Siasconset offer the ultimate romantic getaway, with a stunning view. $$$$

⋆ **Wauwinet, Nantucket.** Very likely Nantucket's best-kept and best-equipped inn, with great lawn and immediate proximity to ocean and harbor beaches, historic Wauwinet is *the* all-out option on the island. $$$$

## Children's Activities

⋆ **Sandwich** has the **Thornton W. Burgess Museum,** namesake of the creator of Peter Cottontail and great for small children; **Heritage Plantation,** with superb old cars and grounds to run around on; and the **Green Briar Nature Center and Jam Kitchen,** with walking trails, "Smiling Pool" pond, natural-history exhibits, and Peter Cottontail's great-great-grandchildren.

⋆ **Pirate's Cove** in South Yarmouth far outdoes the average miniature golf course, architecturally speaking. Give in and pick up a putter at least once—you may have a good time even if your kids aren't alongside you.

★ **Whale watches** out of Provincetown can be a tremendous thrill—genuinely exciting and pleasantly educational.

★ **Flying Horses Carousel** on Martha's Vineyard isn't just for kids, you know. But we wouldn't recommended it for anyone who's too old to *feel* like a kid. Across the street, the **Game Room** has *the* authentic seaside arcade game, Skee-ball.

★ **Vineyard Playhouse** and the **Actors Theater of Nantucket** both have entertaining children's events in July and August.

# FESTIVALS AND SEASONAL EVENTS

The Massachusetts Office of Travel & Tourism (☞ Visitor Information *in* the Gold Guide) offers events listings and a whale-watch guide for the entire state. Also see the events calendar in *Cape-Week*, an arts and entertainment supplement published every Friday in the *Cape Cod Times*.

**WINTER**

➤ EARLY DEC.: Many Cape and island towns do up the Christmas season in grand style. The best-known celebration is the Nantucket **Christmas Stroll** (☎ 508/228–1700), which takes place the first Saturday of the month. Carolers and musicians entertain strollers as they walk the festive cobblestone streets and sample shops' wares and seasonal refreshments. Activities include theatrical performances, art exhibitions, crafts sales, and a tour of historic homes. To avoid the throngs, visit on one of the surrounding weekends (festivities begin the day after Thanksgiving and last through New Year's Eve). Various Cape towns also have strolls; call the Cape Cod Chamber of Commerce for a free brochure (☎ 508/362–3225).

On Martha's Vineyard, **Christmas in Edgartown** (☎ 508/693–0085), the second weekend of December, includes tours of historic homes, teas,

carriage rides, a parade, caroling, and other entertainment. Vineyard Haven packs a chowder contest, church bazaars, plays, and concerts into its **Come Home for the Holidays** festival (☎ 508/693–0085).

Falmouth's **Christmas by the Sea** (☎ 508/548–8500 or 800/526–8532), the first full weekend of December, includes lighting ceremonies at the Village Green, caroling at Nobska Light in Woods Hole, an antiques show, a house tour, church fairs, and a parade.

➤ DEC. 4–6, 12–13: Ashumet Holly and Wildlife Sanctuary, in East Falmouth, holds its annual **Holly Days** (☎ 508/563–6390), where you can pick up cut holly, wreaths, greens.

➤ DEC.: Chatham's **Main Street Open House** (☎ 508/945–0342) takes place the following weekend, with hayrides, caroling, a book-and-author tea, and more, culminating in a dinner dance at the grand Chatham Bars Inn. The open house is part of a monthlong celebration beginning just after Thanksgiving and ending with a lavish **First Night** celebration, including fireworks over Oyster Pond, on New Year's Eve.

**SPRING**

➤ LATE-APR.: Nantucket's four-day **Daffodil Festival**

(☎ 508/228–1700) celebrates spring with a flower show, shop-window displays, and a procession of antique cars adorned with daffodils that ends in tailgate picnics at Siasconset. For about five weeks from mid-April to mid-May, literally millions of daffodils bloom along Nantucket roadsides and in private gardens.

➤ MAY 9: As this year's spring production, **Opera New England of Cape Cod** (☎ 508/775–3974) performs Gounod's *Romeo and Juliet* at Sandwich High School (✉ Quaker Meetinghouse Rd.).

➤ MID-MAY: In Hyannis, **Cape Cod Maritime Week** (☎ 508/362–3828) celebrates Cape Cod's maritime history.

➤ MAY 15–17: The Green Briar Nature Center and Jam Kitchen in Sandwich holds its annual **Green Briar Herb Festival** (☎ 508/888–6870), where you can pick up perennials, wildflowers dug out of their garden, plus dozens of herb varieties.

**SUMMER**

➤ JUNE–AUG.: The Cape and islands are all busy with **summer theater, town band concerts,** and **arts-and-crafts fairs.**

➤ EARLY JUNE: **Hyannis Harbor Festival** (☎ 508/775–2201 or 800/449–6647) is a two-day celebration that includes the blessing of the fleet, boat races, entertainment, and food. **Cape Cod Antique Dealers Association Annual Antiques Show** at Sandwich's Heritage Plantation (☎ 508/888–3300) is attended by 50 dealers in fine 18th- and 19th-century English and American furniture, folk art, Sandwich glass, jewelry, paintings, and quilts.

➤ MID- TO LATE JUNE: **Cape Heritage '98** (☎ 508/362–3225) is a weeklong celebration of the Cape's history and culture hosted by museums, historical societies, theater groups, libraries, and the Cape Cod National Seashore. Falmouth welcomes the **Soundfest Chamber Music Festival** (☎ 508/548–2290), featuring the Colorado Quartet and guest artists. The all-instrumental program will include works by Beethoven, Mozart, and Haydn. In Edgartown's **A Taste of the Vineyard** (☎ 508/627–4440), ticket holders sample treats provided by local restaurants, caterers, and wine merchants. It's a formal evening, with dancing and an auction. **Nantucket Harborfest** (☎ 508/228–1700) is three days of boat and children's parades, a blessing of the fleet, a clambake, and culinary and watersports competitions.

**The Blessing of the Fleet** (☎ 508/487–3424) in Provincetown is the culmination of a weekend of festivities, including a banquet and dance. On the last Sunday of the month, a parade ends at the wharf, where fishermen and their families and friends pile onto their boats and form a proces-

sion. The bishop stands on the dock and blesses the boats with holy water as they pass by. **The Portuguese Festival** coincides with the Blessing of the Fleet, with traditional foods, dances, games, and events honoring the Portuguese heritage in Provincetown.

➤ JULY 4 WEEKEND: **The Mashpee Powwow** (☎ 508/477–0208) brings together Wampanoag Indians from North and South America for three days of dance contests, drumming, a fireball game, and a clambake, plus the crowning of the Mashpee Wampanoag Indian princess on the final night. **Fireworks displays** are still a part of July 4th celebrations in several Cape towns and on Nantucket.

➤ EARLY JULY: The Cape Cod Symphony Orchestra plays the first of its three **Sounds of Summer Pops Concert** (☎ 508/362–1111) on the Hyannis Village Green; the two others are in Orleans and Mashpee (☞ *below*).

➤ EARLY JULY–EARLY AUG.: **Martha's Vineyard Chamber Music Society** (☎ 508/696–8055) performs 10 concerts over five weekends in Edgartown and Chilmark.

➤ MID-JULY: **Edgartown Regatta** (☎ 508/627–4361), started in 1923, is three days of yacht racing around Martha's Vineyard. In East Sandwich, the **Cape Cod Antiquarian Book Fair** (☎ 508/888–2331), welcomes more than 70 dealers in old and rare books.

➤ JULY 25: The 90-member Cape Cod Symphony Orchestra sets up

at the Mashpee Commons for the second of its three annual **Sounds of Summer Pops Concert** (☎ 508/362–1111).

➤ LATE JULY: **The Barnstable County Fair** (✉ Rte. 151, E. Falmouth, ☎ 508/563–3200), begun in 1844, is Cape Cod's biggest event. The seven-day affair features livestock and food judging; horse, pony, and oxen pulls and shows; arts-and-crafts demonstrations; carnival rides; lots of food; and appearances by the formerly famous, bent on comebacks, such as Mary Wilson, Helen Reddy, and the Box Tops.

➤ AUG.: The **Cape & Islands Chamber Music Festival** (☎ 508/255–9509) is three weeks of top-caliber performances, including a jazz night, at various locations in August. On Nantucket, an **Annual House Tour** is held by the garden club, and a **Sandcastle Contest** at Jetties Beach results in some amazing sculptures (☎ 508/228–1700 for both events). On Martha's Vineyard, **fireworks** (☎ 508/693–0085) explode over the ocean while the town watches from the Oak Bluffs village green and the town band plays on the gazebo.

➤ EARLY AUG.: At the **Possible Dreams Auction** (☎ 508/693–7900), held the first Monday of August in Edgartown as a benefit for Martha's Vineyard Community Services, guest auctioneer Art Buchwald sells off such starry prizes as a sail with Walter Cronkite, lunch with Carly Simon, and a bike ride with John F. Kennedy Jr. The Otis Air Base holds a free two-day **open house and air show** (☎ 508/968–4090 or 508/968–4003) with exhibitions that may include precision flying by teams of the Air Force's Thunderbirds or the Navy's Blue Angels.

➤ MID-AUG.: The **Falmouth Road Race** (✉ Box 732, Falmouth 02541, ☎ 508/540–7000) is a world-class race covering 7.1 mi of coast from Woods Hole to Falmouth Heights; apply the previous fall or winter.

➤ AUG. 16: The Boston Pops Esplanade Orchestra wows the crowds with its annual **Pops By the Sea** concert (☎ 508/790–2787), held at the Hyannis Village Green at 5 PM. Each year, a guest conductor strikes up the band.

➤ AUG. 20–22: **Martha's Vineyard Agricultural Fair** (☎ 508/693–4343) is pure Americana, with livestock and food judging, animal shows, a carnival, plus evening musical entertainment.

➤ AUG. 22: The Cape Cod Symphony Orchestra comes to Eldridge Field Park in Orleans for its **Sounds of Summer Pops Concert** (☎ 508/362–1111).

➤ LATE AUG.: **Sails Around Cape Cod** (☎ 508/430–1165) is a two-day 155-nautical-mi race to Harwich Port and back via Provincetown and the canal. The Osterville Historical Society holds its annual **antique show** (☎ 508/428–5861).

AUTUMN

➤ SEPT. 4–6: The annual **New England Jazz Festival** (☎ 508/477–2580) takes place at the Boch Center for the Performing Arts in Mashpee.

➤ SEPT.: **The 29th Annual Bourne Scallop Fest** (☎ 508/759–6000), the weekend after Labor Day, attracts thousands of people to Buzzards Bay for three days of fried scallops.

➤ SEPT. 11–20: The **Harwich Cranberry Festival** (☎ 508/430–2811) features a country-western jamboree, an arts-and-crafts show, fireworks, pancake breakfasts, an antique-car show, and much more.

➤ MID-SEPT.: **Tivoli Day** (☎ 508/693–0085), an end-of-summer celebration in Oak Bluffs on Martha's Vineyard, features a street fair with live entertainment and crafts.

➤ MID-SEPT.–MID-OCT.: Locals go crazy over the monthlong **Martha's Vineyard Striped Bass and Bluefish Derby** (✉ Box 2101, Edgartown 02539, ☎ 508/693–0728), one of the East Coast's premier fishing contests (it raises money for the island's conservation projects), offering more than $100,000 in prizes for albacore, bluefish, striped bass, and bonito catches.

➤ LATE SEPT.: **Bird Carvers' Exhibit and Auction** (☎ 508/896–3867) at the Cape Cod

Museum of Natural History in Brewster showcases local and regional bird carvers' art over three days.

➤ OCT.: **Nantucket Cranberry Harvest Festival** (☎ 508/228–1700) is a three-day celebration, including bog tours, a cookery contest, and a crafts exhibition.

➤ MID-OCT.: **Seafest,** (☎ 508/945–0719) an annual tribute to the maritime industry, marks one of the few times that Chatham Light (☎ 508/945–0719) is open to the public.

➤ OCT. 25: For their annual fall production, **Opera New England of Cape Cod** (☎ 508/775–3974) performs Puccini's *Tosca,* at Sandwich High School (✉ Quaker Meetinghouse Rd.).

➤ THANKSGIVING EVE: **Provincetown Festival of Lights** (☎ 508/487–3424), commemorating the Pilgrims' landing, begins with the lighting of 5,000 white and gold bulbs draped over the Pilgrim Monument. The lights are lit nightly until Little Christmas (January 7). A performance of the "Hallelujah Chorus" accompanies the lighting, and the monument museum offers an open house and tours. Other events include a re-enactment of the first landing of the Pilgrims on Provincetown shores, dramatic readings of the Mayflower Compact (which was signed in Provincetown harbor), fireworks, and an abundance of Thanksgiving dinner celebrations.

# 2 Cape Cod

*Continually shaped by ocean currents, this windswept land of sandy beaches and dunes has an amazing natural beauty. Everyone comes for the seaside, yet crimson cranberry bogs, forests of birch and beech, and grassy meadows and marshlands gracing the interior are equally splendid. Local history is fascinating, whale-watching provides an exhilarating experience of the natural world, cycling trails lace the landscape, shops purvey everything from the useful to the kitsch, and you can dine on simple, as-fresh-as-it-comes seafood or delicious, thoughtfully prepared cuisine—or most anything in between.*

**A**PATTI PAGE SONG FROM THE 1950S PROMISES that "If you're fond of sand dunes and salty air, quaint little villages here and there, you're sure to fall in love with old Cape Cod." The tourism boom since the '50s certainly proved her right. So popular did the Cape become that today parts of it have lost the charm that brought everyone here in the first place. Although traditional associations of weathered, shingled cottages, long dune-backed beaches, and fog-enshrouded lighthouses still hold, the serenity of the landscape has in many places been crowded out by sub-urbanization—the usual housing developments, condominium complexes, and strip malls built to serve an expanding population. In summer, large crowds make you work for the tranquillity once met at every turn.

Updated by
Dorothy
Antczak and
Alan W.
Petrucelli

Dining
updated by
Seth Rolbein

Still, much of the Cape remains compellingly beautiful and unspoiled. Even at the height of the season, there won't be crowds at the less-traveled nature preserves and historic old villages. Off-season, many beaches are dream material for solitary walkers, and life returns to a small-town hum.

In 1961 the Cape Cod National Seashore was established to preserve virtually all of the eastern shoreline in its natural state for all time, and in 1990 the Cape Cod Commission was created to put a stop to the unchecked development of years past. For the sake of its economy, largely based on the area's continued appeal to tourists, and for the sake of preserving a landscape just as dear to most of the people who live here, Cape Cod has moved toward responsible management of its principal resources.

Separated from the Massachusetts mainland by the 17.4-mi Cape Cod Canal—at 480 ft, the world's widest sea-level canal—and linked to it by three bridges, the Cape is always likened in shape to an outstretched arm bent at the elbow, its Provincetown fist turned back toward the mainland. (The Cape "winds around to face itself," is how the writer Philip Hamburger has put it.) Cape Cod Bay rests within the arm's embrace. The open Atlantic lies to the east. And Nantucket Sound washes the southern shore. Being surrounded by all this water has its cost: Tides regularly eat away at the land, sometimes at an alarming rate. In *Cape Cod,* Henry David Thoreau described the Atlantic-side beach—which he walked from end to end on several trips in the mid-19th century—as "the edge of a continent wasting before the assaults of the ocean." Through the years, many lighthouses—some built hundreds of feet from water's edge—have fallen into the sea, and others are now in danger of being lost (Highland Light was recently moved inch by inch, and Nauset Light is now being similarly pushed back from its perch). Billingsgate Island off Wellfleet, which once held a number of cottages and a lighthouse, today is a bare sandbar occupied only by resting birds. And the once continuous Chatham bar was breached by storms in the winter of 1987. Sand eroded from one shore often ends up on another—Provincetown and Monomoy Island, for instance, continue to grow. Nevertheless, the U.S. Geological Survey estimates that "at some distant time—not for many generations, however—Cape Cod will be nothing more than a few low sandy islands surrounded by shoals."

The Cape's Atlantic coast is notorious for its shoals, shifting sandbanks that wrecked more than 3,000 ships in 300 years of recorded history. Nicknamed "the graveyard of ships," the area once had 13 lifesaving stations from Monomoy to Provincetown. They were manned by a crew of surfmen who drilled in lifesaving techniques during the day and took turns walking the beach at night in all kinds of weather, watching for ships in distress. Their motto: "You have to go out, but you don't have to come back." In the service's 43-year history, hundreds of victims were rescued, but only twice were surfmen's lives lost. Thoreau's stories of shipwrecks and those in Henry Beston's *The Outermost House* fascinate with their revelations of the awesome power of the sea, the tragedy of lives lost in icy waters, and the bravery of the men of the U.S. Life Saving Service. When the tide is low on the Cape, you can sometimes see the skeletons of wrecked ships emerge briefly, still caught in the treacherous sands where they lie buried.

Cape Cod is only about 70 mi from end to end, and you can make a cursory circuit of it in about two days. But it is really a place for relaxing—for swimming and sunning, for fishing, boating, and playing golf or tennis, for attending the theater, hunting for antiques, and making the rounds of art galleries, for buying lobster and fish fresh from the boat, or for taking leisurely walks, bike rides, or drives along pretty country roads that continue to hold out against modernity.

# Pleasures and Pastimes

## Beaches
Cape Cod has more than 150 ocean and freshwater beaches, with something for just about every taste. Bay-side beaches generally have cold water, carried down from Maine and Canada, and gentle waves. South-side beaches, on Nantucket Sound, have rolling surf and are warmed by the Gulf Stream. Ocean beaches on the Cape Cod National Seashore again have cold water, serious surf, and are by and large superior—wide, long, sandy, and dune-backed, with great views. They're also contiguous: you can walk from Eastham to Provincetown practically without leaving sand. This almost always ensures privacy, if you stroll far enough away from crowds. From late July through August, Wellfleet beaches are sometimes troubled with red algae in the water, which, while harmless, can be annoying; check with the Seashore on conditions. To mitigate the crowd factor, arrive either early in the morning or later in the afternoon—those long-light hours when the water is warmest. Parking lots tend to fill up by 10 AM.

In season, you will have to pay for access to public beaches, and restricted beaches are open to residents and visitors with permits. The latest official take on seasonal boundaries: between the last June weekend before July 1 and Labor Day. Plan accordingly.

## Bicycling
Biking on the Cape will satisfy avid and occasional cyclists alike. There are plenty of flat back roads, as well as a number of well-developed and scenic bike trails. The Cape's premier bike path, the **Cape Cod Rail Trail**, offers a scenic ride through the area. Following the paved right-of-way of the old Penn Central Railroad, it is about 30 mi long, stretching from South Dennis to South Wellfleet, with an extension underway. The Rail Trail passes salt marshes, cranberry bogs, ponds, and Nickerson State Park, which has its own path. Along the way you can veer off to spend an hour or two on the beach, or stop for lunch. The terrain is easy to moderate. The Cape Cod National Seashore maintains three bicycle trails. **Head of the Meadow Trail** is 2 mi of easy cycling between dunes and salt marshes from High Head Road, off Route 6A in North Truro, to the Head of the Meadow Beach parking lot. **Province Lands Trail** is a 5¼-mi loop off the Beech Forest parking lot on Race Point Road in Provincetown, with spurs to Herring Cove and Race Point beaches and to Bennett Pond. The paths wind up and down hills amid dunes, marshes, woods, and ponds and offer spectacular views—on

# Cape Cod *(Boxes Refer to Detail Maps)*

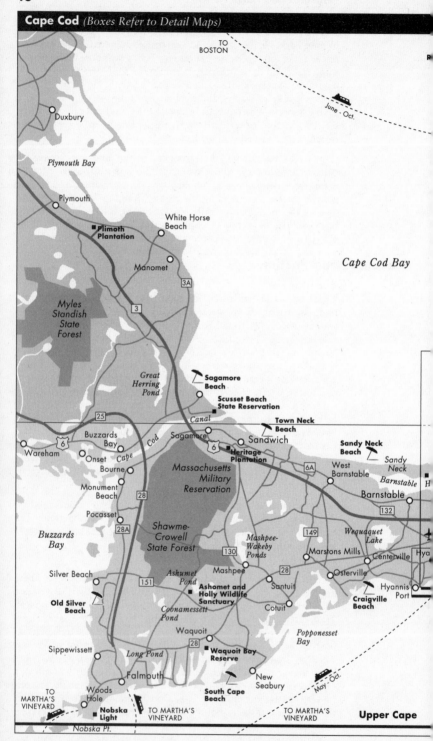

TO
BOSTON

June - Oct.

Duxbury

*Plymouth Bay*

Plymouth

White Horse
Beach

■ **Plimoth
Plantation**

Manomet

3A

*Cape Cod Bay*

3

*Myles
Standish
State
Forest*

*Great
Herring
Pond*

▲ **Sagamore
Beach**

**Scusset Beach
State Reservation**

25

*Canal*

**Town Neck
Beach**

Buzzards
Bay

*Cod*

Sagamore

Sandwich

**Sandy Neck
Beach**

6

Wareham

Onset

*Cape*

Bourne

6

■ **Heritage
Plantation**

*Sandy
Neck*

West
Barnstable

6A

*Barnstable*

H

Monument
Beach

*Massachusetts
Military
Reservation*

28

Barnstable

132

Pocasset

28A

*Shawme-
Crowell
State Forest*

*Wequaquet
Lake*

149

*Buzzards
Bay*

130

*Mashpee-
Wakeby
Ponds*

Marstons Mills

Centerville

Hya

Silver Beach

151

*Ashumet
Pond*

Mashpee

28

Osterville

**Old Silver
Beach**

■ **Ashomet and
Holly Wildlife
Sanctuary**

Santuit

Cotuit

Hyannis
Port

▲ **Craigville
Beach**

*Coonamessett
Pond*

Sippewissett

Waquoit

28

*Long Pond*

■ **Waquoit Bay
Reserve**

*Popponesset
Bay*

Falmouth

New
Seabury

May - Oct.

TO
MARTHA'S
VINEYARD

Woods
Hole

■ **Nobska
Light**

TO MARTHA'S
VINEYARD

**South Cape
Beach**

TO MARTHA'S
VINEYARD

**Upper Cape**

*Nobska Pt.*

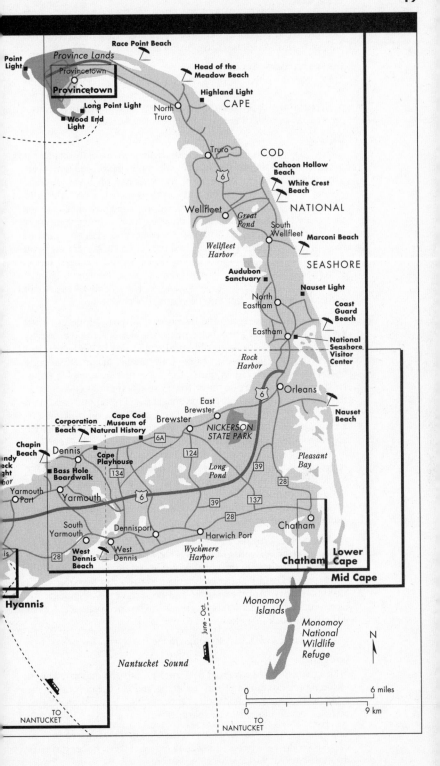

Point
Light ■

*Province Lands*

Race Point Beach ☂

Head of the
Meadow Beach ☂

Provincetown

**Provincetown**

Long Point Light ■

■ Wood End
Light

North
Truro

Highland Light ■

**CAPE**

Truro

6

**COD**

Cahoon Hollow
Beach ☂

White Crest
Beach ☂

Wellfleet

*Great
Pond*

**NATIONAL**

South
Wellfleet

Marconi Beach ☂

*Wellfleet
Harbor*

Audubon
Sanctuary ■

**SEASHORE**

North
Eastham

Nauset Light ■

Coast
Guard
Beach ☂

Eastham

National
Seashore
Visitor
Center

*Rock
Harbor*

6

Orleans ☂

Nauset
Beach

East
Brewster

Corporation
Beach ☂

Cape Cod
Museum of
Natural History ■

Brewster

**NICKERSON
STATE PARK**

Chapin
Beach

Dennis

6A

*Long
Pond*

*Pleasant
Bay*

Cape
Playhouse ■

124

39

■ Bass Hole
Boardwalk

andy
eck
ght
bor

134

Yarmouth
Port

**Yarmouth**

6

28

South
Yarmouth

Dennisport

39

137

28

Chatham

is

28

West
Dennis
Beach ☂

West
Dennis

Harwich Port

*Wychmere
Harbor*

**Lower
Cape**

**Chatham**

**Mid Cape**

**Hyannis**

June - Oct.

*Monomoy
Islands*

*Monomoy
National
Wildlife
Refuge*

N

*Nantucket Sound*

0 _____ 6 miles
0 _____ 9 km

TO
NANTUCKET

TO
NANTUCKET

an exceptionally clear day, you can see the Boston skyline. There's a picnic grove at Pilgrim Spring. Nickerson State Park in Brewster has 8 mi of trails through forest (trail map available).

## Conservation Areas

The Cape is crowded with conservation areas; a walk, cruise, or canoe trip through one of the many protected beaches, ponds, or salt marshes is a delightful escape, and there's no better way to experience all of the natural wonder offered by this unique landscape.

## Dining

Cape Cod is rightly famous for its seafood, and until recently that meant eating cheaply. Fish and shellfish are still in abundance, and until over-fishing and pollution drove so many fishermen off their grounds, catches were so great that the fruits of the sea were a verifiable bar-gain. Colonial jailers chafed at laws that restricted how much lobster they could feed their prisoners, all the while looking down their noses at what they considered a trash fish with an exoskeleton unworthy of their own labors. Three hundred and fifty years later, there's a—you name it—Lobster Pot-Trap-Claw-Bowl-Net every few hundred yards—even though you don't eat lobster from a bowl and only the cleverest fisherman could catch a lobster with a net. These restaurants are still authentic, but they're not the cheap eats they used to be.

Raw shellfish on the rocks and finfish served baked, broiled, or fried are typical of the Cape's seafood offerings. The local shellfish is invariably fresh, kept that way by both public health officials and local pride. Up-scale vacationers have brought a requisite number of correspondingly upscale restaurants; Italian dishes and Asian touches are most popu-lar right now. But you don't need to be a stodgy traditionalist to agree that the fry-o-lators in many home-style seafood restaurants and clam shacks impart the real lusciousness to classic Cape Cod cooking. For a truly authentic summertime Cape experience, nothing beats simple standbys. The quintessential summer lunch is a lobster roll—very light lobster salad with practically no mayonnaise on a plain white frank-furter roll—a bag of Cape Cod–brand potato chips, and a soda. One of the best meals you may ever have on the Cape might just consist of nothing more than a "Catch of the Day" dinner and a drive to a sun-set beach: a plate of steamers or mussels, a six-pack of Sam Adams, and a special someone.

At dinner time, dress is largely casual. One older Cape Codder takes this edict so seriously that he chops off any ties he sees, and adorns his cabin in Chatham with the remains. But in certain rooms—like Chill-ingsworth or Dan'l Webster or Chatham Bars—a sports coat is virtu-ally mandatory. Another recent and important development: An increasing number of restaurants are remaining open year-round, even if they scale back in winter.

For price ranges *see* Chart 1 (A) *in* On the Road with Fodor's.

## Lodging

With a tourism-based economy, the Cape naturally abounds in lodgings, including self-contained luxury resorts, grand old oceanfront hotels, chain hotels, mom-and-pop motels, antiques-filled bed-and-breakfasts, cottages, condominiums, and apartments.

Choosing where to stay depends on the kind of vacation you have in mind. If you love the beach, think about whether you'd rather stay near the dune-backed National Seashore, where waters are coldest, or near the warmer south shore waters. The Seashore is less developed and great for walking, where Mid Cape and Falmouth beaches are generally, but not always, more circumscribed and crowded with families.

Sandwich and other towns along the north-shore Route 6A historic district have quiet, traditional villages with old-Cape atmosphere and charming B&Bs with gardens. If you want more action head for the Mid Cape. Hyannis is the center of it all, with a busy Main Street, active nightlife, and some fine warm-water beaches.

For the austere Cape of dunes and sea, try the beach cottages of the sparsely developed Lower Cape between Wellfleet and Provincetown. P-town itself is something completely different, in summer a frantic wall-to-wall jumble of shops and houses bursting with a large contingent of lesbians and gay men, and a hopping nighttime scene. Staying in town makes getting to everything by foot or bike possible.

For price ranges *see* Chart 2 *in* On the Road with Fodor's.

## Nightlife and the Arts

Since before the turn of the century, creative people have been drawn to Cape Cod summers, and their legacy and ongoing contribution is a thriving arts scene. In addition to professional theater, which offers top-name talent in season, almost every town has a community theater that provides quality entertainment—often mixing local players with visiting pros—throughout the year (☞ Pleasures and Pastimes *in* Chapter 1). The Cape also gets its share of music stars, from pop to classical, along with local groups ranging from barbershop quartets to Bach chorales to early music or chamber ensembles, often playing at school auditoriums or town halls.

Provincetown has a long history as an art colony and remains an important center. Its many galleries exhibit Cape and non-Cape artists, and the Fine Arts Work Center has launched the careers of many well-known award-winning authors. Wellfleet and Orleans have emerged as vibrant art centers. Both towns have attracted craftspeople who sell through a number of unique and sophisticated shops.

Nighttime on Cape Cod can be very special in many ways. In less developed areas, the stars are amazingly bright and make beach walks even more wondrous. Lighthouse beacons cutting shafts through the night sky have a fascination impossible to resist. If you're up *really* late, or really early, head for Chatham Light or a Seashore beach to catch a terrific sunrise.

Another Cape twist on nightlife: Many daytime activities, such as fishing, take on a completely different aspect at night. Scuba enthusiasts might consider night diving: Colors are more vivid by flashlight, much sea life is phosphorescent or bioluminescent, and nocturnal species come out to play. It's important to know the tides and safe locations—check with a dive shop.

## Outdoor Activities and Sports

On the water, canoeing and sea-kayaking are great around the bay's marshy inlets. Fishing on the Cape is both a profession and a pastime. Hundreds of freshwater ponds are good for perch, pickerel, and trout; surf casting off beaches and deep-sea fishing for blues, bass, and flounder is also popular. You'll need a license for freshwater fishing, available for a nominal fee at most tackle and bait shops. Bird-watchers have a variety of habitats to choose from, and the Cape is a great place to come for spring and fall migrants. The Cape is one of the world's finest whale-watching spots; a whale breaching alongside your boat is truly an impressive sight. *See* Outdoor Activities and Sports *in* Provincetown, *below.*

The Cape Cod National Seashore has nine walking trails through varied terrain. In winter, ponds and shallow flooded cranberry bogs sometimes freeze hard enough for skating. Check conditions with the local fire department before venturing onto unfamiliar territory.

## Shopping

Like it or not, shopping is an important part of a Cape Cod vacation. On rainy days, enclosed malls and factory outlets are mobbed. Favorite pastimes include antiquing, bookstore browsing, and gallery hopping. Throughout the Cape you'll find weavers, candle makers, glassblowers, papermakers, and potters, as well as artists working in metal, enamel and wood. You'll also find an inordinate number of shops specializing in country crafts, from straw dolls to handmade Christmas-tree ornaments. And you'll run across plenty of junk that'll make you wish conspicuous consumption would go the way of the break in the Chatham bar.

You'll find an auction going on somewhere on the Cape all year long from country-barn types to the internationally known Eldred's auctions. Though the high-end auctions deal in very fine antiques, they also include some lower-priced merchandise, and often yield interesting Cape pieces, such as old sea chests, at good prices.

## Town Band Concerts

Traditional New England town band concerts are held weekly each summer in many Cape towns—take along chairs, blankets, sweaters, and a picnic supper if you like, and go early to get a good spot.

# Exploring Cape Cod

There are essentially two ways to visit the Cape: Either station yourself in one town and take excursions from there, or stay in a string of towns as you make your way from one end to the other. Neither is right or wrong, it all depends on your idea of a good trip.

## Cape Geography, Highways, and Byways

The Cape consists of 15 towns, each broken up into villages, which is where things can get complicated. The town of Barnstable, for example, consists of Barnstable, West Barnstable, Cotuit, Marstons Mills, Osterville, Centerville, and Hyannis. Likewise the terms Upper Cape and Lower Cape may be confusing. They derive from sailing days and latitudes increasing as one sails south. **Upper Cape**—think upper arm, as in the shape of the Cape—refers to the towns of Bourne, Falmouth, Mashpee, and Sandwich. **Mid Cape** includes Barnstable, Yarmouth, and Dennis. **Lower Cape** covers Brewster, Harwich, Chatham, Orleans, Eastham, Wellfleet, Truro, and Provincetown. **Outer Cape,** as in outer reaches, is essentially synonymous with Lower Cape. Technically it includes only Wellfleet, Truro, and Provincetown.

There are three major roads on the Cape. Route 6 is the fastest way to get from the mainland to Orleans. Route 6A winds along the north shore through scenic towns, while Route 28 dips south through the overdeveloped parts of the Cape. Generally speaking, if you want to avoid malls, heavy traffic, and tacky motels, avoid Route 28 from Falmouth to Chatham. Past Orleans on the way out to Provincetown, the roadside clutter of much of Route 6 belies the beauty of what surrounds it.

*Numbers in the text correspond to numbers in the margin and on the Upper Cape, Mid Cape, Hyannis, Lower Cape, Chatham, and Provincetown maps.*

## Great Itineraries

The itineraries below are divided geographically to present the best of the Upper, Mid, and Lower Cape section by section. If you have three days, decide on one area to explore and divide your days into museum or historical outings and outdoor activities—with beaches and lakes all around you'll probably want to spend half of each day in the fresh air. In five days, it would be reasonable to see two parts of the Cape or the length of it if you don't get the car crazies. You could otherwise take all of your time—two weeks, two months—relaxing and getting to know one area. There isn't a better place than the Cape to just unwind—period.

### THE UPPER CAPE

It's astonishing that so close to the mainland the lovely old town of **Sandwich** ① is one of the best examples of the Cape of yesteryear. Visit the Hoxie House, the Sandwich Glass Museum, or **Heritage Plantation**'s ② beautiful grounds and old car collection, have a picnic lunch on the banks of Shawme Pond, or spend the afternoon at the old-timey Green Briar Nature Center and Jam Kitchen walking the nature trails and picking out homemade jam to take with you. Here and all along the bay shore, you can explore salt marshes full of sea and bird life. If history's your thing, take a step back and start your Upper Cape trip at **Plimoth Plantation,** about 22 mi northwest of the Sagamore Bridge on Route 3A. The reconstructed settlement, where actors dress in period clothing and carry out the activities typical of earlier days, relates a very tangible sense of the Cape's history.

To reach the Upper Cape's south shore, take Route 28A south through some lovely little towns with detours to beautiful white-sand beaches. This is the way to **Falmouth** and to **Woods Hole** ⑥, the center for international marine research and the year-round ferry port for Martha's Vineyard. An aquarium in town has regional sea life exhibits, and there are several shops and museums. On the way out of town, the view from **Nobska Light** is breathtaking. In ⊞ **Falmouth** ④, stroll around the village green, look into some of the historic houses, and stop at the Waquoit Bay National Estuarine Research Reserve for a walk along the estuary and barrier beach. Inland, on the eastern extent of the Upper Cape, the Wampanoag Indian township of **Mashpee** ⑧ is one of the best places to learn about the Cape's Native American heritage.

### THE MID CAPE

The crowded belly of the Cape is a center of activity unlike any other here, excepting perhaps Provincetown at the height of its summer crush. **Hyannis** ⑬ is the hub of Cape Cod, with plenty to do and see. Take a cruise around the harbor, or go on a deep-sea fishing trip. There are shops and restaurants along Main Street, and plenty of amusements for the kids. Kennedy fans will want to spend time at the JFK Museum. End the day with a concert at the Cape Cod Melody Tent, or have dinner aboard the Cape Cod Dinner Train. **Barnstable** ⑩, the county seat, has plenty of its own history and the wonderful Sandy Neck Beach to keep you occupied. Scenic Route 6A passes through **Yarmouth Port** ⑱ and **Dennis** ㉔. There are beaches and salt marshes, museums, walking trails, and old graveyards all along this route if you feel like stopping. In Dennis there are historic houses to tour, the Cape Museum of Fine Art will introduce you to the work of Cape-associated artists, the Cape Cod Rail Trail provides a premier path for bicyclists, and the **Bass Hole Boardwalk** ⑳ makes for a beautiful stroll. Toward the end of the day, head for Scargo Hill and climb 30-ft **Scargo Tower** ㉕ stone tower to watch the sun set. At night you could catch a film at the Cape Cinema, on the grounds of the Cape Playhouse. As for the south shore, there are plenty of activities for kids, from **Centerville** ⑫ to **South Dennis.**

### THE LOWER CAPE

The Lower Cape stretches from Brewster and Harwich out to Provincetown. **Brewster** ㉙ has something for everyone—antiques shops, museums, an old gristmill open to the public, Bassett's Wild Animal Farm for kids, freshwater ponds for swimming or fishing, the beach and tidal flats to explore when the water is low, and miles of biking and hiking trails through **Nickerson State Park** ㉜. Don't miss the **Cape Cod Museum of Natural History** ㉛, which will take a couple of hours to explore. Main Street in the handsome town of **Chatham** ㉞ is perfect for strolling, shopping, and dining. A trip to the **Monomoy Islands** is a must for bird-watchers. Back in town, you can watch glassblowing in process at the Chatham Glass Company, visit the Old Atwood House and Railroad Museums, and drive over to take in the view from Chatham Light.

On the way north from Chatham, take the back road to **Orleans** ㊶, stopping for the view at Rock Harbor. You might want to allow time for a good long bike ride on the Cape Cod Rail Trail, or for an afternoon relaxing at Coast Guard Beach. **Eastham** ㊷ is the next stop on the way up the arm, where the **Ft. Hill Area** ㊸ has the historic Penniman House Museum and some wonderful walks along the adjacent trails. Stop at the National Seashore's **Salt Pond Visitor Center** ㊹ for some interesting info about the area, then take the bike trails to Coast Guard Beach and Nauset Light, with a view of the Three Sisters Lighthouses, now settled in a small park area.

**Wellfleet** ㊺ follows, with pleasant shops along Main Street, dune-backed ocean beaches, and marshes great for canoeing and kayaking. Historic **Marconi Station** ㊼ was the landing point for the transatlantic telegraph early this century, and its White Cedar Swamp trail is quite beautiful. In Wellfleet, visit galleries, browse through great bookstores, take a dip in the bay or the ocean, rent a canoe or kayak and tour the marshes, and have dinner at one of the fine area restaurants. Don't leave without visiting the **Massachusetts Audubon Wellfleet Bay Sanctuary** ㊻, especially if you are a bird-watcher. It's a fantastic place to catch a sunset over the bay. **Truro** ㊽ solidly represents the quietude of this end of the Cape, at least until you reach the seasonally bustling **Provincetown** ㊼ out at the tip. Catch a whale-watch boat from P-town—the Cape ranks fourth in the world for sighting whales, which is an incredible experience. Take a trolley tour in town or bike through the National Seashore on its miles of trails. Climb the **Pilgrim Monument** ㊾ for a spectacular view of the area. Visit the museums and shops and art galleries, or spend the afternoon swimming and sunning on one of the beaches. To escape the crowds, walk across the breakwater to Long Point. Rent a sailboat, or take a parasail ride over the bay. Then choose from an abundance of restaurants and the Cape's wildest nightlife.

## When to Tour Cape Cod

Summer is the most popular season on Cape Cod—from mid-June through the end of August, everything is open and happening and busy. These are the months for maritime history celebrations, Fourth of July fireworks, county fairs, and all manner of summer-only activities, like summer theater and town band concerts. Fall has come into its own on the Cape in recent years, and extends the "season" with Halloween and Thanksgiving events, and all sorts of harvest celebrations. Winter has its own charm, and Christmas strolls and First Nights also make for good getaways (☞ "The Cape's Six Seasons" Close-Up box, *below*). In spring, everything is in bloom, and there are open houses and garden tours.

# THE UPPER CAPE

Especially considering the difference between towns on its north and south shores, the Upper Cape contains enough diversity to make generalizations about its character close to impossible. It just happens to be the closest part of the Cape to the mainland. There are four major towns. **Sandwich** on the north shore is the Cape's oldest town. Centered inland, **Mashpee** is a long-standing Native American township whose Indian-owned land is governed by local Wampanoags. **Falmouth,** the Cape's second most-populous town, is still green and historic, if seemingly overrun with strip malls. **Woods Hole,** one port from which to ferry over to Martha's Vineyard, is an international center of marine and biological scientific research. Along the west coast you'll find wooded areas ending in secluded coves, and on the south coast, long-established seaside resort communities.

The towns are listed as you would approach them, first from the Sagamore Bridge along Route 6A to Sandwich, then from the Bourne Bridge down Route 28 to Falmouth and Woods Hole.

## Sagamore to Sandwich

A pleasant place for a walk or a bike ride near the Sagamore bridge, the **Scusset Beach State Reservation** encompasses 490 acres near the canal, with a cold-water beach on the bay. Its pier is a popular fishing spot; other activities include biking, hiking, picnicking, and tent and RV camping on its 103 sites (some wooded), three of which are accessible to people with disabilities. There are cold showers on the beach and hot showers in the campground. ⊠ *140 Scusset Beach Rd., off Rte. 3 at the Sagamore Bridge rotary, Sandwich 02532,* ☎ *508/ 888–0859.* ▨ *$2 parking fee.* ⊙ *Daily 8–8.*

At **Pairpoint Crystal,** you can watch richly colored lead crystal being handblown in the factory, as it has been since 1837. The shop sells candlesticks, vases, stemware, sun catchers, ornamental cup plates (used to hold the cup in the days when tea was poured into saucers to drink), and reproductions of original Boston and Sandwich glass pieces. ⊠ *851 Rte. 6A, Sagamore,* ☎ *508/888–2344 or 800/899–0953.* ▨ *Free.* ⊙ *Mon.–Sat. 9–6, Sun. 10–6; demonstrations weekdays 9–4:30.*

### Dining

**$$** ✕ **The Bridge.** Aptly named, as the restaurant stands virtually in the shadow of the Sagamore Bridge, this spot bills itself as "New England and Italian dining." For years, its very plain, traditional menu was in the hands of talented owner Helen Trout, who would boast "nobody can cook a bottom round like Helen." Today, however, the menu no longer really rests on solid culinary footing. The heaping eggplant, chicken, or veal parmigiana for lunch is fine, but the finesse needed to pull off a Catalan alternative to bouillabaisse for dinner is not quite there. ⊠ *Rte. 6A, Sagamore,* ☎ *508/888–8144. D, DC, MC, V.*

**$$** ✕ **Sagamore Inn.** This local favorite—owned by the Pagliarani fam-
★ ily since 1963—keeps old Cape Cod traditions alive. Pressed-tin walls and ceilings, fans, old wooden booths and tables, lace curtains, and white linen add up to a dining room so casual you almost overlook its elegance. All the food here is homemade (except the stuffed quahogs), very good, and traditional at the same time. The seafood platter is a knockout, the chicken potpie substantial, and the prime rib huge. Luncheon specials attract regulars from all over town, and the service is friendly as family. A cozy, handsome old bar is under the same roof. ⊠ *1131 Rte. 6A, Sagamore,* ☎ *508/888–9707. Reservations not accepted. AE, MC, V. Closed Tues. and Dec.–Mar.*

# Sandwich

★ ❶   *3 mi east of the Sagamore Bridge, 11 mi west of Barnstable.*

The oldest town on Cape Cod, Sandwich was established in 1637 by some of the Plymouth Pilgrims and incorporated on March 6, 1638. Today it is a well-preserved, quintessential New England village that wears its history proudly. Driving through town past the white-columned town hall, the gristmill on Shawme Pond, the spindlelike spire of the First Church of Christ, and the 18th- and 19th-century homes that line the streets is like driving back in time—you'll feel like you should be holding a horse's reins rather than the steering wheel of a car. When you get to Main Street, park the car and get out for a stroll. Start on Main Street, look at old houses, stop at a museum or two, and work your way to the delightful Shawme Pond. While you walk, look for etched Sandwich glass from the old factory on front doors—there probably aren't two identical glass panels in town.

Unlike other Cape towns, whose deepwater ports opened the doors to prosperity in the whaling days, Sandwich was an industrial town for much of the 19th century. The main industry was the production of vividly colored glass, called Sandwich glass, which is now sought by collectors. The glass was made in the Boston and Sandwich Glass Company's factory here from 1825 until 1888, when competition with glassmakers in the Midwest—and finally a union strike—closed it. The **Sandwich Glass Museum** contains relics of the town's early history, a diorama showing how the factory looked in its heyday, and an outstanding collection of blown and pressed glass. Glassmaking demonstrations are held in the summer. ✉ *129 Main St.,* ☎ *508/888–0251.* 🎟 *$3.50.* ☉ *Apr.–Oct., daily 9:30–4:30; Nov.–Dec. and Feb.–Mar., Wed.–Sun. 9:30–4.*

A lovely place for a stroll or a picnic is the park around **Shawme Pond,** where children seem always to be fishing and the ducks and swans love to be fed, though posted signs warn you not to indulge them. Across the way—it's a perfect backdrop for this setting—is the spired, white, 1848 **First Church of Christ,** inspired by a design by London architect Christopher Wren. ✉ *Water and Grove Sts.*

Where the pond drains over its dam, a little wooden bridge leads over a watercourse to the waterwheel-powered **Dexter Gristmill,** built in 1654. In season, the miller demonstrates and talks about the mill's operation and sells the ground corn. ☎ *508/888–4910.* 🎟 *$1.50; combination ticket with Hoxie House (☞ below), $2.50.* ☉ *July–Labor Day, Mon.–Sat. 10–4:45, Sun. 1–4:45; Memorial Day–June and Labor Day–mid-Oct., Tues.–Sat. 10–4:45.*

The **Old Town Cemetery,** along Shawme Pond, is a classic, undulating, New England graveyard. Stop in for a peaceful moment and trace the genealogy of old Sandwich. ✉ *Grove St.*

---

NEED A
BREAK?

Stop in at the delightful **Dunbar Tea Shop** for a delicious treat either at tables outside by a shady hillside or indoors in a former billiards room/carriage house, now converted into a country cottage with paneled walls, a beamed ceiling, and assorted antiques and kitschy doodads. Lunch—everything from a smoked fish platter with salad (fisherman's lunch) to quiche or a salmon tart—English cream tea, and tasty sweets are served from 11 to 4:30 daily, year-round. There's also a gift shop with British tea, specialty foods, and home decorating items. ✉ *1 Water St.,* ☎ *508/833-2485.*

---

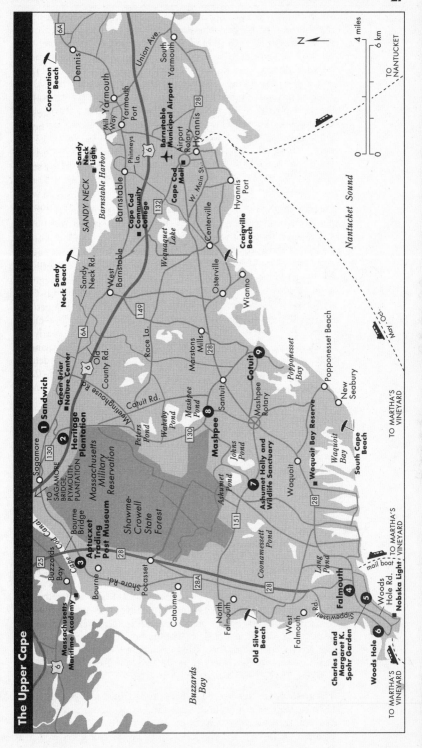

# The Upper Cape

TO NANTUCKET

*Nantucket Sound*

TO NANTUCKET

N

4 miles

6 km

Corporation Beach

Dennis

6A

Union Ave

South Yarmouth

Mill Yarmouth Way

Yarmouth Port

6

28

Barnstable Municipal Airport

Hyannis

Phinneys La.

Airport Rotary

Sandy Neck Light

SANDY NECK

Barnstable Harbor

Barnstable

Cape Cod Mall

W. Main St.

Hyannis Port

Cape Cod Community College

132

Wequaquet Lake

Centerville

Craigville Beach

Sandy Neck Beach

Sandy Neck Rd.

West Barnstable

Osterville

Wianno

6A

149

Race La.

Marstons Mills

28

Poponesset Beach

Poponesset Bay

**9** Cotuit

New Seabury

Meetinghouse Rd.

Old County Rd.

6

**1** Sandwich

Sagamore

130

**2**

Heritage Plantation

Green Briar Nature Center

Catuit Rd.

Wakeby Pond

Mashpee Pond

Santuit

Mashpee Rotary

**8** Mashpee

Massachusetts Military Reservation

Peters Pond

130

Waquoit Bay Reserve

Waquoit Bay

South Cape Beach

TO MARTHA'S VINEYARD

TO SAGAMORE BRIDGE PLYMOUTH PLANTATION

Buzzards Bay

Bourne Bridge

25

**3**

Aptucxet Trading Post Museum

28

Shawme-Crowell State Forest

Johns Pond

Ashumet Pond

**7** Ashumet Holly and Wildlife Sanctuary

Waquoit

28

Massachusetts Maritime Academy

6

Cape Cod Canal

Bourne

Pocasset

Shore Rd.

Cataumet

28A

Coonamessett Pond

151

Long Pond

**4** Falmouth

Siders Pond

TO MARTHA'S VINEYARD

mail boat

Woods Hole Rd.

**5** Nobska Light

Buzzards Bay

North Falmouth

West Falmouth

Old Silver Beach

Charles D. and Margaret K. Spohr Garden

**6** Woods Hole

TO MARTHA'S VINEYARD

Small children and nostalgic adults will enjoy a visit with their pals at
the **Thornton W. Burgess Museum,** dedicated to the Sandwich native
whose tales of Peter Cottontail, Reddy Fox, and a host of other creatures of the Old Briar Patch have been part of children's bedtimes for
decades. Burgess, an avid conservationist, made his characters behave
true to their species to educate children as he entertained them. A storytelling session, featuring the live animal that the Burgess story is about,
is held regularly in July and August. The many displays crowded into
the small old house include some of Burgess's 170 books. The small
gift shop carries puppets, Burgess books, and Pairpoint Crystal cup plates
decorated with Burgess characters. Out back, benches overlook Shawme
Pond and the ducks and swans bathing along its shore. ⊠ *4 Water St.,*
☎ *508/888–4668.* ⌨ *$2 suggested donation.* ◷ *Mon.–Sat. 10–4, Sun.
1–4; closed Sun. and Mon. Jan.–Mar.*

Overlooking Shawme Pond is the **Hoxie House,** a truly old saltbox remarkable in that it has been virtually unaltered since it was built in
1675—even though it was lived in until the 1950s, it was never mod-
ernized with electricity or plumbing. It has been furnished to reflect
daily life in the Colonial period, with some pieces on loan from the
Museum of Fine Arts in Boston. Highlights are diamond-shaped lead-
glass windows and a collection of antique textile machines. ⊠ *18
Water St.,* ☎ *508/888–1173.* ⌨ *$1.50; combination ticket with Dex-
ter Gristmill (☞ above), $2.50.* ◷ *Mid-June–mid-Oct., Mon.–Sat. 10–
5, Sun. 1–5.*

The **Yesteryears Doll Museum** is housed in the 1833 First Parish Meetinghouse. Its enormous collection includes antique German, French,
and Chinese dolls in bisque, china, wax, and other media; Henry VIII
and his wives, elegantly clothed in velvets and brocades; samurai war-
riors; and Balinese shadow puppets. The museum also has some wonderful miniature sets, such as a toy millinery shop with display cases,
hatboxes, even ladies trying on hats; period German kitchens, complete with pewter, brass, copper, and tin implements; and an elaborately
detailed four-story late-Victorian dollhouse with a wedding feast going
on. The gift shop sells antique dolls, books, and toys, as well as doll
costumes and the patterns with which to make them. ⊠ *143 Main St.,*
☎ *508/888–1711.* ⌨ *$3.50.* ◷ *Mid-May–Oct., Mon.–Sat. 10–3:30.*

★ ❷  On 76 beautifully landscaped acres overlooking the upper end of
Shawme Pond is the **Heritage Plantation,** an extraordinary complex
of museum buildings, gardens, and a café. In 1967 pharmaceuticals
magnate Josiah K. Lilly III purchased the estate and turned it into a
nonprofit museum. The Shaker Round Barn showcases classic and historic cars—including a 1930 yellow-and-green Duesenberg built for
Gary Cooper, a 1919 Pierce-Arrow, and a 1911 Stanley Steamer—as
well as art exhibitions. The Military Museum houses antique firearms,
a collection of 2,000 hand-painted miniature soldiers, military uniforms,
and Native American arts. The Art Museum boasts an extensive Cur-
rier & Ives collection, Americana (including a mechanical-bank collection), antique toys such as a 1920 Hubley Royal Circus, and a
working 1912 Coney Island–style carousel.

The grounds are crisscrossed by paths and planted with daylily, hosta,
heather, herb, fruit-tree, and other gardens. Rhododendron enthusiasts will recognize the name of onetime estate owner and hybridizer
Charles O. Dexter; the rhododendrons are in full glory mid-May
through mid-June. Daylilies reach their peak mid-July through early
August. In summer, evening concerts are held in the gardens. The center of the plantation is about ¾ mi on foot from the in-town end of

Shawme Pond. ⊠ *Grove and Pine Sts.,* ☎ *508/888–3300.* 🎟 *$8.* ☉ *Mid-May–late Oct., daily 10–5.*

Take a walk to Town Neck Beach on the **Sandwich Boardwalk,** built over a salt marsh, creek, and low dunes (there is good birding here). In 1991 Hurricane Bob and an October nor'easter destroyed the previous boardwalk. It's a testament to the town's solidarity that individuals and businesses donated planks to rebuild it, which volunteers then installed. The donors' names, jokes (GET OFF OUR BOARD), thoughts (SIMPLIFY/THOREAU), and memorials to lovers, grandparents, and boats are inscribed on the planks. The long sweep of Cape Cod Bay stretches out around the beach at the end of the walk, where a platform provides fine views, especially at sunset. With stone jetties, dunes and waving grasses, and the entrance to the canal in the foreground, look out toward Sandy Neck, Wellfleet, and Provincetown, or toward the white cliffs beyond Sagamore. The sandy strip on this mostly rocky beach is near the rugosa rose–patched dunes; the flowers have a delicious fragrance. If you have a canoe, the creeks running through the salt marsh would make for great paddling. From town cross Route 6A on Jarves Street, and at its end turn left, then right on the mile-plus trip to the boardwalk parking lot.

OFF THE BEATEN PATH
**GREEN BRIAR NATURE CENTER AND JAM KITCHEN** – Is it the soothing pond-side setting, or its simple earthiness—who's to say? The Green Briar Nature Center and Jam Kitchen, owned and operated by the Thornton Burgess Society, is as solid a symbol of the old Cape as you could find. You'll pass a wildflower garden on your way in, and the delightful Smiling Pool sparkles out back. Birds flit about the grounds, and great smells waft from vintage stoves in the Jam Kitchen, where you can watch as jams and pickles are made according to Ida Putnam's recipes, used here since 1903. (Sun-cooked fruit preserves are especially superb.) Come weekdays mid-April through mid-December to see jam being made in the Jam Kitchen—you can even take a jam-making class. The nature center has classes for adults and children, as well as walks, lectures, and an annual May herb festival, where you can take home dozens of different herbs, wildflowers, and perennials. And the Briar Patch Conservation Area behind the building has nature trails to explore—take a walk and visit with Peter Cottontail, Grandfather Frog, and other creatures that inspired the beloved Thornton Burgess characters. ⊠ *6 Discovery Hill Rd., off Rte. 6A,* ☎ *508/888–6870.* 🎟 *$1 suggested donation.* ☉ *Mon.–Sat. 10–4, Sun. 1–4; Jan.–Mar., closed Sun.–Mon.*

☾ At the **Sandwich Fish Hatchery,** you'll see more than 200,000 brook, brown, and rainbow trout at various stages of development, being raised to stock the state's ponds. The mesh over the raceways is to keep kingfishers and herons from a free lunch. You can buy feed for 25¢ and watch the fish jump for it. There's a modest interpretive center. ⊠ *Main St., Rte. 6A,* ☎ *508/888–0008.* 🎟 *Free.* ☉ *Daily 9–3.*

OFF THE BEATEN PATH
**PLYMOUTH** – Back across the Sagamore Bridge on the mainland, Plymouth is the settlement site the Pilgrims chose in December 1620 after scouting locations from their base at Provincetown. In addition to a number of historical museums, you can view **Plymouth Rock** (reputed first footfall) and tour *Mayflower II,* a 1957-built replica of the *Mayflower,* as well as **Plimoth Plantation**—a painstaking reconstruction of the original settlement, from the thatched roofs, cramped quarters, and open fireplaces to the long-horned livestock. At the plantation, actors in period costume speak Jacobean English as they walk the grounds and carry on the

daily life of the 17th century. Feel free to engage them in conversation about their life, and expect curious looks if you ask them about anything beyond 1627. A crafts center features demonstrations of early techniques of making pottery, baskets, furniture, and woven goods. Self-guided tours start with a 12-minute film. ✉ *Warren Ave., Rte. 3A, about 22 mi northwest of the Sagamore Bridge via Rte. 3 or the coastal Rte. 3A,* ☎ *508/746-1622.* ▭ *$15 Plantation only; $5.75 Mayflower II only; $18.50 combination Plantation/Mayflower II.* ◷ *Plantation: Apr.–Nov., daily 9–5; Mayflower II: Apr.–June and Sept.–Nov., daily 9–5.*

## Dining and Lodging

**$–$$** ✕ **The Beach House.** In a ramshackle old house near the Barnstable
★ line, the Beach House serves seafood, with a rib or two thrown in. The menu lists things they *don't* do ("valet parking, schmooze too much, sing happy birthday very well") as well as the things they do ("cook your seafood right, keep your beer cold, always say 'no problem' "). The menu, happily, gets it just about right. The portions of everything are generous (except the lobster, which is boiled and no bigger than the chicks most establishments pass off with their stock surf dinners), and the kitchen stays open until 11 PM. Live entertainment is in the bar on summer weekends. ✉ *674 Rte. 6A, near the turnoff to Sandy Neck,* ☎ *508/362–6403. Reservations not accepted. AE, D, MC, V.*

**$** ✕ **Marshland Restaurant and Bakery.** This tiny coffee shop is the sure antidote to the overspending you're susceptible to if you eat all your meals out while you're on the Cape. For breakfast try an Italian omelet, a rich mix of Italian sausage, fresh vegetables, and cheese. The lunch specials—grilled chicken club sandwich, lobster salad, turkey Reuben, a daily quiche, and the like—are the best choices midday, and for dinner the prime rib does not quit. It's not a particularly comfortable spot, and certainly not fancy—but it is good, reliable roadside cooking. ✉ *Rte. 6A,* ☎ *508/888–9824. No credit cards. No lunch or dinner Sun., no dinner Mon.*

**$$–$$$$** ✕▥ **Dan'l Webster Inn.** Built in 1971 on the site of a 17th-century inn, the Dan'l Webster is essentially a contemporary hotel with old New England friendliness and hospitality. In the restaurant, the menu and wine list are thoughtful, interesting, and up-to-date. Some hydroponic vegetables and fish come from the local, family-run D. W. Aquafarm. Standout dishes include striped bass crusted with cashews and macadamia nuts, accompanied by mango sauce. Choose from one of four intimate dining rooms, a garden room, or the less formal tavern serving more casual fare. Guest rooms, in the main inn and wings or in two nearby historic houses with four suites each, have floral fabrics and fine reproduction mahogany and cherry furnishings, including some canopy beds, creating a generally colonial look. Some suites have fireplaces or whirlpools, and one has a baby grand piano. Guests have access to nearby golf courses and a local health club; pets are not permitted. ✉ *149 Main St., 02563,* ☎ *508/888–3622 or 800/444–3566,* ⅻ *508/888–5156. 47 rooms, 9 suites. 2 restaurants, bar, air-conditioning, no-smoking rooms, room service, pool. AE, D, DC, MC, V.*

**$$** ✕▥ **Inn at Sandwich Center.** Directly across from the Sandwich Glass Museum, this house is listed on the National Register of Historic Places. All of the five rooms have private baths, Laura Ashley comforters and bedding, and hooked rugs, as well as such amenities as terry cloth robes; three have fireplaces. Each room is named after the color whose decor it boasts (white, green, yellow, beige, and blue). Particularly lovely is the Blue Room, which has a two-poster bed, comfortable rockers, a chintz chaise, and a library of travel books crammed in the room's original crooks and nannies and built-in cabinets—great material to peruse on the private deck overlooking the lush gardens. Owner Eliane Thomas's breakfast specialties include rhubarb muffins and French crepes with raspberry sauce, served by candlelight in the antiques-furnished "Keeping Room," which has a fireplace and original 1750 beehive oven. Expect handmade chocolates on your pillow at night, and sherry at the end of a long day. ⊠ *118 Tupper Rd., 02563,* ☎ *508/888–6958 or 800/249–6949,* ⨎ *508/833–2770. 5 rooms. No smoking. AE, D, MC, V. CP.*

$$ ⊞ **Wingscorton Farm.** This is, perhaps, the Upper Cape's best-kept se-
★ cret: an enchanting oasis on a working farm with ducks, chickens (you can help gather eggs), sheep, a donkey, and a pair of llamas. Built in 1763, the main house, once a stop on the Underground Railroad, has a dining room with one of the largest fireplaces in New England—the hearth alone is 9 ft long! Oriental rugs, wing chairs, and a TV are in the paneled library/den. The main house has three second-floor suites, each with a fireplace, braided rugs topping wide-plank floors, wainscoting, and, adjoining the main guest room, a smaller bedroom, once the house's original "birthing rooms." (Today, they hold twin beds.) The property also includes a detached cottage and, better, the stone carriage house, that offers a fully equipped kitchen, a living room with a pull-out queen-size sofa and a wood-burning stove; a spiral staircase leads to a loft (with a queen-size bed) and an oversize sundeck. Traditional clambakes are prepared year-round by visiting members of Martha's Vineyard's Wampanoag tribe, and a private bay beach is a five-minute walk away. ⊠ *11 Wing Blvd., 02537,* ☎ *508/888–0534. 4 suites, 1 carriage house, 1 2-bedroom cottage. AE, MC, V. BP.*

$ ⊞ **Captain Ezra Nye House.** In the heart of town, Elaine and Harry Dickson's 1829 house belonged to one of the town's past notables, one of a few local Nyes whose legacy you'll encounter. It isn't crammed with state-of-the-period pieces or overly precious from a museumlike restoration, but it is brimming with the Dickson's hospitality—they cook full breakfasts and offer suggestions for activities by day or night—for which there is no substitute. ⊠ *152 Main St., 02563,* ☎ *508/888–6142 or 800/388–2278,* ⨎ *508/833–2897. 6 rooms, 1 suite. No smoking. AE, D, MC, V. BP.*

$ ⊞ **Dunbar House.** Stepping into the Dunbar House is like stepping through time. Built in 1741 by descendants of Sandwich founder John Dillingham, the inn, though it has been modernized, retains all its original charm. The three rooms—each named for English lakes—meld history and hospitality; the Ennerdale, for instance, has a fine selection of antiques, a four-poster queen-size bed, pine-plank floors, and lovely views of Shawme Pond. The Loweswater and Buttermere have queen or twin beds, private baths, and water views, as well. If you're here in summer, you can savor owners Mike and Mary Bell's hearty homemade breakfasts, with their wonderful scones, on the outdoor patio overlooking the pretty gardens. You'll receive a voucher good for a complimentary tea at the Dunbar Tea Shop (☞ Exploring, *above*) next door. ⊠ *1 Water St., 02563,* ☎ *508/833–2485,* ⨎ *508/833–4713. 3 rooms. No smoking. MC, V. BP.*

$ ⊞ **Earl of Sandwich Motor Manor.** Single-story Tudor-style buildings are arranged in a U around a duck pond and wooded lawn set with lawn chairs. Rooms in the main building (1966) and the newer buildings (1981–83) have rather somber decor, with dark exposed beams on white ceilings, dark paneled walls, quarry-tile floors with Oriental throw rugs, and chenille bedspreads, but they are of good size and have large Tudor-style windows and small-tiled baths. The innkeepers, Kathy and Brian Clifford, are friendly and helpful. Pets are permitted with advance notice. ⊠ *378 Rte. 6A, 02537,* ☎ *508/888–1415 or 800/ 442–3275,* FAX *508/833–1039. 24 rooms. Air-conditioning, no-smoking rooms. AE, D, DC, MC, V. CP.*

$ 🖬 **Sandwich Lodge & Resort.** Set amid 10 rolling acres, this glorified motel—emphasis on glorified—offers a variety of rooms, all nicely decorated in soft mauve and navy. Suites (the deluxe are the newest), like efficiencies, are equipped with gleaming kitchens with two-burner stoves, microwave ovens, coffeemakers, and refrigerators; some even have two-person whirlpool tubs. The grounds include private volleyball and shuffleboard courts, game rooms, hot tub, and two of the nicest (read: cleanest) pools on the Cape. The on-site restaurant serves two meals a day (ample portions, great prices) in a homey atmosphere. Pets are allowed, with advance notice, for a $15 fee. ⊠ *54 Rte. 6A, Box 1038, 02653,* ☎ *508/888–2275 or 800/282–5353,* FAX *508/888– 8102. 41 rooms, 17 suites, 5 efficiencies. Restaurant, bar, no-smoking rooms, indoor-outdoor pool, hot tub. AE, D, MC, V. CP.*

$ 🖬 **The Summer House.** A skip from the center of town but set back enough for a spell of quietude, this charming 1835 Greek Revival is surrounded by beautiful gardens. Plenty of windows (most with the original wavy bubble-glass), antique furniture, hand-stitched quilts, painted hardwood floors, and light airy fabrics conjure the breezy feeling of an old-fashioned summer home. There are working fireplaces in all but one of the large rooms. The common areas include a sitting room and a sunporch. A sumptuous breakfast is served in the sunny breakfast room, once a formal parlor. ⊠ *158 Main St., 02563,* ☎ *508/888–4991 or 800/241–3609. 5 rooms share 4 baths. Bikes. No smoking. AE, D, MC, V. BP.*

$ ⚠ **Shawme-Crowell State Forest.** Less than a mile from the Cape Cod Canal, this 742-acre state forest is a good base for local biking and hiking, and campers get free day use of Scusset Beach. Open-air campfires are allowed at the 285 wooded tent and RV (no hookups) campsites. ⊠ *Rte. 130, 02563,* ☎ *508/888–0351.*

## Nightlife and the Arts

**Bobby Byrne's Pub** (⊠ Rte. 6A, ☎ 508/888–6088) offers a comfortable pub atmosphere, a jukebox, and good light and full menus.

**Heritage Plantation** (☞ *above*) sponsors summer jazz and other concerts in its gardens from June to mid-September; bring chairs or blankets. Most concerts are free with admission to the complex.

**Opera New England of Cape Cod** has two performances a year, in spring and fall, at **Sandwich High School** (⊠ Quaker Meetinghouse Rd.) by Opera Northeast, a New York City–based touring company, under the direction of Donald Westwood. Upcoming productions are Gounod's *Romeo and Juliet*, to be performed on May 9, 1998, and Puccini's *Tosca*, to be performed on October 25, 1998. For more information or to make reservations, call 508/775–3974.

**Rof-Mar Diplomat Club** (⊠ Popple Bottom Rd., ☎ 508/428–8111; reservations required), a function room, has ballroom dinner dances and dancing year-round. There's a large dance floor and seating on outdoor porches overlooking Lawrence Pond.

**Town band concerts** are held Thursday evenings in late June through late August starting at 7:30. ⊠ *Bandstand, Henry T. Wing Elementary School, Rte. 130,* ☎ *508/888–5281.*

**Sandwich Auction House** (☞ *below*) has weekly sales.

## Shopping

**Brown Jug** (⊠ 155 Main St., at Jarves St., Sandwich, ☎ 508/833–1088) specializes in antique glass, such as Sandwich glass and Tiffany iridescent glassware, as well as Staffordshire china.

**Cape Cod Factory Outlet Mall** (✉ Factory Outlet Rd., Exit 1 off Rte. 6, Sagamore, ☎ 508/888–8417) has a food court and more than 20 outlets, including Corning/Revere, Carter's, Bass, Bugle Boy, Van Heusen, Izod, London Fog, and Champion/Hanes.

**Chocolate House Fudge and Gift Shop** (✉ 11 Cranberry Hwy., Sagamore, ☎ 508/888–7065), just over the Sagamore Bridge on the Cape side, sells creamy fudge in 12 flavors, hand-dipped chocolates and truffles, and saltwater taffy and penny candy. The gift shop has cranberry glass and other Cape items.

**Christmas Tree Shops** (✉ Exit 1 off Rte. 6, Sagamore, ☎ 508/888–7010; other locations at Falmouth, Hyannis, Yarmouth Port, West Yarmouth, West Dennis, and Orleans) sell everything from furniture to clothes to kitchen goods.

**Home for the Holidays** (✉ 154 Main St., ☎ 508/888–4388), set in an old home, is a lovely place to browse for decorations, gifts, and hand-crafted cards for nearly every holiday or special occasion—baby gifts, Christmas ornaments and papers, goblin lights for Halloween—as well as elegant glassware and china.

**Horsefeathers** (✉ 454 Rte. 6A, E. Sandwich, ☎ 508/888–5298) sells antique linens, lace, vintage baby and children's clothing, and Victoriana such as valentines.

**H. Richard Strand** (✉ Town Hall Sq., Sandwich, ☎ 508/888–3230), in an 1800 home, displays very fine pre-1840 and Victorian antique furniture, paintings, and American glass.

**Sandwich Auction House** (✉ 15 Tupper Rd., Sandwich, ☎ 508/888–1926) is a great place to spend part of a Wednesday night (Saturday in the off-season) —come sometime during the day to preview the items for sale. It's a local institution, with weekly sales; in addition, there are specialty sales every five to six weeks that offer up the cream of the crop of antiques received in that period. Specialty sales are also held for collections of modern rugs, silver, or toys.

**Titcomb's Bookshop** (✉ 432 Rte. 6A, E. Sandwich, ☎ 508/888–2331) has used, rare, and new books, including a large collection of Cape and nautical titles and Americana, as well as an extensive selection of quality new children's books.

*En Route* **Route 6A** heads east from Sandwich, passing through the oldest settlements on the Cape. It is part of the Old King's Highway historic district and is therefore protected from development. Driving this quiet section of the north shore on a countrylike road brings you close to the Cape's pleasures. In autumn the foliage along the road is bright—maples with their feet wet in ponds and marshes put on a good display. Along 6A just east of Sandwich center, you can stop to watch cranberries being harvested in flooded bogs. If you're heading to Orleans and you're not in a hurry, this is a lovely route to take.

# Bourne to Falmouth

*A 15-mi stretch along Rte. 28.*

Several villages along Route 28 between the Bourne Bridge and Falmouth might not offer a tremendous amount of interest in their own right, but they are worthy of a stop on the way to larger towns. Route 28 from the Bourne Bridge south to Falmouth is overly commercial in many areas, while Route 28A between Pocasset and West Falmouth is more scenic, with side roads leading to attractive beaches and small harbors. The area is steeped in history. Indian names serve as reminders of the true first settlers here, and houses that date from the 18th century line the streets.

# SUMMERTIME:
# A WEEK AT A GLANCE

**E**VERY SUMMER the Cape's social and cultural calendar fills up with regularly scheduled weekly events. Before setting out, phone in advance, as even longstanding events change.

➤ MONDAY: West Yarmouth's summertime **town band concerts** (☎ 508/778–1008) take place at 7:30 at Mattacheese Middle School.

➤ TUESDAY: **Tales of Cape Cod** (☎ 508/362–8927), Barnstable's historical society, sponsors **slide-illustrated lectures.** In Woods Hole, **free guided village walking tours** begin at 4 at the Bradley House Museum (☎ 508/548–7270). The 2,500-acre **Waquoit Bay National Estuarine Research Reserve** (☎ 508/457–0495) in **Waquoit** holds evening talks on environmental, historical, and artistic subjects; you can bring a picnic dinner. **Town band concerts** in Harwich take place at 7:30 in Brooks Park (☎ 508/432–1600). The Nau-Sets (☎ 508/394–0257) hold **weekly square dances** (year-round) in Dennis.

➤ WEDNESDAY: **Hyannis's** Cape Cod Melody Tent (☎ 508/775–9100) hosts a morning **children's theater** series from 11–12:30, for ages 3–11. Also in Hyannis, **town band concerts** (☎ 508/362–5230) are held at 7:30 on the Village Green. The **Cape Cod Museum of Natural History** (☎ 508/896–3867) in Brewster hosts stargazing sessions and "Live at the Museum" world music concerts at 7:30. The bidding begins at 6:30 each week at the **Sandwich Auction House** (☎ 508/888–1926).

➤ THURSDAY: Morning **performances of fairy tales and folktales for children** are given at Brewster's First Parish Church (☎ 508/896–5577). In Bourne, **town band concerts** begin at 7 in Buzzards Bay Park (☎ 508/759–6000). Sandwich's **town band concerts** get underway at 7:30 at the bandstand at the Henry T. Wing Elementary School

(☎ 508/888–5281). Falmouth's **town band concerts** start at 8 in Marina Park (☎ 508/548–8500). Also in Falmouth, the **Nimrod Inn** (☎ 508/540–4132) offers the Big Band and jazz sounds of Stage Door Canteen (year-round). At Provincetown's Old Harbor Station (✉ Race Point Beach, no phone), you can see **reenactments of an old-fashioned lifesaving procedure.**

➤ FRIDAY: Look for **Molly Benjamin's fishing column** in the *Cape Cod Times.* The Cape Playhouse (☎ 508/385–3911) puts on morning **children's theater** shows. Chatham's **town band concerts** (☎ 508/945–5199) begin at 8. You may see 500 fox-trotting on the roped-off dance floor; there are special dances for children and singalongs for all. It's **Rock Night** at the Charles Moore Arena (☎ 508/255–2971) in Orleans. From 8 to 10 PM, kids 9–14 ice-skate to DJ-spun rock.

➤ SATURDAY: You can catch professional and local folk and blues on the first Saturday of each month, at the Liberty Folk Society's **Benefit Coffeehouse** (☎ 508/428–1053), in Marstons Mills. During Wellfleet's **Gallery Crawl,** walk from gallery to gallery meeting artists and checking out the works in their just-opened shows. **Merlyn Auctions** (☎ 508/432–5863) in North Harwich offers moderate to inexpensive antique and used furniture. Do a turn around the Cape's largest dance floor at **Betsy's Ballroom** (☎ 508/362–9538) in South Yarmouth.

➤ SUNDAY: Between 4:30 and 7:30, tours are given of the newly relocated and relit **Nauset Light** (☎ 508/240–2612). At 8 PM you can hear a concert at Wellfleet's Congregational Church, with its 1873 organ. The Ashumet Holly and Wildlife Sanctuary sponsors all-day **tours to Cuttyhunk Island** (☎ 508/563–6390), which leave Falmouth Harbor at 9 AM and return at 5.

**❸** A monument to the birth of commerce in the New World, the **Aptucxet Trading Post Museum** was erected on the foundation of the original post archaeologically excavated in the 1920s. Here, in 1627, Plimoth Plantation leaders established a way station between the Indian encampment at Great Herring Pond 3 mi to the northeast, Dutch colonists in New Amsterdam (New York) to the south, and English colonists on Cape Cod Bay. Before the canal was built, the Manomet River connected Herring Pond with Buzzards Bay (no, scavengers don't frequent it—it was misnamed for the migrating osprey that do), and a short portage connected the pond to Scusset River, which met Cape Cod Bay. The Indians traded furs; the Dutch traded linen cloth, metal tools, glass beads, sugar, and other staples; and the Pilgrims traded wool cloth, clay beads, sassafras, and tobacco (which they imported from Virginia). Wampum was the medium of exchange.

Inside are 17th-century cooking utensils hanging from the original brick hearth, beaver and otter skins, furniture, and other artifacts such as Indian arrowheads, tools, and tomahawks. Also on the grounds are a gift shop in a Dutch-style windmill, a saltworks, herb and wildflower gardens, a picnic area overlooking the canal, and a small Victorian railroad station built for the sole use of President Grover Cleveland, who had a summer home in Bourne. ⊠ *24 Aptucxet Rd., Bourne,* ☎ *508/ 759–9487.* ⊡ *$2.50.* ☉ *May–Columbus Day, Tues.–Sat. 10–5, Sun. 2–5, and Mon. holidays; also open Mon. July–Aug. 10–5.*

The **Massachusetts Military Reservation,** consisting of the Camp Edwards Army National Guard and Reserve Training Site, the Coast Guard Air Station Cape Cod, the Otis Air National Guard Base (with F-15 fighter planes), and the Cape Cod Air Station (with a PAVE PAWS radar station, a sophisticated system designed to track military satellites and nuclear missiles) sprawls across 23,000 acres of land in the upper portion of the Shawme-Crowell State Forest.

More often referred to as **Otis Air Base,** the reservation holds a free two-day open house and air show, usually on the first weekend in August, with exhibitions that may include precision flying by teams of the Air Force's Thunderbirds or the Navy's Blue Angels. Other highlights may be performances by the Air Force Band, the Air Force honor guard and drill team, the National Guard's Equestrian Unit, or the Army's Golden Knights Parachute Team. Military planes are on display, and concession stands serve the crowds. ⊠ *Entrances at the so-called Otis rotary on Rte. 28, Rte. 130 in Forestdale, and Rte. 151 in Falmouth.* ☎ *508/ 968–4090 or 508/968–4003 for dates and information.*

**☾** A break for energetic kids pent up in a car for too many miles, **Adventure Isle** has minibikes, a giant slide, a Ferris wheel, children's rides, an arcade, bumper boats, batting cages, a double gyroscope (for those with strong stomachs), and a miniature golf course built in 1997. ⊠ *Rte. 28, Bourne, 2 mi south of Bourne Bridge,* ☎ *508/759–2636 or 800/535–2787.* ⊡ *$1.75–$4.50 per ride.* ☉ *Mid-Mar.–Nov., daily 9 AM–11 PM.*

## Dining and Lodging

**$$** ✕ **Stir Crazy.** One of just a few authentic Asian restaurants on the Cape— owner Bopha Samms hails from Cambodia—Stir Crazy fits the bill when you reach your inevitable limit of seafood and Yankee cooking. Overlook the anonymous location (beside self-storage lockers, in a building that also houses a real estate office)—the real spice is where it belongs: in the food. Dishes blur the ethnic line with lively Cambodian-Thai-Vietnamese influences. Look for *nhem shross* (an appetizer of vegetables and shrimp) and beef *lock lack* (sirloin tips on a bed of watercress). ⊠ *626 MacArthur Blvd., Pocasset,* ☎ *508/564–6464. Reservations not accepted. MC, V. No smoking. Closed Mon. No lunch Sun.*

$ ✕ **My Tinman.** "The family diner with a heart" is one of the only real diners (architecturally speaking) left on the Cape. The outside is covered in silver aluminum; inside, the pink booths share space with *Wizard of Oz* photos and paraphernalia. As you'd expect, it's diner food all the way, and it's all good, especially the meat loaf. Most dishes are named after the film: Lionhearted sandwiches, Scarecrow Garden Salad—you get the picture. Breakfast is served all day (it's open from 5 AM to 2 PM). ⊠ *MacArthur Blvd., near the MMR rotary, Pocasset, no phone. Reservations not accepted. No credit cards.*

$$$ ⌂ **Inn at West Falmouth.** This luxurious, 1898 estate house inn is in a secluded area of an exclusive village. A mixture of contemporary and antique furnishings and polished hardwood floors set an elegant but relaxed mood. Guest rooms have queen-size beds, some with canopies, Italian marble bathrooms with whirlpool tubs, phones, hair dryers, and wall safes; some have wood-burning fireplaces and private decks. Beyond the French doors, off the pale pink and green breakfast and sun room, a patio spilling over with lush potted plants and trees overlooks woods, gardens, and a tennis court, and leads to the small pool and deck. ⊠ *Off Blacksmith Shop Rd., Box 1208, W. Falmouth 02574,* ☎ *508/540–7696 or 800/397–7696,* FAX *508/548–6974. 6 rooms. Pool, tennis court. No smoking. AE, MC, V. CP.*

$$$ ⌂ **Sea Crest.** Location and amenities are strong draws at this conference center and resort, whose eight buildings sprawl along one end of beautiful Old Silver Beach. Recently renovated from floor to ceiling, rooms are crisp and clean, done in dark blues and pastels. Many rooms have ocean views, and some have gas-log fireplaces. A number of lodging packages are available. ⊠ *350 Quaker Rd., N. Falmouth 02556,* ☎ *508/540–9400 or 800/225–3110,* FAX *508/548–0556. 266 rooms, 8 suites. Restaurant, deli, piano bar, room service, indoor-outdoor pool, hot tub, sauna, putting green, 2 tennis courts, health club, shuffleboard, beach access, video games, children's programs. AE, DC, MC, V.*

$ ⌂ **Wood Duck Inn.** Maureen and Dick Jason literally open their 1848 house to guests: You can watch TV in the Jasons' private living room, and if you don't want to use the cooking facilities in your suite, you can use the house kitchen. Such friendliness and accessibility add to the pleasure of staying here, though it's the rooms that make it tough to leave. Both the Garden and the Treetops suites are light and airy and tastefully decorated with antiques, handmade quilts, and (what else?) decoys; each has a private entrance, sparkling kitchenette, TV, pull-out sofa in the living area, separate bedroom and private bath stocked with amenities such as bath crystals and oils and plush terry robes. The private decks offer idyllic views overlooking conservation lands and cranberry bogs. (Winter guests take note: the bog is ideal for ice-skating or ice sailing.) There's also a small, two-story detached house nearby called the Duckling, which houses a suite with a private deck, fully furnished kitchen, and private bath. ⊠ *167 Palmer Ave., Box 305, 02540,* ☎ *508/564–6404. 5 rooms, 1 cottage. No smoking. AE, MC, V. BP.*

## Nightlife and the Arts

The **Army Corps of Engineers** (☎ 508/759–4431), which maintains the Cape Cod Canal via a field office in Bournedale, offers free daily programs in summer, including slide shows about the canal, sing-alongs, night walks, and storytelling around campfires at the Bourne Scenic Park and Scusset Beach State Park.

**Sea Crest Resort** (⊠ 350 Quaker Rd., N. Falmouth, ☎ 508/540–9400) has a summer and holiday-weekend schedule of nightly entertainment on the outdoor terrace, including dancing to country, Top 40, reggae, and jazz bands and a big-band DJ, as well as karaoke and comedy.

**The Wharf** (⊠ Grand Ave. S, Falmouth Heights, ☎ 508/548–0777) is the Upper Cape's hot beachfront club, with dancing in the nightclub and outdoor deck to high-energy rock and Top 40 bands in season. Saturday afternoon beach parties with entertainment and activities (volleyball tournaments, raft races, and the like) add to the fun. Keep up your energy with selections from the deli or the upstairs restaurant's full or light menu.

In Bourne, **town band concerts** are held Thursday evenings in July and August starting at 7 in **Buzzards Bay Park** (⊠ Main St., ☎ 508/759–6000).

## Outdoor Activities and Sports

### BEACH

**Old Silver Beach** in North Falmouth is a long, beautiful crescent of soft white sand, with the Sea Crest resort hotel at one end. It is especially good for small children because a sandbar keeps it shallow at one end and creates tidal pools full of crabs and minnows. There are lifeguards, rest rooms, showers, and a snack bar.

### BIKING

An easy, straight trail stretches on either side of the **Cape Cod Canal,** 6½ mi on the south side, 7 mi on the north, with views of the bridges and ship traffic on the canal. Directly across the street from the canal bike path on the mainland, **P & M Cycles** (⊠ 29 Main St., Buzzards Bay, ☎ 508/759–2830) rents a large selection of mountain and hybrid bikes at reasonable rates.

### HIKING AND WALKING

The **Army Corps of Engineers** sponsors guided walks, bike trips, and hikes. Outside its field office's visitor center, on a bank of the canal with an excellent view, are picnic tables, access to the canal bike path, a herring run through which the fish travel on their spawning run in May, and short self-guided walking trails through woodland. The visitor center is on the mainland side of the canal on Route 6 in Bournedale, between the bridges. ☎ *508/759–4431 or 508/759–5991 for tides, weather, and special events.*

### HORSEBACK RIDING

**Haland Stables** (⊠ Rte. 28A, W. Falmouth, ☎ 508/540–2552) offers lessons and trail rides, by reservation.

### ICE-SKATING AND ROLLER-SKATING

**John Gallo Ice Arena** (⊠ 231 Sandwich Rd., Bourne, ☎ 508/759–8904) is the place to go for ice-skating fall through spring. Hours vary widely; call for details.

### SCUBA DIVING

Rentals, instruction, group dives, and information are available through **Aquarius Diving Center** (⊠ 3239 Cranberry Hwy., Buzzards Bay, ☎ 508/759–3483).

### TENNIS

**Ballymeade Country Club** (⊠ Rte. 151, N. Falmouth, ☎ 508/457–7620) has six Har-Tru and four hard courts, lessons, clinics, ball machines, a grass croquet court, and a pro shop (☎ 508/457–7620) that accepts court-time reservations from mid-June through mid-October.

## Shopping

**Tanger Outlet Center** (⊠ Rte. 6 at the Buzzards Bay Rotary, Bourne, ☎ 207/874–4775) sells women's, men's, and children's apparel from brand name manufacturers and designers such as Izod, Nine West, Levi's, and Liz Claiborne.

*En Route*   Before crossing the Bourne Bridge on your way to the Cape, the towns of Wareham, Onset, and Buzzards Bay have some unusual attractions and adventures for road-weary travelers.

**The Massachusetts Maritime Academy** (⊠ Taylor's Point, Buzzards Bay 02532), founded in 1891, is the oldest such academy in the country. Future members of the Merchant Marine receive their training at its 55-acre campus in Buzzards Bay. The library has nautical paintings and scale models of ships from the 18th century to the present and is open to the public at no charge (hours vary widely; call 508/830–5000, ext. 1201). For a 20- to 30-minute tour of the academy weekdays at 10 and 2, call 48 hours in advance (ext. 1102).

☾   A wet and wild adventure for the kids awaits at the **Water Wizz Water Park,** with a 50-ft-high water slide complete with tunnels and dips, a river ride, a kiddies' water park, three tube rides, two enclosed water mat slides, a kiddy slide, a pool, miniature golf, and food. The enclosed Black Wizard water slide descends 75 ft in darkness. ⊠ *Rtes. 6 and 28, Wareham, 2 mi west of the Bourne Bridge,* ☎ *508/295–3255.* ⌨ *$18.* ⊙ *Memorial Day–mid-June, weekends 11–4; mid-June–Labor Day, daily 10–6:30.*

# Falmouth

❹   *15 mi south of the Bourne Bridge, 4 mi north of Woods Hole.*

The Cape's second-largest town, Falmouth was settled in 1660 by Congregationalists from Barnstable who had been ostracized from their church and deprived of voting privileges and other civil rights for being sympathizers with the Quakers (then the victims of severe repression). The **Village Green,** added to the National Register of Historic Places in 1996, was used as a militia training field in the 18th century and a grazing ground for horses in the early 19th. Today it is flanked by attractive old homes, some built by sea captains, and the 1856 **Congregational Church,** built on the timbers of its 1796 predecessor, with a bell made by Paul Revere. The cheery inscription reads: "The living to the church I call, and to the grave I summon all."

Two museums that represent life in colonial Cape Cod are maintained by the **Falmouth Historical Society.** The 1790 **Julia Wood House** retains wonderful architectural details—a widow's walk, wide-board floors, lead-glass windows, a Colonial kitchen with wide hearth—and is otherwise full of antique embroideries, baby shoes and clothes, toys and dolls, portraits, furniture, and the trappings of an authentically equipped doctor's office, from the house's onetime owner. Out back, the Hallett Barn Museum displays antique farm implements, a 19th-century horse-drawn sleigh, and other interesting items. The smaller **Conant House** next door, a 1794 half Cape, has military memorabilia, whaling items, scrimshaw, sailors' valentines, and a genealogical and historical research library. There's also a collection of books, portraits, and things relating to Katharine Lee Bates, the native daughter who wrote "America the Beautiful." Guides give tours of the museums, and a pretty, formal garden with a gazebo and flagstone paths is adjacent. Free walking tours of the town are available in season, as are historical trolley tours; call for prices and times. ⊠ *Palmer Ave.,* ☎ *508/548–4857.* ⌨ *3.* ⊙ *Mid-June–mid-Sept., weekdays 2–5.*

The 1812 white Cape house at 16 Main Street marks the **birthplace of Katharine Lee Bates.** Now owned by the historical society, it is no longer open to the public (though the society hopes to eventually turn the building into a museum). A plaque out front commemorates Bates's birth, in 1859.

NEED A
BREAK? **Peach Tree Circle Farm** includes a bakery, a farm stand, and a cheery tearoom where inexpensive lunches of soups, sandwiches, salads, and a few entrées like quiche and chicken potpie are served year-round amid the smells of baking bread and herbs hung to dry. ⊠ *881 Old Palmer Ave.,* ☎ *508/548–4006.*

❺ The **Charles D. and Margaret K. Spohr Garden,** a private garden of 3 planted acres on Oyster Pond, is a pretty, peaceful place that the generous owner invites the public to enjoy. The springtime explosion of more than 700,000 daffodils gives way in turn to the tulips, azaleas, magnolias, flowering crabs, rhododendrons, lilies, and climbing hydrangeas that carry garden goers into summer. A collection of old millstones, bronze church bells, and ships' anchors is interspersed in the landscape. ⊠ *Fells Rd.,* ☛ *Free.* ☉ *Daily sunrise–sunset.*

☾ The spacious facility of the **Cape Cod Children's Museum** welcomes children of all ages with interactive play, science exhibits, a 30-ft pirate play ship, a portable planetarium, and other playtime activities for toddlers and parents most mornings. ⊠ *Falmouth Mall, Rte. 28,* ☎ *508/457–4667.* ☛ *$3.* ☉ *Mon.–Sat. 10–5, Sun. noon–5.*

☾ Another good place for kids on a rainy day, the **Leary Family Amusement Center** has video-game rooms and candlepin bowling. ⊠ *Town Hall Sq., Rte. 28,* ☎ *508/540–4877.*

## Dining and Lodging

$$$ ✕ **Regatta of Falmouth-by-the-Sea.** One of the Cape's nicest dining
★ rooms offers beautiful views of Nantucket Sound and Martha's Vineyard and a menu that's constantly evolving, becoming less traditional and at the same time more appealing. The dining room is modern and subdued, with soft colors, lamplight, and an intimate, romantic ambience. Continental and Asian cuisines have been colliding on the menu lately, resulting in dishes such as the sautéed shellfish sampler: scallops, mussels, lobster, and shrimp with a sensational curried lobster sauce over Asian greens. Three menus are served each night: lighter fare, regular fare, and a three-course early dinner menu. The Regatta of Cotuit (☞ Dining *in* Mashpee, *below*) is a sibling. ⊠ *217 Clinton Ave., Falmouth Harbor,* ☎ *508/548–5400. AE, MC, V. Closed Oct.– Memorial Day. No lunch Sat.*

$$ ✕ **Flying Bridge.** This place is more of a dining-entertainment megaplex than a restaurant, strictly speaking. Find a seat on one of the decks overlooking the harbor, and settle in for classic bar food, which is also to say classic Cape food: seafood and burgers, steak and pasta, lots of battered and fried appetizers, salads, and specials. A great place to meet and be met, the lounges have live music Friday through Sunday. ⊠ *220 Scranton Ave.,* ☎ *508/548–2700. Reservations not accepted. AE, DC, MC, V.*

$$ ✕ **Quarterdeck Restaurant.** Part bar, part restaurant—but all Cape Cod—this spot is across the street from Falmouth's town hall, so the lunch talk tends to focus on local politics. The stained glass is not old and authentic, but the huge whaling harpoons certainly are. Low ceilings and rough-hewn beams suit the menu, which features hearty sandwiches like Reubens and grilled chorizo, and then specials at night. The swordfish kabob, skewered with mushrooms, onion, and green pep-

per and served over jasmine rice, is especially good. ⊠ *164 Main St.,* ☎ *508/548–9900. AE, D, DC, MC, V.*

**$$** ✕🏠 **Coonamessett Inn.** Delightful and old-fashioned, with plenty of
★ art and hanging plants and wood all around, this is one of the best—
and oldest (since 1953)—of many "quaint" inn-restaurants on the
Cape. The main dining room is lovely, and the smaller garden room
looks out over a beautiful pond and garden. The menu, not surpris-
ingly, is also traditional, maybe too much so for more adventuresome
palates. A low-fat dish is regularly offered, billed as heart-healthy. If
you're staying the night, one- or two-bedroom suites are in five build-
ings ranged around a broad, landscaped lawn that spills down to a scenic
wooded pond—a tranquil country setting. Three suites directly over-
look the pond. Rooms are casually decorated, with bleached wood or
pine paneling, New England antiques or reproductions, and upholstered
chairs and couches. A large collection of Cape artist Ralph Cahoon's
work is displayed throughout the inn. ⊠ *311 Gifford St., at Jones St.,*
*02540,* ☎ *508/548–2300,* 🅵🅰🆇 *508/540–9831. 25 suites, 1 cottage.*
*Restaurant, bar. AE, MC, V. CP.*

**$$** 🏠 **Capt. Tom Lawrence House.** Steps from downtown yet set back
from the street enough to feel secluded, this pretty white house with a
cupola and black shutters is surrounded by flowers, bushes, and a lawn
shaded by old maple trees. Built in 1861 for a whaling captain, the intimate B&B offers romantic rooms with antique and painted furniture,
French country wallpaper, soft colors, and thick carpeting. The beds,
all queen-size canopy or king-size, have firm mattresses, Laura Ashley or Ralph Lauren linens, and down comforters in winter. A new efficiency apartment is bright and spacious, with a fully equipped, eat-in
kitchen, and a private entrance. The large common room has a piano
and a fireplace. The full breakfast is lavish and delicious. ⊠ *75 Locust*
*St., 02540,* ☎ *508/540–1445 or 800/266–8139,* 🅵🅰🆇 *508/457–1790.*
*6 rooms, 1 efficiency. Air-conditioning. No smoking. MC, V. Closed*
*Jan. BP.*

**$$** 🏠 **Mostly Hall.** Set in a landscaped park far back from the street and
★ separated from it by tall bushes, trees, and a wrought-iron fence, Car-
oline and Jim Lloyd's inn looks very much like a private estate. (The
house got its name when a young turn-of-the century visitor exclaimed
to his mother, "Look! It's mostly hall!") The imposing 1849 house has
a wraparound porch and a dramatic widow's walk fitted as a guest
den with travel library and television. Accommodations are in corner
rooms, with leafy views through shuttered casement windows, read-
ing areas, antique pieces and reproduction queen-size canopy beds, flo-
ral wallpapers, wall-to-wall carpeting, and Oriental accent rugs.
First-floor rooms have 13-ft ceilings, though most baths are small. The
yard in back, with Adirondack chairs set around lush gardens, is a lovely
place to relax. ⊠ *27 Main St., 02540,* ☎ *508/548–3786 or 800/682–*
*0565,* 🅵🅰🆇 *508/457–1572. 6 rooms. Air-conditioning, bicycles, library.*
*No smoking. AE, D, MC, V. Closed Jan. BP.*

**$$** 🏠 **Palmer House Inn.** This turn-of-the-century Queen Anne home is
set on a tree-lined street in the historic heart of Falmouth, just a short
stroll past the village green and the hustle and bustle of Main Street.
The interior is slightly suffocating—dark hardwoods, heavy period furniture, endless lace, and ornate stained-glass windows. The Victorian
theme is carried into those of the 12 rooms not done in a floral motif,
although to a much lesser degree. Each room has an antique, wicker
or a four-poster bed, ceiling fan, and private bath. Your best bet: the
Tower Room, a third-floor showplace with a view across the treetops

into Woods Hole and beyond—make up for it. There's also a private cottage, set off in the backyard, with a Jacuzzi. A candlelight breakfast is served and afternoon refreshments are available. ⊠ *81 Palmer Ave., 02540,* ☎ *508/548–1230 or 800/472–2632,* ℻ *508/540–1878. 12 rooms, 1 suite. Air-conditioning, bicycles. No smoking. AE, D, DC, MC, V. BP.*

**$$** 🏨 **Quality Inn.** Across from a pond ½ mi outside Falmouth center, this group of three buildings has large rooms with plush carpeting, contemporary pastel decor, and wood-tone furniture; suites come with king-size beds. The large pool area is bright and nicely arranged with patio furniture and greenery. The restaurant is open for breakfast in summer and dinner year-round. Discount packages are available, and children under 18 stay free. ⊠ *291 Jones Rd., 02540,* ☎ *508/540–2000,* ℻ *508/548–2712. 98 rooms, 5 suites. Restaurant, bar, no-smoking rooms, room service, indoor pool, sauna, video games. AE, D, DC, MC, V.*

**$$** 🏨 **Wildflower Inn.** You'll find here many instances of what innkeep-★ ers Phil and Donna Stone call their "old made new again" decorating style: Tables are constructed from early-1900s pedal sewing machine bases, and the living room–breakfast area's sideboard was once a '20s Hotpoint electric stove. Each of the inn's five rooms, all with private baths and two with whirlpool tubs, are also innovatively decorated. In what was once a stable, the fully furnished Loft Cottage has a spiral staircase that winds up to the bedroom. In the kitchen, Donna whips up edible concoctions using the wildflowers she grows out back. (Her floral feasts have been featured on the PBS series, *Country Inn Cooking.*) The wraparound porch serves as the summer's breakfast nook; one morning the five-course breakfast might include sunflower crepes, the next, calendula corn muffins. ⊠ *167 Palmer Ave., 02540,* ☎ ℻ *508/548–9524 or 800/294–5459. 5 rooms, 1 cottage. Air-conditioning. No smoking. AE, MC, V. BP.*

**$** 🏨 **Admiralty Resort.** Rooms and suites at this large roadside motel outside Falmouth center have contemporary decor, with mauve carpeting, pastel Formica, and light oak, and come with wet bar and coffeemaker. Standard rooms have two queen-size beds or one queen and one Murphy bed. King Jacuzzi rooms have king-size beds and whirlpool tubs in the bedroom. Town house suites have cathedral ceilings with skylights, two baths (one with whirlpool), a loft with a king-size bed, a living room with a sofa bed and a queen-size or king-size bed. There is beach access, and golf and other packages are available. Children under 12 stay free. ⊠ *51 Teaticket Hwy., Rte. 28, 02540,* ☎ *508/548–4240,* ℻ *508/457–0535. 70 rooms, 28 suites. Restaurant, bar, indoor-outdoor pools, hot tub, children's programs, playground. AE, D, DC, MC, V.*

**$** 🏨 **Inn on the Sound.** Despite its location smack-dab on car-choked Grand Avenue, this inn, perched on a bluff overlooking Vineyard Sound and offering glimpses of the island itself, is an oasis of tranquillity. Every room, from the living room—with its oversize boulder fireplace, oversize windows, and overstuffed chairs–to each of the 10 guest rooms (all named for various Falmouth area landmarks) face the water. David Ross owns the inn with his sister Renee, an interior decorator who has decorated each guest room with a certain flair. Every room has a queen-size bed and simple, contemporary-casual furnishings: natural oak tables, unbleached cottons, and ceiling fans. Common areas include the art-laden living room, a bistrolike breakfast room, and a 40-ft deck with even more stunning water views. ⊠ *313 Grand Ave., Falmouth Heights 02540,* ☎ *508/457–9666 or 800/564-9668* ℻ *508/ 457–9631. 10 rooms. Beach. No smoking. AE, D, MC, V. BP.*

## Nightlife and the Arts

The **College Light Opera Company** (✉ Highfield Theatre, off Depot Ave., ☎ 508/548–0668), founded in 1969, presents music and theater majors from Oberlin and other colleges performing nine musicals or operettas, each running one week during the summer. The company includes more than 30 singers and an 18-piece orchestra. Be forewarned: When the shows are good, they are mildly enjoyable. When they are bad—and it does happen—they are dreadful.

**Coonamessett Inn** (✉ 311 Gifford St., ☎ 508/548–2300) has dancing to soft piano, jazz trios, or other music in its lounge on weekends year-round.

The **Flying Bridge** restaurant (✉ 220 Scranton Ave., ☎ 508/548–2700) on Falmouth Harbor has live entertainment in its lounges on Friday evenings during the summer, plus steel bands on the outdoor deck on summer afternoons.

**Nimrod Inn** (✉ 100 Dillingham Ave., ☎ 508/540–4132) offers the Big Band and jazz sounds of Stage Door Canteen, one of the Cape's best bands, every Thursday evening year-round.

Falmouth's **town band concerts** are held on Thursday evenings starting at 8 in summer (✉ Marina Park, Scranton Ave., ☎ 508/548–8500 or 800/526–8532).

## Outdoor Activities and Sports

### BIKING

The **Shining Sea Trail** is an easy 3½-mi route between Locust Street, Falmouth, and the Woods Hole ferry parking lot. It follows the coast, giving views of Vineyard Sound and dipping into oak and pine woods; a detour onto Church Street takes you to Nobska Light. A brochure is available at the trailheads.

**Corner Cycle** (✉ 115 Palmer Ave., Falmouth, ☎ 508/540–4195) rents tandem and trail bikes by the hour, day, or week. They also do on-site repairs.

**Holiday Cycles** (✉ 465 Grand Ave., Falmouth Heights, ☎ 508/540–3549) has surrey, tandem, and other unusual bikes.

### FISHING

The Cape Cod Canal is a good place to fish, from the service road on either side, for blues, flounder, herring, mackerel, and striped bass seasonally making their way through the passage (April–November). The Army Corps of Engineers has a **canal fishing hot line** (☎ 508/759–5991). Freshwater ponds are good for perch, pickerel, trout, and more; the required license (along with rental gear) is available at tackle shops, such as **Eastman's Sport & Tackle** (✉ 150 Main St., Falmouth, ☎ 508/548–6900). **Patriot Boats** (✉ Falmouth Harbor, ☎ 508/548–2626 or 800/734–0088 in MA) has deep-sea fishing from party or charter boats. Midweek you can most likely get a spot on the party boat without reservations; reserve ahead for weekend trips.

### ICE-SKATING

Fall through spring, ice-skating is available at the **Falmouth Ice Arena** (✉ Palmer Ave., Falmouth, ☎ 508/548–9083).

### TENNIS AND RACQUETBALL

**Falmouth Sports Center** (✉ 33 Highfield Dr., Falmouth, ☎ 508/548–7433) is a huge facility with three all-weather tennis, six indoor tennis, and three racquetball–handball courts, as well as steam rooms and saunas, free weights, and physical and massage therapist services. Day and short-term rates are available.

Public tennis courts abound: Falmouth has more than 20.

## Shopping

**Howlingbird** (✉ 91 Palmer Ave., Falmouth, ☎ 508/540–3787) carries detailed, hand-silk-screened, marine-theme T-shirts and sweatshirts, plus silver and shell jewelry, hand-painted cards, and silk-screened hats and handbags.

**Maxwell & Co.** (✉ 200 Main St., Falmouth, ☎ 508/540–8752) has traditional men's and women's clothing with flair from European and American designers, handmade French shoes and boots, and leather goods and accessories.

# Woods Hole

❻ *4 mi southwest of Falmouth, 19 mi south of the Bourne Bridge.*

The village of Woods Hole dangles at the Cape's southwestern tip. Well known as a center for international marine research, it is home to several major scientific institutions. The National Marine Fisheries Service was here first, established in 1871 to study fish management and conservation. In 1888 the Marine Biological Laboratory (MBL), a center for research and education in marine biology, moved in across the street. Then in 1930 the Woods Hole Oceanographic Institution (WHOI) arrived, and the U.S. Geological Survey's Branch of Marine Geology followed suit in the 1960s.

Most of the year, Woods Hole is a peaceful community of intellectuals quietly going about their work. In summer, however, the basically one-street village overflows with the thousands of scientists and graduate students who come from around the globe either to participate in summer studies at MBL, WHOI, or the National Academy of Sciences conference center, or to work on independent research projects. A handful of waterside cafés and shops along Water Street compete for the most bicycles stacked up at the door.

What accounts for this incredible concentration of scientific minds is, first, the variety and abundance of marine life in Woods Hole's unpolluted waters, and the natural deepwater port. Second is the opportunity for easy interchange of ideas and information and the stimulation of daily lectures and discussions (many open to the public) by important scientists. Third, the pooling of resources among the various institutions makes for economies that benefit each while allowing all access to highly sophisticated equipment.

A good example of this joining of forces is the **MBL–WHOI Library,** one of the best collections of biological, ecological, and oceanographic literature in the world. And on top of its access to more than 200 computer databases and the Internet, the library subscribes to more than 5,000 scientific journals in 40 languages, with complete collections of most from their first issues. During World War II, the librarian arranged with a German subscription agency to have German periodicals sent to neutral Switzerland, to be stored until the end of the war. Thus the library's German collections are uninterrupted, where even many German institutions' are incomplete. All journals are always accessible, because they cannot be checked out and because the library is open 24 hours a day. The Rare Books Room contains photographs, monographs, and prints, as well as journal collections that date from 1665.

Unless you are a scientific researcher, the only way you'll get to see the library is by taking the **Marine Biological Laboratory tour** (☎ 508/548–3705; call for reservations and meeting instructions at least a week in advance if possible). The 1½-hour tours are led by retired scientists (mid-June–August, weekdays at 1, 2, and 3 PM) and include an introductory slide show, as well as stops at the library, the marine resources center (where living sea creatures collected each day are kept), and one of the many working research labs.

WHOI is the largest independent private oceanographic laboratory in the world. Several buildings in the village of Woods Hole house its shore-based facilities; others are on a 200-acre campus nearby. During World War II its research focused on underwater explosives, submarine detection, and the development of antifouling paint. Today its $87 million annual budget helps to operate many specialized laboratories with state-of-the-art equipment. A graduate program is offered jointly with MIT, in addition to undergraduate and postdoctoral studies. WHOI's several research vessels roam the world's waters. Its staff led the successful U.S.–French search for the *Titanic* (found about 400 mi off Newfoundland) in 1985.

The Oceanographic Institution (⊠ 86 Water St.) is not open to the public, but you can learn about it at the small **WHOI Exhibit Center,** with videos and exhibits on the institution and its various projects, including research vessels. ⊠ *15 School St.,* ☎ *508/289–2663.* ☞ *$2 suggested donation.* ⊙ *Memorial Day–Labor Day, Mon.– Sat. 10–4:30, Sun. noon–4:30; Apr. and Nov., Fri.–Sat. 10–4:30, Sun. noon–4:30; May, Sept., and Oct., Tues.–Sat. 10–4:30, Sun. noon–4:30.*

The **National Marine Fisheries Service Aquarium** displays 16 tanks of regional fish and shellfish. There are magnifying glasses and a dissecting scope for kids to use to examine marine life, and several hands-on pools with banded lobsters, crabs, snails, sea stars, and other creatures. The star attractions are two harbor seals, which can be seen in the outdoor pool near the entrance in summer. ⊠ *Corner of Albatross and Water Sts.,* ☎ *508/495–2267.* ☞ *Free.* ⊙ *Late June–mid-Sept., daily 10–4; mid-Sept.–late June, weekdays 10–4.*

Free guided walking tours of the village are available Tuesdays at 4 in July and August from the **Bradley House Museum,** across from the Martha's Vineyard ferry parking lot. In the archives of the historical collection you'll find old ships' logs, postcards, newspaper articles, maps, diaries and photographs, more than 200 tapes of oral history provided by local residents, and a 100-volume library on maritime history. The three-building museum complex houses paintings, a restored Woods Hole Spritsail boat, boat models, and a model of the town as it looked in the 1890s. ⊠ *573 Woods Hole Rd.,* ☎ *508/548–7270.* ☞ *Donations accepted.* ⊙ *Museum: mid-June–late Sept., Tues.–Sat. 10–4; archives: year-round, Tues. and Thurs. 10–2.*

The 1888 **Episcopal Church of the Messiah,** a stone church with a conical steeple and a small medicinal herb garden in the shape of a Celtic cross, is a good place for some quiet time. The garden is enclosed by a holly hedge and has a bench for meditation. Inscriptions on either side of the carved gate read ENTER IN HOPE and DEPART IN PEACE. ⊠ *Church St.,* ☎ *508/548–2145.* ☞ *Free.* ⊙ *Daily sunrise–sunset.*

**Nobska Light,** a 42-ft cast-iron tower lined with brick, was built in 1876 with a stationary light. It shows red to indicate dangerous waters or white for safe passage. Since the light was automated in 1985, the adjacent keeper's quarters have been the headquarters of the Coast Guard group commander—a fitting passing of the torch from one safeguarder of ships to another. The lighthouse is not open to the public except during special tours. Still, it's an impressive sight, as are the spectacular views from its base of the nearby Elizabeth Islands and of Martha's Vineyard, across Vineyard Sound. ⊠ *Church St.*

## Dining and Lodging

**$$** ✕ **The Fish Monger's Café.** To say that this is the best restaurant in Wood's
**★** Hole sounds like faint praise in the scheme of Cape and island restau-
rants, but the menu is ambitious—particularly the inventive daily spe-
cials that take a new look at traditional seafood. The fried calamari
appetizer is light and perfectly fried and served with a hot pepper
sauce, and the hearty *bruschetta* (broiled bread slices) with ricotta, fresh
herbs, basil leaves, and olives is a treat. Many grilled seafood dishes
come with tropical fruit sauces or glazes—mango and cilantro sauce
over grilled salmon is particularly delectable. The main dining room
overlooks the water, and there is a bar attached to the restaurant. Break-
fast and lunch are also served. All in all, it's one of those places that
manages to be both relaxed and attentive at the same time. ⊠ *25 Water
St.,* ☎ *508/540-5376. Reservations not accepted. AE, MC, V. Closed
Dec.–mid-Feb. and Tues. off-season.*

**$$** ▥ **Woods Hole Passage.** Recent "wash ashores" Todd and Robin
Norman have tastefully converted this century-old carriage house and
barn into a romantic showcase. There's nothing traditional about their
style—rich earth tones and muted natural hues prevail. The large common room, with its dusky rose walls and yellow pine floor, is sparsely
furnished with a smattering of clay pots, books, and a floor-to-ceiling
abstract painting that instantly says "relax." There are five rooms; one
(the smallest) is in the main house, the others are in the restored barn.
The ultra-feminine Angel Room has lace curtains, eyelet bedding on a
queen-size rattan bed, and a ceiling fan for warm nights. The home-
cooked breakfast is served on nice days on the flagstone patio amid
lush gardens. The inn is close to the Martha's Vineyard ferry stop, and
the Normans sell (at cost) tickets to avoid lines at the dock. Early-risers
can request a "breakfast-in-a-bag" to take with them. Afternoon refreshments are served. ⊠ *186 Woods Hole Rd., 02540,* ☎ *508/548–
9575 or 800/790–8976,* ℻ *508/540–4771. 5 rooms. Air-conditioning,
tennis court, bicycles. No smoking. AE, D, DC, MC, V. BP.*

**$** ▥ **Marlborough House.** This gem is somewhat hard to find—look for
the white picket fence and small black-and-white sign. A classic Cape
house, built in 1942, it has five rooms in the main house, each with
private bath and each named for its predominate color. Whether you
choose the Green (or Yellow, Rose, Blue, or Pink) Room, expect to find
the same basic touches: wicker beds, lace curtains, hooked rugs, fresh
flowers, floral print bedcoverings and pillows. Lavender-scented sheets
are nice if you don't mind that the smell permeates the rooms. There's
a detached cottage out back, next to the kidney-shape, in-ground pool.
The Wisteria Cottage, small in scale and with redbrick floors, has a
queen-size bed and a sitting room. ⊠ *320 Woods Hole Rd., 02543,*
☎ *508/548–6218 or 800/320–2322,* ℻ *508/457–7519. 5 rooms, 1
cottage. Pool. No smoking. AE, MC, V.*

## Nightlife and the Arts

**Dome Restaurant** (⊠ Woods Hole Rd., ☎ 508/548–0800) has dancing to a jazz band in its lounge, under an authentic Bucky Fuller
geodesic dome, on Sunday in July and August; there is easy-listening
piano on Friday and Saturday in season.

**Woods Hole Folk Music Society** (⊠ Community Hall, Water St., ☎ 508/
540–0320) presents professional and local folk and blues in a no-
smoking, no-alcohol environment, with refreshments available during
intermission. There are concerts of nationally known performers the
first and third Sundays of the month from October to April, as well
as on the first Sunday of May.

# East Falmouth and Waquoit

*3 mi northeast of Falmouth.*

**❼** Like Heritage Plantation in Sandwich, **Ashumet Holly and Wildlife Sanctuary,** a 45-acre tract of woodland, shady groves, meadows, and hiking trails, was purchased and donated by Josiah K. Lilly III to preserve local land. Overseen by the Massachusetts Audubon Society, the reserve is true to its name and its original plantsman, Wilfred Wheeler, with its 1,000-plus holly trees and shrubs, comprised of 65 American, Asian, and European varieties. Grassy Pond is home to numerous turtles and frogs, and in summer, 35 nesting pairs of barn swallows live in open rafters of the barn—the sanctuary director can show them to you. Tours to nearby Cuttyhunk Island (on a 50-ft sailing vessel) leave Falmouth Harbor every Sunday from mid-July to mid-October. In December they hold a three-day holly sale. Self-guided-tour maps are available, and downstairs in the nature center there is a small gift shop with handcrafted items. ⊠ *286 Ashumet Rd., E. Falmouth,* ☎ *508/563–6390.* ☞ *$3.* ☉ *Trails: daily sunrise–sunset; office: Tues.–Sat. 9–4.*

**Waquoit Bay National Estuarine Research Reserve** encompasses 2,500 acres of estuary and barrier beach around the bay, making it a good birding spot. **South Cape Beach** is part of the reserve, where interpretive walks are held or you can sun on the sands. **Flat Pond Trail** runs through a variety of habitats, including fresh- and saltwater marshes. **Washburn Island** is accessible by boat (your own) or by Saturday-morning tours (call to reserve), with 330 acres of pine barrens and trails, 10 wilderness campsites (permit required), and swimming. The visitor center at **reserve headquarters,** a 23-acre estate, is staffed in summer by volunteers and has guided field trips. A Tuesday-evening series is held midsummer (bring picnics) on environmental, historical, and artistic subjects; call for program updates. ⊠ *Rte. 28, 3 mi west of Mashpee Rotary, Waquoit,* ☎ *508/457–0495.*

**Tony Andrews Farm and Produce Stand** has pick-your-own strawberries (mid-June–early July), peas, beans, corn, and tomatoes (late June–late August), as well as other produce on the stand. In the fall you can pick your own pumpkins. ⊠ *398 Old Meeting House Rd., E. Falmouth,* ☎ *508/548–5257.*

# Mashpee

**❽** *9 mi east of Falmouth, 10 mi south of Sandwich.*

Mashpee is one of two Massachusetts towns (the other is Gay Head, now also known by its original Wampanoag name, Aquinnah, on Martha's Vineyard) that has municipally governed, as well as Native-American governed, areas. In 1660 the missionary Reverend Richard Bourne gave a 16-square-mi parcel of land to the Wampanoags (what irony—an outsider "giving" natives their own land). Known as the Mashpee Plantation and governed by two local sachems, it was the first Indian reservation in the United States. In 1870 Mashpee was founded as a town. Just over a hundred years later, in 1974 the Mashpee Wampanoag Tribal Council was formed to continue their government with a Chief, Supreme Sachem, Medicine Man, and Clan Mothers. More than 600 residents are descended from the original Wampanoags, and some continue to observe their ancient traditions.

Housed in a 1793 half Cape, the **Wampanoag Indian Museum** is a small and somewhat disappointing museum on the history and culture of the tribe. Exhibits include baskets, weapons, hunting and fishing tools, clothing, arrowheads, and a small diorama depicting a scene from an early settlement. If you're curious, don't hesitate to ask questions of the

Wampanoag staff. There is also a herring run on site, a small tidal stream that runs from the ocean into a pond. In spring, thick schools of herring swim to the calm safety of the pond to spawn. Indians observed this cycle and would net great numbers of herring as the fish made their way back to the sea. ⊠ *Rte. 130,* ☎ *508/477–1536.* ⊡ *Donations accepted.* ⊙ *Call for hours.*

The town landing gives a spectacular view of the interconnecting **Mashpee and Wakeby ponds,** the Cape's largest freshwater expanse, popular for swimming, fishing, and boating.

The **Old Indian Meeting House** was originally built on Santuit Pond in 1684, and later moved to its present site on Route 28. The oldest standing church on Cape Cod, the Meeting House is still used by the Mashpee tribe for worship and meetings. In summer, memorials and religious events are held, some incorporating traditional Wampanoag services. ⊠ *Rte. 28,* ☎ *508/477–1536.* ⊙ *By appointment only.*

The **Old Indian Burial Ground,** near the meeting house, is an 18th-century cemetery with interesting headstones typical of the period, carved with scenes and symbols, and inscribed with witty sayings. A perfect place for bird-watching, fishing or canoeing, the **Mashpee River Woodlands** comprises 391 acres of conservation land along the Mashpee River. More than 8 mi of trails meander through the marshlands and pine forests. Park on Quinaquisset Avenue, River Road, or Mashpee Neck Road (where there is a public landing for canoe access).

### Around Mashpee

Several towns surrounding Mashpee are worth a visit if time allows. **New Seabury** and Popponesset are resort–residential communities that are actually divisions of Mashpee itself. New Seabury has one of the best golf courses on the Cape.

**❾** **Cotuit** is a small, picturesque New England town with several historic attractions. Once called Cotuit Port, the town was formed around seven Crocker family homesteads.

**★** The **Cahoon Museum of American Art** is in one of the old Crocker family buildings in town, a 1775 Georgian Colonial farmhouse that was once a tavern and an overnight way station for travelers on the Hyannis–Sandwich Stagecoach line. Its several rooms display selections from the permanent collection of American primitive paintings by Ralph and Martha Cahoon along with other 19th- and early 20th-century art. Special exhibitions, classes, talks, and demonstrations are held throughout the summer. A gift shop sells cards, prints, and collectibles based on works from the museum's permanent collection and special exhibitions. ⊠ *4676 Rte. 28, Cotuit,* ☎ *508/428–7581.* ⊡ *Donations accepted.* ⊙ *Apr.–Jan., Tues.–Sat. 10–4.*

The first motor-driven fire fighting apparatus on Cape Cod, a 1916 Model T Chemical Fire Engine, is housed in the **Santuit–Cotuit Historical Society Museum,** behind the **Samuel Dottridge Homestead,** which itself dates from the early 1800s. ⊠ *1148 Main St., Cotuit,* ☎ *508/428–0461.* ⊙ *Mid-June–mid-Oct., Thurs.–Sun. 2:30–5.*

### Dining and Lodging

**$$$$** ✕ **Regatta of Cotuit.** Like its sister restaurant in Falmouth (☞ *above*), the Cotuit Regatta serves elegant interpretations of local foods in an equally elegant setting. Lately the menu has been reaching for a wider global cuisine, and a younger audience, but it is still much more haute than hot. In a restored Colonial stagecoach inn, Regatta is fitted with wood and brass and decorated with Oriental carpets. In the dining room, classic yet original fare comes in the form of pâtés of rabbit, veal, and venison, and a signature seared loin of lamb with cabernet sauce, surrounded by chèvre, spinach, and pine nuts. Three menus are served each night: lighter fare, regular fare, and a three-course early dinner menu. There is also a cozy taproom with its own bar menu. ⊠ *Rte. 28, Cotuit,* ☎ *508/428–5715. AE, MC, V. No lunch.*

**$$$** ✕ **Popponesset Inn.** You'll see a wedding or two at the Popponesset Inn every weekend in summer, and for good reason. The site is picture-perfect with old Cape Cod saltbox houses, lovely Nantucket Sound in the background, and perfect evening light. The restaurant has a lounge, Poppy's, and a number of small dining rooms, some with skylights and others looking over the water through glass walls. Best of all, sit outside, either under the tent or at one of the umbrella tables. The food may not be exceptional—traditional offerings include baked stuffed lobster, grilled swordfish, and steak—but the atmosphere is. There's dancing to a band on weekends. ⊠ *Shore Dr.,* ☎ *508/477–1100 or 508/477–8258. AE, DC, MC, V. Closed Nov.–Mar. and Mon. and Tues. Labor Day–Apr. No lunch Labor Day–mid-June.*

**$$** ✕ **The Flume.** The Cape and islands belonged to the Wampanoags for
★ centuries before the Pilgrims arrived, and the Flume, owned for 25 years by Wampanoag elder (and author) Earl Mills, expresses that cultural tradition in great food and atmosphere—Indian artifacts adorn the dining room. The food concentrates on a few New England staples—the chowder is considered to be among the best on the Cape. Traditional codfish cakes and beans, Yankee pot roast, and Indian pudding are done to perfection. The Sunday roast turkey is a genuine treat, as are Aunt Jan's desserts. In spring and summer, seafood specials are featured, especially soft-shell crabs. And if you have any questions about Mashpee's remarkable Wampanoag history, Earl will happily provide the answers. ⊠ *Lake Ave., off Rte. 130,* ☎ *508/477–1456. MC, V. Closed Thanksgiving–Easter. No lunch Oct. 12–Thanksgiving.*

**$$$$** ✕🖭 **New Seabury Resort and Conference Center.** On a 2,000-acre point
★ surrounded by the waters of Nantucket Sound, this self-contained resort community rents apartments in some of its 13 "villages." For example, Maushop Village is a gray-shingled oceanfront complex of buildings set among narrow lanes of crushed seashells, with rugosa roses trailing over white picket fences and trellises; the interiors attractively mix Cape-style and modern furnishings. Town house units in Sea Quarters have solariums with whirlpools and gas-log fireplaces. All units have full kitchens and washer-dryers. Among the amenities (many of them seasonal) are fine oceanfront dining, a lounge and restaurant overlooking the fairways, a vast tennis facility, a 3-mi private beach and an oceanfront pool, miles of jogging trails through the woods, and Popponesset Marketplace. Golf and other packages are available. ⊠ *Box 550, New Seabury 02649,* ☎ *508/477–9400 or 800/999–9033,* FAX *508/477–9790. 167 1 or 2-bedroom units. 5 restaurants, 2 pools, 2 18-hole golf courses, miniature golf, 16 tennis courts, health club, jogging, soccer, beach, windsurfing, boating, jet skiing, bicycles, pro shops, children's programs. AE, DC, MC, V.*

## Nightlife and the Arts

**Bobby Byrne's Pub** (⊠ Mashpee Commons, Rtes. 28 and 151, ☎ 508/477–0600)—celebrating its 25th anniversary in 1998—offers a comfortable pub atmosphere, an outdoor café, a jukebox, and good light and full menus.

The **Boch Center for the Performing Arts** (☎ 508/477–2580) presents top-name performers year-round, and is home to the Annual New England Jazz Festival, held Labor Day weekend. Plans are underway to complete a complex of buildings by 1999. The new facilities, to be set in an idyllic 10-acre tract of countryside in Mashpee, will include an amphitheater that will accommodate 2,000 people. Until the complex is completed, performances are held at Mashpee Commons or in the auditoriums of local high schools.

The 90-member **Cape Cod Symphony Orchestra** (☎ 508/362–1111) comes to Mashpee on Saturday, July 25, 1998, for its annual Sounds of Summer Pops Concert (they also perform once a summer in Hyannis and Orleans). The performance takes place at the Mashpee Commons.

### Outdoor Activities and Sports

BEACH

**South Cape Beach** in Mashpee is a 2½-mi-long state and town beach on warm Nantucket and Vineyard sounds, accessible via Great Neck Road south from the Mashpee rotary. Wide, sandy, and pebbly in parts, with low dunes and marshland, it's a beach where you can walk to get a bit of privacy. Its only services are portable toilets. A hiking trail loops through marsh areas and ponds, linking it to the **Waquoit Bay Reserve** (☞ East Falmouth and Waquoit, *above*).

GOLF

**New Seabury Country Club** (✉ Shore Dr., New Seabury, ☎ 508/477–9110) has a superior 18-hole, par-72 championship layout on water and an 18-hole, par-70 course. Both are open to the public September–May (space-available basis, proper attire required).

### Shopping

**Mashpee Commons** (✉ Junction of Rtes. 28 and 151, Mashpee, ☎ 508/477–5400) has about 50 restaurants, art galleries, and shops (including Starbucks and the Gap) in an attractive village square setting. There's also a multi-screen movie theater, and free outdoor entertainment in summer.

**Popponesset Marketplace** (✉ 2½ mi south of Rte. 28 from Mashpee rotary, New Seabury, ☎ 508/477–9111), open late spring to early fall, has 20 shops (boutique clothing, antiques), eating places (a raw bar, pizza, ribs, Ben & Jerry's), miniature golf, and weekend entertainment (bands, fashion or puppet shows, sing-alongs) by the sea.

# THE MID CAPE

The designation Mid Cape is one that actually makes sense. It comprises the central section of the Cape and the major towns of Barnstable, the commercial hub of **Hyannis,** and **Hyannis Port,** a well-groomed enclave and site of the Kennedy compound. You'll find some of the most heavily populated and touristed areas Mid Cape, along with a number of historic districts, and scenic back roads.

**Barnstable** is first in line, with its villages of **West Barnstable, Osterville, Centerville, Marstons Mills,** and Hyannis. The **Yarmouths** on both routes 6A and 28 follow. Then come the **Dennis** townships on both the bay and sound sides.

## Barnstable

🔟 *4 mi west of Yarmouth, 4 mi north of Hyannis.*

With more than 43,000 year-round residents, Barnstable is the largest town on the Cape. It's also the second oldest—it was founded in 1639, two years after Sandwich, its lone predecessor. You'll get a feeling for its age in **Barnstable Village,** a lovely area of large old homes dominated by the Barnstable County Superior Courthouse, behind which is a complex of government buildings best avoided—the county covers

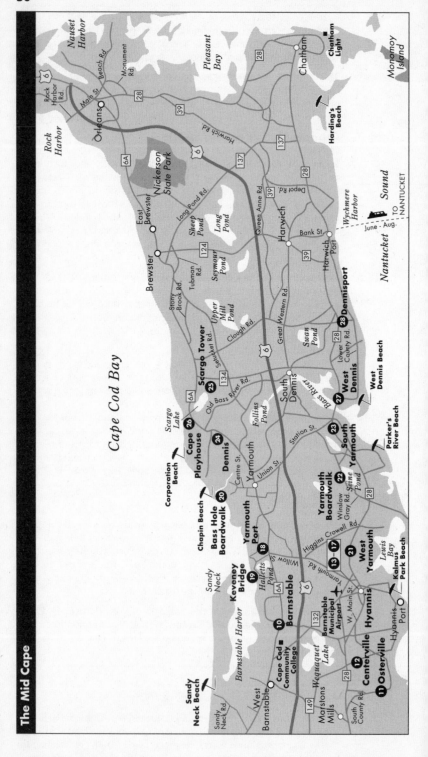

the entire Cape. The Village Hall is home to the Barnstable Comedy Club, one of the oldest community theater groups in the country. The Cape Cod Conservatory of Music and Arts and Cape Cod Community College are also in the municipality.

The **Olde Colonial Courthouse,** built in 1772 as the colony's second courthouse, is the home of the historical society **Tales of Cape Cod** (⊠ Rte. 6A, ☎ 508/362–8927), which is restoring it to serve as a museum. A series of slide-illustrated lectures, given by guest speakers, is held Tuesday nights in July and August. From the courthouse, take a peek across the street at the old-fashioned English gardens at **St. Mary's Episcopal Church.**

Set in a 1644 building listed on the National Register of Historic Places, the **Sturgis Library** was established in 1863. Its holdings date from the 17th century and include hundreds of maps and land charts, the definitive collection of Cape Cod genealogical material ($5 daily fee for nonresidents), and an extensive maritime history collection. ⊠ *3090 Rte. 6A,* ☎ *508/362–6636.* ◑ *Mon. and Thurs. 10–2, Tues.– Wed. 1–9, Fri. 1–5, Sat. 10–4.*

Barnstable's maritime past is on display at the **Trayser Museum Complex.** Listed on the National Register of Historic Places, the red-painted brick main building houses a small collection of maritime exhibits— telescopes, captains' shaving boxes, items brought back from voyages, ship models and paintings—as well as ivory, Sandwich glass, and Indian arrowheads. The downstairs re-creates the way the building looked in 1856, when it served as a customs house—don't miss the ornate bronze Corinthian-order columns. A restored customs-keeper's office with an original safe and a view of the harbor are on the second floor. Also on the grounds are a jail, circa 1690, with two cells bearing former inmates' graffiti, and a carriage house with early tools, fishing implements, and a 19th-century horse-drawn hearse. ⊠ *Rte. 6A,* ☎ *508/362–2092.* ▨ *$2 suggested donation.* ◑ *Mid-June–mid-Oct., Tues.–Sun. 1:30–4:30.*

If you are interested in Cape history, the **Nickerson Memorial Room** at **Cape Cod Community College** has the largest collection of information on Cape Cod, including books, records, ships' logs, oral-history tapes, photographs, films, and more. It also has materials on the islands of Martha's Vineyard and Nantucket. ⊠ *Rte. 132, W. Barnstable,* ☎ *508/362–2131, ext. 4445.* ◑ *Mon., Wed., Fri. 8:30–4; Tues. 8:30–3.*

★ Hovering above Barnstable Harbor and the 4,000-acre **Great Salt Marsh, Sandy Neck Beach** stretches some 6 mi across a peninsula that ends at **Sandy Neck Light.** The beach is one of the Cape's most beautiful—dunes, sand, and sea spread endlessly east, west, and north. The marsh used to be harvested for salt hay; now it is a haven for birds, which are out and about in greatest number morning and evening, at low tide, and during spring and fall migration. The lighthouse, standing just a few feet from eroding shoreline at the tip of the neck, has been out of commission since 1952. It was built in 1857 to replace an 1827 light, and it used to run on acetylene gas. It is now privately owned. The main beach at Sandy Neck has lifeguards, a snack bar, rest rooms, and showers. ⊠ *Sandy Neck Rd., W. Barnstable, off Rte. 6A just over the West Barnstable line.* ▨ *$10 parking fee, Memorial Day–Labor Day.* ◑ *Daily 9–9, but staffed only until 5 PM.*

## Dining and Lodging

$$$  ✕ **Mattakeese Wharf.** Route 6A goes through a lot of changes as it winds its way up the arm of the Cape, and around Barnstable, it gets downright residential, almost suburban. Head north at the traffic light, and the road eventually ends at Mattakeese Wharf, a traditional seaside haunt that's a favorite of ladies who lunch, and who also demand a screened-in view, preferably through glass. The high points of this place are the sea breeze and chowder. The bouillabaisse, with a savory saffron broth, is extremely tasty, and seafood specials are very traditional but good, particularly herb-grilled tuna, or broiled scrod for purists. ✉ *271 Millway, Barnstable Harbor,* ☎ *508/362–4511. AE, DC, MC, V. Closed Nov.–Apr.*

$$   ⊞ **Ashley Manor.** Set back from the Old King's Highway by high privet hedges and a wide lawn, this fine B&B is just a walk from the village. The inn preserves its antique wide-board floors and has open-hearth fireplaces, one with a beehive oven, in the living room, the dining room, and the keeping room. Antique and country furnishings, Oriental rugs, and brass and crystal details create an elegant atmosphere. All rooms but one have a working fireplace or woodstove. Breakfast is served on the backyard terrace, which looks onto fruit trees, a gazebo, and the tennis court, or in the formal dining room with china and crystal. There's also a country-Colonial one-room cottage. ✉ *3660 Main St., Rte. 6A, Box 856, 02630,* ☎ *508/362–8044 or 888/535–2246,* ⅏ *508/362–9927. 2 rooms, 3 suites, 1 cottage. Air-conditioning, tennis court, croquet, bicycles. No smoking. D, MC, V. BP.*

$$   ⊞ **Beechwood Inn.** Debbie and Ken Traugot's yellow and pale green
★    1853 Queen Anne Victorian is trimmed with gingerbread, wrapped by a wide porch with wicker furniture and a glider swing, and shaded by old beech trees. While the parlor is pure mahogany-and-red-velvet Victorian, guest rooms are decorated with antiques in lighter, earlier Victorian styles (including Eastlake). Two have fireplaces, and one has a water view. Bathrooms have pedestal sinks and antique plumbing and lighting fixtures. Breakfast is served in the dining room, which has a pressed-tin ceiling, fireplace, and lace covered tables set with hurricane lamps, fresh flowers, china, and crystal. Afternoon tea and homemade snacks are also available. ✉ *2839 Main St., Rte. 6A, 02630,* ☎ *508/ 362–6618 or 800/609–6618,* ⅏ *508/362–0298. 6 rooms. Refrigerators, bicycles. No smoking. AE, MC, V. BP.*

$$   ⊞ **Cobb's Cove.** Proprietors Henri-Jean and Evelyn Chester built this inn in 1974 for people who share their fondness for nature and relaxation. Lush gardens, with a dazzling collection of bird feeders and fountains, surround the house—ask for a tour. The 12-by-12-ft timber-cut interior is Colonial rustic, with exposed wood, rough burlap walls, and heavy wooden doors dotted (just like they did in the olden days) with nail heads. Guest rooms are large; each has a dressing area and a private bath with whirlpool tub and robes. The two top-floor rooms have glass walls, and from the sitting areas you can enjoy the spectacular view of Cape Cod Bay—on a good day, you can see all the way to Provincetown! For an extra charge, you can request in advance a three- to five-course dinner served in the rustic dining room, dominated by a Count Rumford–designed fireplace. Historic Barnstable village, beaches, even whale-watching boats in the marina, are a short stroll away. ✉ *31 Powder Hill Rd., Box 208, 02630,* ☎ ⅏ *508/362–9356. 6 rooms. No smoking. D, MC, V. BP.*

$$   ⊞ **Heaven on High.** There's no more appropriately named B&B on the
★    Cape. Deanna and Gib Katten's 11-year-old haven is indeed heaven, nestled high on a hill, on one of the Cape's oldest roads, overlooking dunes, Great Salt Marsh, and the Bay at Sandy Neck. The house's decor

is a meld of California beach house and Cape Cod comfort—light, airy, and breezy. The Great Room is filled with overstuffed chairs and couches, a fireplace, TV, and natural oak floors. Sliding glass doors lead to the living room, where you can relax in solitude, and also to the deck that runs the full length of the house, offering unobstructed panoramic views of sand and surf. Each of the three rooms is named for the collection which it houses; the Silhouettes and Mirrors Room, for instance, has silhouettes and mirrors of all types, as well as a queen-size bed, fireplace, tiled bath with two vanity sinks, and a large private deck, accessible through sliding glass doors. The breakfasts are almost angelic. ⊠ *70 High St., Box 346, 02668,* ☎ *508/362–4441 or 800/362–4044,* 𝔽𝔸𝕏 *508/362–4465. 3 rooms. Air-conditioning. No smoking. No credit cards. BP.*

$ ⊞ **Acworth Inn.** Cheryl and Jack Ferrell, self-described "corporate drop-outs," have run Acworth since 1994. Jack will help prepare your sightseeing excursions, while Cheryl sees to fresh flowers and chocolates in your room. Their historic 1860 house, which has been added to throughout the years, has six large rooms; each is decorated in soft pastels, designer linens, and tasteful, hand-painted furniture. Cheryl is also known for her breakfasts (one specialty is homemade granola, with strawberries that come from the strawberry patch out back), cinnamon rolls, and chocolate chip cookies. You can work off the calories by taking advantage of the inn's free loaner bikes, or just doze off the sugar rush in a rocker on the porch. ⊠ *4352 Main St., Rte. 6A, Box 256, Cummaquid 02637,* ☎ *508/362–3330 or 800/362–6363,* 𝔽𝔸𝕏 *508/375–0304. 6 rooms. Bicycles. No smoking. AE, D, MC, V. BP.*

$ ⊞ **Crocker Tavern.** A stately Georgian red clapboard home that dates back to 1754, the Crocker Tavern sits on more than 3 acres of an unspoiled stretch of Route 6A, in walking distance of Barnstable harbor and several historic sites. Jeff and Sue Carlson have lovingly restored it to look (and feel) as it did back when it served as a stagecoach stop and tavern during the Revolutionary War. (The stone mileage marker, with distances to Plymouth and Boston, sits in front of the house.) The five large guest rooms, all with four-poster or canopy beds and sitting areas, are decorated with antique or reproduction Shaker or Colonial sideboards and dressers, eyelet curtains, quilts, Oriental and hooked rugs, and window seats. Two rooms have working fireplaces; three have detached (albeit private) baths (some with claw-foot tubs and antique sinks) and terry robes. Afternoon tea and light snacks are served in the parlor. ⊠ *3095 Main St., Rte. 6A, 02630,* ☎ *508/362–5115 or 800/773–5359,* 𝔽𝔸𝕏 *508/362–5562. 5 rooms. Air-conditioning, bikes. No smoking. MC, V. CP.*

## Nightlife and the Arts

The **Barnstable Comedy Club** (⊠ Village Hall, Rte. 6A, ☎ 508/362–6333), the Cape's oldest amateur theater group (they celebrated their 75th anniversary in 1997), gives much-praised musical and dramatic performances throughout the year. Some folk who appeared here before they made it big: Geena Davis, Frances McDormand, and Kurt Vonnegut, a past president of the BCC.

## Outdoor Activities and Sports

### FISHING

The **Barnstable Harbor Charter Fleet** (⊠ 186 Millway, ☎ 508/362–3908) has fishing trips on 10 sport-fishing vessels on a walk-on basis from spring through fall.

### WHALE-WATCHING

On **Hyannis Whale Watcher Cruises** out of Barnstable Harbor, a naturalist narrates and comments on whale sightings and the natural history of Cape Cod Bay. Trips last four hours, there are concessions on board, and in July and August sunset whale-watches depart at 5 PM. ⊠ *Millway, Barnstable,* ☎ *508/362–6088 or 800/287–0374.* ☞ *$23. Tours Apr.–Oct.*

### Shopping

**Black's Handweaving Shop** (⊠ 597 Rte. 6A, ☎ 508/362–3955), in a barnlike shop with working looms, makes beautiful handwoven goods in traditional and jacquard weaves.

**The Crystal Pineapple** (⊠ 1540 Rte. 6A, ☎ 508/362–3128 or 800/462–4009) has cranberry glass and many collectibles lines, including Dept. 56, Snowbabies, Swarovski crystal, and Disney Classics.

**Salt & Chestnut** (⊠ 651 Rte. 6A, ☎ 508/362–6085) has antique and custom-designed weather vanes displayed indoors and in the yard—a fun place to browse.

**Whippletree** (⊠ Rte. 6A, ☎ 508/362–3320) is a large barn, decorated for each season and filled with country gift items and a year-round Christmas section. Offerings include German nutcrackers, from Prussian soldiers to Casanovas.

## Osterville and Centerville

*4 mi west of Hyannis, 6 mi southwest of Barnstable.*

⓫ A wealthy Barnstable enclave southwest of the town center, **Osterville** is lined with elegant waterfront houses, some of which are large "cottages" built in the 19th century, when the area became popular with a monied set. There is upscale shopping downtown.

An 1824 sea captain's house is the setting for the **Osterville Historical Society Museum,** which displays antiques, dolls, and exhibits on Osterville's history. Two boat museums, each showcasing various sailing vessels, as well as the Cammett House (dating from the early 18th century) are also on the property. ⊠ *155 West Bay Rd., Osterville,* ☎ *508/428–5861.* ☞ *$2.* ☺ *Mid-June–Oct., Thurs.–Sun. 1:30–4:30.*

⓬ **Centerville** was once a busy seafaring town, which is evident from the 50 or so shipbuilders' and sea captains' houses along its quiet, tree-shaded streets. Offering the pleasures of both sheltered ocean beaches on Nantucket Sound, such as **Craigville Beach** (☞ *below*), and freshwater swimming in Lake Wequaquet, it has been a popular vacation spot since the mid-19th century. Shoot Flying Hill Road, named by the Indians, is the highest point of land on the Cape, with panoramic views to Plymouth and Provincetown to the north, and to Falmouth and Hyannis to the south.

Set in a 19th-century house, the **Centerville Historical Society Museum** exhibits furnished period rooms, Sandwich glass, miniature carvings of birds by Anthony Elmer Crowell, models of ships, marine artifacts, military uniforms and artifacts, antique tools, perfume bottles (dating from 1760 to 1920), 300 quilts and costumes (from 1650 to 1950), and a research library. Each summer there is a special costume exhibit and one other select exhibit. ⊠ *513 Main St.,* ☎ *508/775–0331.* ☞ *$2.50.* ☺ *Mid-June–mid-Sept., Wed.–Sun. 1:30–4:30; or by appointment.*

The **1856 Country Store** sells penny candy, except that these days, the candy costs at least 25 pennies. The store also sells newspapers, coffee, crafts, jams, and all kinds of gadgets and toys. You can sip your coffee—and take a political stance—by choosing one of the wooden benches out front, marked DEMOCRAT and REPUBLICAN. ⊠ *555 Main St.,* ☎ *508/775–1856.*

| NEED A BREAK? | Sample the homemade offerings at **Four Seas Ice Cream,** a tradition with generations of summer visitors. It's open Memorial Day–Labor Day. ⊠ *360 S. Main St.,* ☎ *508/775-1394.* |

## Dining and Lodging

**$$$** ✕ **East Bay Lodge.** Quality and attentive service are the defining factors here, along with a creative handling of old favorites on the menu. The main dining room (part of it is a beautiful glassed-in veranda) is simple and comfortable. Favorite appetizers like wild mushroom tart and chilled poached salmon seem inspired every time. The signature Chateaubriand Boquetière—perfectly done with an outstanding side plate of potato gratin, tomato confit, and asparagus with bordelaise sauce—realistically serves two. The wine list—immense and pricey—is heavy on heavy-hitting classics, with lots of big reds that are unusual for the Cape's seasonal visitors; there are also the more typical summer whites. Many nights jazz or cabaret piano is played in the lounge. ⊠ *199 East Bay Rd., Osterville,* ☎ *508/428–5200. AE, D, DC, MC, V. No lunch Labor Day–Memorial Day.*

**$$** ⌂ **Inn at Fernbrook.** Part of a parcel originally landscaped by Frederick Law Olmsted, designer of New York's Central Park and Boston's Emerald Necklace, the inn has duck ponds, a heart-shaped sweetheart rose garden, and exotic trees; a right-of-way grants access to the eponymous fern-rimmed brook. The house itself is an 1881 Queen Anne Victorian mansion on the National Register of Historic Places—a beauty from its turreted exterior to the fine woodwork and furnishings within. Most rooms have garden views; some have bay-window sitting areas, canopy beds (all beds are queen- or king-size), and working fireplaces. Breakfast is friendly and delicious. ⊠ *481 Main St., Centerville 02632,* ☎ *508/775–4334,* 𝔽𝔸𝕏 *508/778–4455. 4 rooms, 1 2-bedroom suite, 1 cottage (no kitchen). No smoking. AE, D, MC, V. BP.*

**$$** ⌂ **Tradewinds Inn.** Cape Cod conjures up images of sand and surf, and for the past 28 years, the Tradewinds Inn has offered both—and a lot more. The 6-acre property, overlooking the Atlantic Ocean, Lake Elizabeth, and the quaint village of Craigville, offers oceanview rooms, all with private balconies or patios, *thisclose* to the water. (There are also three efficiencies with fully equipped kitchenettes, but skip the less attractive non-water-view rooms in the detached cottages.) The furnishings are simple but appealing: whitewashed furniture, geometric print wallpapers, and tasteful, if somewhat unspectacular, prints dotting the walls. Just for guests, there's a small, sandy private beach across from the L-shaped main lodge. The wicker-festooned, knotty-pine paneled lobby has a fireplace and a cocktail bar. ⊠ *780 Craigville Beach Rd., Craigville 02636,* ☎ *508/775–0365,* 𝔽𝔸𝕏 *508/790–1404. 28 rooms, 4 suites, 3 efficiencies. Lobby lounge, putting green, beach. AE, MC, V. Closed Nov.–Apr. CP.*

## Nightlife and the Arts

**Benefit Coffeehouse** (⊠ Liberty Hall, Main St., Marstons Mills, ☎ 508/428–1053), of the Liberty Folk Society, presents professional and local folk and blues in a no-smoking, no-alcohol environment, with refreshments available during intermission. Benefit concerts—helping various causes, including children suffering from cancer—are held on the first Saturday of each month.

**East Bay Lodge** (⊠ East Bay Rd., Osterville, ☎ 508/428–5200) has dancing to jazz and to music of the '60s, '70s, and '80s, as well as cabaret-style piano and occasional vocals in its lounge several nights a week year-round.

## Outdoor Activities and Sports

BEACH

**Craigville Beach** is a long, wide strand that is extremely popular with the collegiate crowd (it's also known by its nickname, Muscle Beach). It has lifeguards, showers, rest rooms, and food nearby.

HORSEBACK RIDING

**Holly Hill Farm** (⊠ 240 Flint St., Marstons Mills, ☎ 508/428–2621) offers instruction and day camps but no trail rides.

## Shopping

**Oak & Ivory** (⊠ 1112 Main St., Osterville, ☎ 508/428–9425) specializes in Nantucket lightship baskets made on the premises, as well as gold miniature baskets and scrimshaw.

# Hyannis

**⓭** *11 mi east of Mashpee, 23 mi east of the Bourne Bridge.*

Hyannis was named for the Native American Sachem Iyanno, who sold the area for £20 and two pairs of pants. He would have sold it for far more had there been any indication that Hyannis would become known as the "home port of Cape Cod," or that the Kennedys would have pitched so many tents here. Hyannis is effectively the transportation center of the Cape. A bustling year-round hub of activity, it has the Cape's largest concentration of businesses, shops, malls, hotels and motels, restaurants, and entertainment hot spots. Hyannis's Main Street is lined with book and gift shops, jewelers, clothing stores, summerwear and T-shirt shops, and ice-cream and candy stores. And there are plenty of fun and fancy eateries.

Perhaps best known for its association with the Kennedy clan, the Hyannis area was also a vacation spot for President Ulysses S. Grant in 1874, and later for President Grover Cleveland.

Hyannis is making an effort to preserve its historical connection with the sea. By 1840 more than 200 shipmasters had established homes in the Hyannis–Hyannisport area. Aselton Park (at the intersection of South and Ocean streets) and the Village Green on Main Street are the sites of events celebrating this history, and Aselton Park marks the beginning of the scenic Walkway to the Sea. The project to continue the walkway to the dock area should be completed by the end of 1998.

Three parallel streets run through the heart of town. Busy, shop-filled Main Street is one-way from east to west; South Street runs from west to east; and North Street is open to two-way traffic. The airport rotary connects with major routes 132, 28, and 6.

**⓮** Located in Main Street's Old Town Hall, the surprisingly sparse **John F. Kennedy Hyannis Museum** explores JFK's Cape years (1934–63) through enlarged and annotated photographs, culled from the archives of the JFK Library near Boston, as well as a seven-minute video. The gift shop sells mugs, T-shirts, and presidential memorabilia. ⊠ *397 Main St.,* ☎ *508/790–3077.* ⊡ *$3.* ⊙ *Mon.–Sat., 10–4, Sun. 1–4.*

**⓯** The **St. Francis Xavier Church** (⊠ 347 South St., ☎ 508/775–0818) is where Rose Kennedy and her family worshiped during their summers on the Cape; the pew that John F. Kennedy used regularly is marked by a plaque.

Beyond the bustling docks where waterfront restaurants draw crowds, and ferries, harbor tour boats, deep-sea fishing vessels come and go,
**⓰** the quiet esplanade by the **John F. Kennedy Memorial** overlooks boat-filled Lewis Bay. JFK loved to sail these waters, and in 1966 the peo-

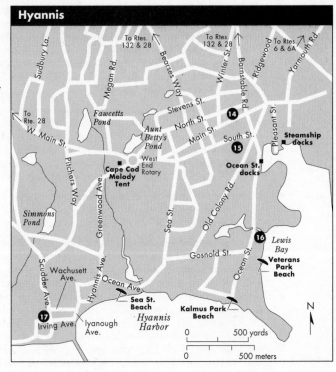

ple of Barnstable erected a plaque and fountain pool here in his memory. Adjacent to the memorial is **Veterans Park,** with a beach, a tree-shaded picnic and barbecue area, and a playground.

Hyannis Port was a mecca for Americans during the Kennedy presidency, when the **Kennedy Compound** became the summer White House. The days of hordes of Secret Service men and swarms of tourists trampling down the bushes are gone, and the area is once again a community of quietly posh estates, though the Kennedy mystique is such that tourists still seek it out. The best way to get a glimpse of the compound is from the water on one of the many harbor tours or cruises.

Joseph P. and Rose Kennedy bought their house here—the largest one, closest to the water—in 1929, as a healthful place to summer with their soon-to-be-nine children. (Son Ted bought the house before his mother's death in 1995.) Sons Jack and Bobby bought neighboring houses in the 1950s. Jack's is the one at the corner of Scudder and Irving, with the 6-ft-high stockade fence on two sides. Bobby's is next to it, with the white fieldstone chimney. Ted bought a home on Squaw Island, a private island connected to the area by a causeway at the end of Scudder Avenue. It now belongs to his ex-wife, Joan. Eunice (Kennedy) and Sargent Shriver have a house near Squaw Island, on Atlantic Avenue.

The compound is relatively self-sufficient in terms of entertainment: Rose Kennedy's former abode (with 14 rooms and nine baths) has a movie theater, a private beach, a boat dock, a swimming pool, a tennis court, and a sports field that was the scene of the famous Kennedy touch-football matches. More recently, Maria Shriver, Caroline Kennedy, and other family members have had their wedding receptions here.

In 1950 the actress Gertrude Lawrence and her husband, producer-manager Richard Aldrich, opened the **Cape Cod Melody Tent** to showcase Broadway musicals and concerts. Today the focus has shifted away from theater to music and stand-up comedy, and the performers who play here, in the round, include Liza Minnelli, Steve and Eydie, Aretha Franklin, Joan Rivers, Vince Gill, Don Rickles, Tony Bennett, and Huey Lewis and the News. The Tent also hosts a Wednesday morning children's theater series in July and August. ⊠ *21 W. Main St., at the West End Rotary,* ☎ *508/775–9100.*

**Cape Cod Storyland Golf** is a 2-acre miniature golf course set up as a mini-Cape Cod, with each of the 18 holes a Cape town. The course winds around small ponds and waterfalls, a full-size working gristmill, and reproductions of historic Cape buildings. There's a $4.50 additional charge for the bumper boats. ⊠ *70 Center St., by the railroad depot,* ☎ *508/778–4339.* ⌷ *$6.* ☺ *Mid-Apr.–Oct., daily 8 AM–midnight.*

Perfect for a rainy day, **Ryan Family Amusement Center** is replete with video-game rooms and that old seaside favorite, Skee-ball. ⊠ *Cape Cod Mall, Rte. 132,* ☎ *508/775–5566.* ☺ *Daily; hours vary.*

The **Cape Cod Scenic Railroad** provides a unique and breathtaking tour of the Cape between Hyannis and Sandwich, accompanied by a running commentary of historical facts and interesting tidbits about old Cape Cod, with stops at the Cape Cod Canal and Sandwich village. In addition to the scenic historical excursions, there is an Ecology Discovery Tour, and an elegant Dinner Train (jackets or ties required for men). ⊠ *Downtown Hyannis Station at Main and Center Sts., 252 Main St.,* ☎ *508/771–3788 or 800/872–4508 (MA only).* ⌷ *Scenic Ride: $11.50, Dinner Ride: $43.95 ($51.30 Saturday).* ☺ *Weekends and holidays in May; June–Oct., Tues.–Sun., and Mon. holidays. Call for schedule.*

OFF THE
BEATEN PATH
**CAPE COD POTATO CHIPS** – Come on weekdays 10–4 for a free factory tour and get free samples of the all-natural chips hand-cooked in kettles in small batches. ⊠ *Independence Park, Rte. 132,* ☎ *508/775–7253.* ☺ *Weekdays 9–5; Sat. in July and August, 10–4.*

## Dining and Lodging

$$$ ✕ **Roadhouse Café.** For more than 16 years, Dave and Melissa
★ Colombo's restaurant has been one the smartest spots for a night out in Hyannis. Candlelight flickers off the white linen tablecloths and dark wood wainscoting. The calamari appetizer is a chef's choice; the codfish chowder, a local favorite. Veal chop is a nightly special, with sauces that change according to the kitchen's whim. All desserts are homemade and well worth the calories, especially the tiramisu. In the more casual bistro and the handsome mahogany bar, you can order from a separate menu, which offers a thin-crust, 16-inch pizza with combinations like arugula, tomatoes, roasted garlic, and goat cheese, as well as burgers, sandwiches, and other lighter fare. Also in the bistro: On Monday nights year-round is the best straight-ahead jazz on Cape Cod, with regulars like pianist Dave McKenna and Lou Colombo (the trumpet-playing father of the chef). There's a piano bar every other night of the week between July 4 and Labor Day, which continues on Friday and Saturday in the off-season. ⊠ *488 South St.,* ☎ *508/775–2386. Reservations essential. AE, D, MC, V. No lunch.*

$$–$$$ ✕ **The Paddock.** For 29 years, the Paddock has been synonymous with
★ excellent, formal dining on the Cape—authentically Victorian with sumptuous upholstery in the main dining room, and old-style wicker on the preferable, breezy summer porch. Its menu manages to be traditional yet deceptively innovative, combining fresh ingredients in novel ways. A salad of duck, apple, sharp Vermont cheddar, and walnuts on baby greens with cider-mustard vinaigrette is but one example. Steak au poivre,

with the promise of five varieties of crushed peppercorns, is masterful; the superb Pacific Rim Chicken is a grilled breast topped with oranges and mangoes, served on mixed greens and Oriental noodles. You can't go wrong with the local shellfish, which turns up in other main course dishes and as a raw-bar appetizer. Terrific brawny reds are on the award-winning wine list. An older crowd enjoys live music in the evenings at the bar, where Manhattans are still a drink of choice. ⊠ *W. Main St. rotary, next to Melody Tent,* ☎ *508/775–7677. AE, DC, MC, V. Closed mid-Nov.–Mar.*

**$$–$$$** ✕ **Penguin Sea Grill.** The fact that a penguin is a bird that swims has symbolic import for this restaurant: Owner-chef Bobby Gold experiments carefully, never losing sight of the fresh grilled seafood that he prepares so well. Baked stuffed lobster (from 1 to 3 pounds) is outstanding, with crabmeat stuffing topped with fresh sea scallops. There are a number of excellent pasta dishes, as well, again with seafood, like pasta Fiore: shrimp, scallops, and lobster with mushrooms, scallions, sherry, and cream on angel-hair pasta. All of the breads and desserts are homemade. The dining room is on two levels, with wood and brick and carvings of sea life galore. ⊠ *331 Main St.,* ☎ *508/775–2023. AE, DC, MC, V. No lunch weekends.*

**$$** ✕ **Harry's.** Homesick for a little bit of the French Quarter on Cape
★ Cod? Harry's feels as though it's transplanted from New Orleans, and serves accordingly—both Cajun and Creole—with great spices and sizable portions. The menu features a number of meal-size sandwiches and blackened local fish that's done to perfection, with a jambalaya that makes you wonder if there isn't a bayou nearby. It's also a prime spot for music: You can hear the blues on Friday and Saturday nights (Wednesday and Thursday as well, Memorial Day through Labor Day). ⊠ *700 Main St.,* ☎ *508/778–4188. Reservations not accepted. AE, DC, MC, V.*

**$$** ✕ **Starbuck's.** No, it's not another branch of the Seattle-based coffee company, but something more like a T. G. I. Friday's with an edge. Walls and rafters are hung with doodads and hoo-has of every description, but un-themed: a tuba here, a couple of mannequins there, a miniature World War I fighter plane from out of nowhere. The menu matches the decor, with selections yanked from all over the planet: ostensibly Asian shrimp deep-fried in a coconut-tempura batter, Buck's beef burrito from the Tex Mex column, and Italian standards like spaghetti and meatballs. There's also an assortment of burgers and Buckwiches. This might not be for everyone, but it's good if you're in the mood for something raucous. ⊠ *645 Rte. 132,* ☎ *508/778–6767. Weekend reservations essential. AE, D, DC, MC, V.*

**$$** ✕ **Up the Creek.** A little more cramped than cozy, with a menu that has changed little in recent years, Up the Creek is a dependable, reasonable spot for an early evening out. Fresh, traditional seafood is what they do best here. The blackboard specials are often the freshest alternatives of the night. Seafood strudel is rich and filling, stuffed with lobster, shrimp, scallops, and crab and doused with a dense sauce. Baked stuffed lobster is also a favorite. ⊠ *36 Old Colony Rd.,* ☎ *508/771–7866. AE, D, DC, MC, V. Closed weeknights Columbus Day–March.*

**$–$$** ✕ **Baxter's Fish N' Chips.** Fried seafood being the Cape staple that it is, you may want to plan ahead for a trip to pay homage to one of the best fry-o-lators around. Fried clams are delicious and generous, cooked up hot to order with french fries and homemade tartar sauce. Outside, a number of picnic tables allow you to lose no time in the sun with lobster, burgers, or something from the excellent raw bar. Indoors, Baxter's Boat House Club, slightly more upscale, serves the same menu, but with a number of very good specials as well. The restaurant is right on Lewis Bay, and it's always been a favorite of boaters and bathers alike. ⊠ *Pleasant St.,* ☎ *508/775–4490. Reservations not accepted. AE, MC, V. Closed weekdays Labor Day–Columbus Day and entirely Columbus Day–Apr.*

**$–$$** ✕ **Sam Diego's.** The bar is busy with people seeing and being seen, and the menu has satisfyingly authentic Mexican tortillas, burritos, and delicious mole *poblano* (chicken with a spicy, bittersweet cocoa sauce). The prop-shop decor is a little cheesy—sombreros and Aztec birds and the like—but the atmosphere is fun and friendly. Especially early evenings, it's popular with families, thanks in part to the all-you-can-eat chili and taco bar—a good option for when the kids need a little change of culinary pace. An interesting dessert, crusty deep-fried ice cream, is served in a giant goblet. Later on, the bar scene picks up steam; dinner is served until midnight. ⊠ *950 Iyanough Rd., Rte. 132,* ☎ *508/ 771–8816. AE, D, MC, V.*

**$** ✕ **Barbyann's.** Sooner or later, your kids are likely to reach their limit of seafood, even fish-and-chips. When it happens, pack everyone off to Barbyann's for an inexpensive family meal—steak and seafood, pizza and burgers, a couple of hardly authentic but kid-pleasing Mexican dishes, and loads of appetizers. Best of all are the Buffalo chicken wings, hot and spicy and filling. There's an outdoor patio with umbrella tables, and "night bird" specials are offered all evening Monday–Thursday for $8.95. ⊠ *120 Airport Rd.,* ☎ *508/775–9795. Reservations not accepted. AE, D, DC, MC, V.*

**$$$$** 🏨 **Breakwaters.** If you were staying any closer to the water, you'd be **★** *in* the water—that's how close these charming weathered gray-shingled cottages are to Nantucket Sound. Privately owned condos that are rented as cottages per day or week, the one-, two-, and three-bedroom efficiencies offer all the comforts of home (except towels and sheets). There's an in-ground heated pool less than 200 ft from the lifeguarded town beach. Each unit has one or two full baths; kitchens with microwave, coffeemaker, refrigerator, toaster and stove; TV/radio and phone (local calls are free); as well as a deck or patio—and most have water views. An added plus: daily (except Sunday) maid service. The Laundromat at the end of the street will wash and fold your dirty duds for a nominal fee. ⊠ *Sea St. Beach, Box 118, 02601,* ☎ ℻ *508/775–6831. 18 cottages (weekly rentals in season). Pool, beach, baby-sitting. No credit cards. Closed mid-Oct.–Apr.*

**$$$** 🏨 **Tara Hyannis Hotel & Resort.** For its beautifully landscaped setting, extensive services and pampering, and superior resort facilities it's hard to beat the Tara. The lobby area is elegant, but the rooms, each with a private balcony, are admittedly a bit dull—standard contemporary hotel style. Views are best from rooms overlooking the golf greens or the courtyard garden. ⊠ *West End Rotary, 02601,* ☎ *508/775– 7775 or 800/843–8272,* ℻ *508/778–6039. 224 rooms. Restaurant, pub, room service, air-conditioning, indoor-outdoor pool, beauty salon, 18-hole golf course, putting green, 2 tennis courts, health spa, children's programs, business services. AE, D, DC, MC, V.*

**$$** 🏨 **Quality Inn Cape Cod.** All the rooms at this business- and family-oriented cinder-block motel just off the highway have white-oak-veneer furnishings, including one king-size or two double beds, a table and chairs or a desk and chair, and a wardrobe. Some king rooms have sofa beds. The quietest rooms are those on the top floor that face the pond and woods; all have free HBO movies. Guests have free use of the nearby Barnstable Athletic Club, and children under 18 stay free. ⊠ *1470 Rte. 132, 02601,* ☎ *508/771–4804 or 800/999–4804,* ℻ *508/790–2336. 104 rooms. Indoor pool, sauna. AE, D, DC, MC, V. CP.*

**$$** 🏠 **Simmons Homestead Inn.** First the bad news: If you can't stand smoke, don't stay here, because smoker Bill Putman readily admits he encourages smoking. Now the good news: The 1820 former sea captain's country estate is a delightful, inviting escape from the world, even though it's only two minutes from congested downtown Hyannis. Each of the 10 rooms (main house) and two suites (detached barn) is named for the animal memorabilia it houses. They have antique, wicker, or canopied four-poster beds and private baths; some have fireplaces and private decks. The Bird Room, the largest, cheeriest room in the menagerie, has an old-fashioned cast-iron tub and a private deck. You can borrow the 10-speed mountain bikes to explore the town, or simply enjoy the garden views and expansive backyard from the wraparound porch. Simmons Pond is a short jaunt away on the property's trail—expect Putnam's dog to tag along. (You can bring your own dog, with advance notice.) ✉ *288 Scudder Ave., Box 578, Hyannis Port 02647,* ☎ *508/ 778–4999 or 800/637–1649,* 🖷 *508/790–1342. 10 rooms, 2 suites. No-smoking rooms, bicycles. AE, D, MC, V. BP.*

**$** 🏠 **Capt. Gosnold Village.** An easy walk to the beach and town, this colony of motel rooms and cottages is ideal for families. Children can ride their bikes around the quiet street, and the pool is fenced in and watched by a lifeguard. In some rooms, walls are attractively paneled with knotty pine; floors are carpeted, and furnishings are colonial or modern, simple, and pleasant. All cottages have decks and gas grills, and receive maid service. ✉ *230 Gosnold St., 02601,* ☎ *508/775–9111,* 🖷 *508/775–8221. 18 cottages (divisible into rooms, efficiencies, and 1- to 3-bedroom cottages). Picnic area, pool, basketball. MC, V. Closed Nov.–mid-Apr.*

**$** 🏠 **HoJo Express Inn.** This centrally located inn—squarely placed on downtown Main Street—is in walking distance to all of Hyannis's major sights. The 36 rooms and three suites, tastefully refurbished in 1992, offer large double beds and typical motel-chain furniture; most, unfortunately, have views of the drab parking lot. There's a value-conscious package plan, which includes complimentary cocktails, discount shopping coupons, and a meal allowance at four nearby restaurants. ✉ *447 Main St., 02601,* ☎ *508/775–3000 or 800/446–4656,* 🖷 *508/771– 1457. 36 rooms, 3 suites. Restaurant, bar, indoor pool, hot tub, meeting room. AE, D, DC, MC, V.*

**$** 🏠 **Hyannis Inn Motel.** The second-oldest motel in Hyannis, this modest, two-story motel, has been family-owned and -run for the past 45 years. It also has a link to the Kennedys (who still worship at St. Francis Xavier Church, a few minutes walk from the motel): During JFK's presidential campaign, the main building served as press headquarters. The main building's immaculate rooms, nicely decorated in shades of dark blue and rose, have double, queen- or king-size beds, cable TV, and direct-dial phones. The newer deluxe rooms—built in 1981 and situated in a separate wing out back—are larger, sunnier and quieter (they don't face Main Street)—and offer queen- or king-size beds, sleeper sofas, walk-in closets, refrigerators; some have whirlpool tubs. The restaurant serves breakfast only. Children 11 and under stay free. ✉ *473 Main St., 02601,* ☎ *508/775–0255 or 800/922–8993.*🖷 *508/ 771–0456. 77 rooms. Restaurant, pub, air-conditioning, pool, sauna. AE, D, MC, V. Closed Dec.–Jan.*

**$** 🏠 **Inn on Sea Street.** This charming, relaxed B&B just a walk from the beach and downtown Hyannis consists of two 19th-century houses furnished with country antiques and lacy fabrics by Lois Nelson, one of the personable innkeepers. Automobile enthusiasts will enjoy the antique cars J. B., her husband, keeps around. Breakfasts are served on china, silver, and crystal in the antiques-and-lace dining room of the 1849 main house or in the glassed-in sunporch. The mansard-roofed house across the street has a common living room, a shared kitchen, large guest rooms with queen-size canopy beds, and one room with a private porch. Out back, a small, charming cottage all in white has its own kitchen. ⊠ *358 Sea St., 02601,* ☎ *508/775–8030,* FAX *508/771– 0878. 9 rooms, 2 with shared bath; 1 cottage. Air-conditioning. AE, D, MC, V. No smoking. Closed Dec.–May. BP.*

$
★ **Sea Breeze Inn.** Owner-innkeeper Patricia Gibney—her Irish accent is almost as lilting as her inn—opened this charming, cedar-shingled seaside B&B 12 years ago. Each of the 14 rooms have antique or canopied beds and are elegantly (but simply) decorated with well-chosen antiques. Larger families may want to stay in one of the three detached cottages— the nicest among these is the Rose Garden, with three bedrooms (two on the second level), two baths, a spacious TV room, a fireplace, even a washer and dryer. If that's booked, try the Honeymoon Cottage, which has a canopy bed and inviting double Jacuzzi. The cottage is fenced in for total privacy—yet another reason romance gets top billing here. Patricia's breakfasts are worth rising early for to enjoy in the dining room or in the outdoor gazebo, surrounded by picture-perfect gardens. ⊠ *397 Sea St., 02601,* ☎ *508/771–7213,* FAX *508/862–0663. 14 rooms, 3 cottages. Air-conditioning, no-smoking cottages, beach. AE, D, MC, V. CP.*

## Nightlife and the Arts

**Bobby Byrne's Pub** (⊠ Rte. 28, ☎ 508/775–1425) offers a comfortable pub atmosphere, a jukebox, and good light and full menus.

The **Boston Pops Esplanade Orchestra** (☎ 508/790–2787) wows the crowds with its annual Pops By the Sea concert, held in August at the Hyannis Village Green. Each year, a guest conductor strikes up the band; recent baton bouncers have included Olympia Dukakis, Joan Kennedy, Mike Wallace, and Art Buchwald.

**Bud's Country Lounge** (⊠ Bearses Way and Rte. 132, ☎ 508/771–2505) has pool tables and features live country music, dancing, and line-dancing lessons year-round.

The **Cape Cod Melody Tent** (⊠ 21 W. Main St., at the West End Rotary, ☎ 508/775-9100) is the Cape's top venue for popular music concerts and comedy shows. It's been an institution since it was founded in 1950 by actress Gertrude Lawrence and her husband, producer-manager Richard Aldrich, to showcase theatrical productions. Performances are held late June–early September in a 2,300-seat theater-in-the-round tent, one of only 10 such venues left in the country. Performers in 1997 included first-name favorites Liza, Aretha, Wynonna, and Steve and Eydie, as well as Tony Bennett, John Denver, Anne Murray, Willie Nelson, Huey Lewis and the News, and Mary Chapin Carpenter. The Tent also hosts a Wednesday morning children's theater series in July and August.

The 90-member **Cape Cod Symphony Orchestra** (☎ 508/362–1111), under the direction of Royston Nash since 1995, plays a Sounds of Summer Pops Concert on the Hyannis Village Green; it's usually on or around the Fourth of July.

**Duval Street Station** (⊠ 477 Yarmouth Rd., ☎ 508/771–7511), in the old Hyannis train station, is the Upper and Mid Cape's only gay club. There's a piano bar on the lower level, and dance bar on the upper floor with music almost as hot as the crowd. Surprisingly, given this is the Cape, expect a bit of attitude with a capital A.

It's noisy. It's smoky. It's crowded. But **The Prodigal Son** (⊠ 10 Ocean St., ☎ 508/771-1337) is also home to some of the best up-and-coming bands and musicians in New England. Their high-octane java also keeps the joint jumpin'. Call for schedule.

**Starbuck's** (✉ Rte. 132, ☎ 508/778–6767; ☞ Dining, *above*) has live acoustic entertainment in its bar many nights year-round.

In July and August, **town band concerts** (☎ 508/362-5230 or 800/449–6647) are held on Wednesday evenings starting at 7:30 on the Village Green on Main Street.

The **Yacht Club,** the sophisticated lounge of the Tara Hyannis Hotel & Resort (✉ West End Rotary, ☎ 508/775–7775 or 800/843–8272), has live pop entertainment nightly in season, and mostly DJs in the off-season. The bar has a wide-screen TV for sports events and a window wall overlooking the pool and golf course.

## Outdoor Activities and Sports

### BEACHES

**Kalmus Park Beach,** in Hyannis Port, is a fine, wide sandy beach with an area set aside for windsurfers and a sheltered area good for children. Located at the south end of Ocean Street, it has a snack bar, rest rooms, showers, and lifeguards.

**Veterans Park** near the end of Ocean Street in Hyannis has a small beach that is especially good for children because it is sheltered from waves and fairly shallow. There are picnic tables, barbecue facilities, showers, rest rooms, and a playground.

### BIKING

**Cascade Motor Lodge** (✉ 201 Main St., ☎ 508/775–9717), by the bus and train station, rents bikes.

### BOATING

**Eastern Mountain Sports** (✉ 1513 Rte. 132, ☎ 508/362–8690) rents kayaks and camping gear.

Sailing tours aboard the catboat **Eventide** (✉ Ocean St. Dock, ☎ 508/775–0222) offer a variety of cruises through Hyannis Harbor and out into Nantucket Sound, including a nature tour and a sunset cruise.

**Hy-Line** (✉ Ocean St. Dock, ☎ 508/778–2600) offers cruises on old-time Maine coastal steamer replicas. The one-hour tours of Hyannis Harbor and Lewis Bay include a view of the Kennedy compound and other special points of interest.

### FISHING

The **Goose Hummock Shop** (✉ Rte. 28, ☎ 508/778–0877) can provide you with the required freshwater license, along with rental gear, including canoes and kayaks.

Fishing trips are operated on a walk-on basis in spring and fall by **Hy-Line** (✉ Ocean St. Dock, ☎ 508/790–0696). Reservations are mandatory during the summer season.

### GOLF

**Tara Hyannis Hotel & Resort** (✉ West End Rotary, ☎ 508/775–7775 or 800/843–8272) has a beautifully landscaped, challenging 18-hole, par-3 course open to nonguests. You also may bump into some famous faces: Many performers from the Cape Cod Melody Tent tee off here while in town.

### HEALTH AND FITNESS CLUBS

**Barnstable Athletic Club** (✉ 55 Attucks La., Independence Park, off Rte. 132, ☎ 508/771–7734) has four racquetball–wallyball courts (wallyball is volleyball played on a racquetball court), a squash court, basketball, day care, an aerobics room, whirlpools, sauna and steam rooms, and cardiovascular and free-weight equipment. Day and short-term memberships are available.

**The Tara Club** (⊠ Tara Hyannis Hotel & Resort, West End Rotary, ☎ 508/775–7775) has two outdoor tennis courts, a fitness club, and indoor and outdoor pools.

ICE-SKATING

**The Kennedy Memorial Skating Rink** (⊠ Bassett La., ☎ 508/790–6346) offers skating spring through fall.

## Shopping

**Cape Cod Mall** (⊠ Between Rtes. 132 and 28, ☎ 508/771–0200), the Cape's largest, has 90 shops, including Macy's, Filene's, Marshall's, Sears, the Gap, Victoria's Secret, a food court and a four-screen movie complex.

A mix of Disneyesque fantasy and Victorian excess, the Hyannis branch of the **Christmas Tree Shops** (⊠ Rte. 132, ☎ 508/778–5521) is the largest on the Cape. Many people come just to gawk at the glass-enclosed clock out front, made in the '20s in Cincinnati.

**Colonial Candle of Cape Cod** (⊠ 232 Main St., ☎ 508/771–3916) has its touristy elements, but the candles are top-notch.

**Hyannis Antique Co-op** (⊠ 500 Main St., ☎ 508/778–0512) has a large selection of jewelry, glassware, porcelain, furniture, dolls, prints, and collectibles at good prices.

**Nantucket Trading Company** (⊠ 354 Main St., ☎ 508/790–3933) sells an incredible array of neat and nifty necessities for your home, including edibles, table linens, cooking gadgets, and kitchen accessories.

# Yarmouth and Yarmouth Port

*21 mi east of the Sagamore Bridge, 4 mi west of Dennis.*

**⑱** Once known as Mattacheese, or "the planting lands," Yarmouth was settled in 1639 by farmers from the Plymouth Bay Colony. **Yarmouth Port** wasn't incorporated as a separate village until 1829. By then the Cape had begun a thriving maritime industry, and men turned to the sea to make their fortunes. Many impressive sea captains' houses—some now B&Bs and museums—still line the streets, and Yarmouth Port has some real old-time stores in town.

For a peek into the past, make a stop at **Hallet's,** a country drugstore preserved as it was in 1889, when the current owner's grandfather, Thacher Hallet, opened it. Hallet served not only as druggist but as postmaster and justice of the peace as well. At the all-marble soda fountain with swivel stools, you can order the secret-recipe ice-cream soda as well as an inexpensive lunch. Above Hallet's Store, the **Thacher Taylor Hallet Museum** displays photographs and memorabilia of Yarmouth Port and the Hallet family. ⊠ *Rte. 6A, Yarmouth Port,* ☎ *508/362–3362.* ☜ *Donations accepted.* ☾ *Call for hours.*

The 1886 **Village Pump,** a black wrought-iron mechanism long used for drawing household water, is topped by a lantern and surrounded by ironwork with cutouts of birds and animals. In front is a stone trough used for watering horses. **Parnassus Book Service** (☞ Shopping, *below*) is right across the street. The 1696 **Old Yarmouth Inn,** near the village pump, is the Cape's oldest inn and a onetime stagecoach stop.

To get a good look at the area's flora, and perhaps some of its fauna, the **Botanical Trails of the Historical Society of Old Yarmouth,** right behind the post office, offer 50 acres of oak and pine woods and a pond, accented by blueberries, lady's slippers, Indian pipes, rhododendrons, holly, and more. Stone markers and arrows point out the 2 mi of trails; you'll find trail maps in the gatehouse mailbox. Just beyond the historical society's trails, the little **Kelley Chapel** was built in 1873 by a father for a daughter grieving over the death of a child. The simple interior is dominated by an iron woodstove and a pump organ. ⊠ *Off Main St., Rte. 6A, Yarmouth Port.* ⊒ *$1 suggested donation.* ☉ *Gatehouse: summer, daily 1–4; trails: during daylight hours year-round.*

Set back from the street, the 1780 **Winslow Crocker House** is an elegantly symmetrical two-story Georgian with 12-over-12 small-pane windows and rich paneling in every room. After Crocker's death, his two sons built a wall dividing the house in half. The house was moved here from West Barnstable in 1936 by Mary Thacher, who donated it—along with her collection of 17th- to 19th-century furniture, pewter, hooked rugs, and ceramics—to the Society for the Preservation of New England Antiquities, which operates it as a museum. ⊠ *250 Main St., Rte. 6A, Yarmouth Port,* ☎ *508/362–4385.* ⊒ *$4.* ☉ *Weekends only, June–Oct. 15. Tours given every hour on the hour 11–4.*

Built in 1840 onto an existing 1740 house for a sea captain in the China trade, then bought by another, who swapped it with a third—this should give you an idea of the proliferation of sea captains hereabouts—the **Captain Bangs Hallet House** is a white Greek Revival building with a hitching post out front and a weeping beech in back. The house and its contents typify the 19th century sea captain's home, with pieces of pewter, china, nautical equipment, antique toys and clothing, and more on display. The kitchen has the original 1740 brick beehive oven and butter churns. ⊠ *11 Strawberry La., off Rte. 6A, Yarmouth Port,* ☎ *508/362–3021.* ⊒ *$3.* ☉ *First Sun. in June–last Sun. in Oct., 1–3:30.*

Purchased in 1640 and established as a prosperous farm in the late 1700s, the **Taylor-Bray Farm** is still a working farm, listed on the National Register of Historic Places. The farm is open to the public in summer, and has picnic tables, walking trails, and a great view of the tidal marsh. ⊠ *Bray Farm Rd., Yarmouth Port,* ☎ *508/398–2231, ext. 292.*

For a scenic loop with little traffic, turn north off Route 6A in town onto Church Street or Thacher Street, then left onto Thacher Shore Road. In fall this route is especially beautiful for its impressive stands of blazing red burning bush. Wooded segments alternate with open views of marsh. Keep bearing right, and at the WATER STREET sign, the dirt road on the right will bring you to a wide-open view of marshland as it meets the bay. Don't drive in too far, or you may get stuck. As you come out, **⑲** a right turn will take you to **Keveney Bridge,** a one-lane wooden bridge over marshy Hallet's Mill Pond, and back to Route 6A.

 One of Yarmouth Port's most beautiful spots is Bass Hole, which stretches from Homer's Dock Road to the salt marsh. **Bass Hole Boardwalk** extends over a marshy creek; amid the salt marshes, vegetated wetlands, and upland woods meander the 2.4-mi **Callery-Darling nature trails.** Gray's Beach is a little crescent of sand with still water good for children. At the end of the boardwalk, benches provide a place to relax and look out over abundant marsh life and, across the creek, the beautiful, sandy shores of Dennis's Chapin Beach. At low tide you can walk out on the flats for almost a mile—a far cry from the 18th-century harbor here that was the site of a schooner shipyard. ⊠ *Trail entrance on Center St. near the Gray's Beach parking lot.*

# THE CAPE'S SIX SEASONS

**S**O MUCH HAS BEEN MADE of the difference between "the season" and "the off-season" that you might think there is a magic switch that flicks Cape Cod on and off. The truth is something more complicated. Far from having only two seasons, much less the traditional four, the Cape and islands really have six identifiable seasons, each distinct and dramatic.

From early April until Memorial Day, it's spring—a season that can be beautiful, bursting with wildflowers and green grass. Locals shake out the kinks and get ready. In the old days, you could smell tar and pitch, as fishermen prepared their nets for another season at sea. Now, you're likelier to smell fresh paint, as B&B owners put on a fresh coat. Provincetown's exotic little gardens, nurturing unusual plants brought home from seaports around the world, are in their glory. And it's a great time to enjoy bird walks, nature hikes, and country drives, along with lower prices.

"The season" is generally thought to start on Memorial Day, but, in fact, the holiday is a quick and busy harbinger, nothing more. After the long weekend, a lull sets in that lasts until the end of June, when schools finally let out. The weather is still iffy—balmy one minute, freezing the next—but everything is open, and not too many summer people have arrived yet. It's the calm before the rush—more anticipation than real hustle.

Then, all at once, comes high tide—from late June through early September, with surges on July 4th and Labor Day. For some, Labor Day to Columbus Day is the most intriguing time. School's back in session, so families are gone. Yet summer often sticks around. And the ocean, not nearly as fickle as the air, holds onto its warmth. Many visitors who are not tied to summer vacations have discovered this "shoulder" season.

Columbus Day to New Year's becomes a time of slow retraction. Under crisp blue skies in the clear autumn light, the lower Cape and islands' cover of heather, gorse, cranberry, bayberry, box berry, and beach plum resembles, in Thoreau's words, "the richest rug imaginable spread over an uneven surface." Oaks and swamp maples burnish into beautiful, subtle tones, and fields of marsh grass that were green just a few weeks ago become tawny, like a miniature African savannah. Thanksgiving is a particularly evocative moment, when the Cape's Pilgrim history comes to the fore.

And then—winter. The mainland gets much more snow than the Cape, but nonetheless, it's a stark, tough time. The landscape has a desolate, moorish quality, which makes the inns and restaurants that do remain open all the more inviting. Many museums and shops close, although a year-round economy has emerged, especially around Hyannis, Falmouth, and Orleans. The Cape's community theater network continues throughout the year, and many golf courses remain open, except when it snows (some open their courses to cross-country skiers). Intimate B&Bs and inns make romantic retreats after a day of ice fishing or pond skating.

Still, the farther toward the fringes you get, the quieter life becomes. If you believe that less can be more, this is the time to discover what others never could have found during the hectic high-season months.

Many towns on the Cape and islands celebrate Christmas in an old-fashioned way, with wandering carolers and bands, theatrical performances, crafts sales, and holiday house tours. Nantucket's Christmas Stroll is the best-known event. The Cape's holiday season extends from Thanksgiving to New Year's, with celebratory activities including First Night celebrations in many towns.

—by Seth Rolbein

## Dining and Lodging

$$$ ✕ **Abbicci.** Unassuming to a fault from the outside, Abbicci's interior tells an entirely different story, with stunning modern decor, explosions of color, and a compelling black slate bar. One of the first to bring northern Italian cooking to Cape Cod, chef-owner Marietta Hickey has remained true to the tradition, and ahead of the crowd. She prepares rich and full-tasting yet heart-healthy fare, thanks to a light touch with olive oil and a watchful eye over the fat content of her dishes. One of the most pleasing is braised rabbit with fresh green beans and tiny onions over garlic mashed potatoes. In summer, fish and a dozen elegant pasta dishes take over. ⊠ 43 Main St., Rte. 6A, ☎ 508/362–3501. AE, D, DC, MC, V.

$$ ✕ **Inaho.** Yuji Watanabe, the chef-owner of the Cape's best Japanese ★ restaurant, makes early morning journeys to Boston's fish markets to shop for the freshest local catch. His selection of sushi and sashimi is vast and artful, and vegetable and seafood tempura come out of the kitchen fluffy and light. If you're a teriyaki lover, you can't do any better than the chicken's beautiful blend of sweet and sour. The serene and simple Japanese garden out back has a traditional goldfish pond. With this kind of authentic ambience, attention to detail, and remarkably high quality, sometimes it's hard to remember you're still on old Cape Cod. ⊠ 157 Main St., Rte. 6A, ☎ 508/362–5522. MC, V. Closed Mon. No lunch.

$–$$ ✕ **Jack's Outback.** Tough to find, tough to forget, this eccentric little ★ serve-yourself-pretty-much-anything-you-want joint is right down the driveway by Inaho, and goes by the motto "Good food, lousy service" (it lives up to its motto in every way). It's the quintessential local hangout. Solid breakfasts give way by midday to thick burgers, freshly concocted sandwiches, pasta salads, homemade soups, and then traditional home-cooked favorites like Yankee pot roast and fried chicken. The owner, Jack Braginton Smith, is a local historian of note. (As far as locals are concerned, if he doesn't insult you on the way in and out, you've been insulted.) Jack's has no liquor license, and you may not BYOB. ⊠ 161 Main St., ☎ 508/362–6690. Reservations not accepted. No credit cards. No dinner Sun. and Mon.

$$ ▥ **Wedgewood Inn.** This handsome Greek Revival building, white with ★ black shutters and fan ornaments on the facade, is on the National Register of Historic Places and dates from 1812. Inside, the sophisticated country decor is a mix of fine Colonial antiques, upholstered wing chairs, Oriental rugs, large Stobart and English sporting prints and maritime paintings, brass accents, handcrafted cherry pencil-post beds, antique quilts, period wallpapers, and Claire Murray hooked rugs on wide-board floors. Two spacious suites have canopy beds, fireplaces, and porches; one has a separate sitting room. Innkeeper Gerrie Graham cooks elegant breakfasts such as Belgian waffles with whipped cream and strawberries, or egg dishes with hollandaise. Milt, her husband, a former Boston Patriot and FBI agent, serves breakfast and helps guests make plans for the day. ⊠ 83 Main St., Rte. 6A, Yarmouth Port 02675, ☎ 508/362–5157 or 508/362–9178. 4 rooms with bath, 2 suites. Airconditioning. No smoking in common areas. AE, DC, MC, V. BP.

$ ▥ **Americana Holiday Motel.** If you want (or need) the convenience ★ of staying on Route 28, this recently renovated family-owned and -operated strip motel is a good choice. All rooms have cable TV, refrigerator, and phone; those in the rear Pine Grove section overlook serene sea pines and one of the motel's three pools (one indoor), rather than traffic snarls. In the off-season, the rates simply can't be beat. ⊠ 99 Rte. 28, W. Yarmouth 02673, ☎ 508/775–5511 or 800/445–4497, FAX 508/790–0597. 149 rooms, 4 suites. Air-conditioning, coffee shop,

*1 indoor and 2 outdoor pools, hot tub, sauna, steam room, putting green, shuffleboard, video games, playground. AE, D, DC, MC, V. Closed Nov.–Mar. CP (off-season only).*

**$**   🏚 **Lane's End Cottage.** Indeed, owner-innkeeper Valerie Butler's house sits at the end of a dirt lane—moved here in 1860 by oxen, to make room for the neighboring church. And what a magical setting it is: a white picket fence, an English cottage garden, an English antique-laced common room/library with a fireplace. There are only three guest rooms, each with private bath (shower only), tasteful appointments, and beds with firm mattresses and feather comforters. The sun-drenched first-floor Terrace Room has French doors that lead to a cobblestone patio overlooking a flower-rimmed garden. Pets are allowed with prior notice. ✉ *268 Main St., Rte. 6A, 02675,* ☎ *508/362–5298. 3 rooms. No smoking. No credit cards. BP.*

**$**   🏚 **Village Inn.** Many of the guests who stay here will say it's just like
★ staying at your grandmother's—provided, of course, your grandmother has the kind of clean, snug rooms found in Esther Hickey's 1795 sea captain's house. Each of the 10 wide pine-plank floored rooms are lessons in history: The Provincetown Room, for instance, served as the house's schoolroom; the original (now unused) light fixtures are still in place. Room sizes and amenities fluctuate; avoid the Wellfleet Room, which is about as big as the oyster that bears its name! (Even the tub is half-size.) Families should note that the Brewster, Truro, and Hyannis rooms connect, and that the first-floor Yarmouth Room is the most spacious, with its own library, bathroom with fireplace, and private entrance. The common rooms have as many books as some public libraries. ✉ *92 Main St., Rte. 6A, Box 1, Yarmouth Port 02675,* ☎ *508/362–3182. 10 rooms. No smoking. MC, V. BP.*

### Nightlife and the Arts

**Oliver's Restaurant** (✉ Rte. 6A, Yarmouth Port, ☎ 508/362–6062) has live acoustic music in its tavern weekends year-round (nightly in summer).

### Shopping

**Cummaquid Fine Arts** (✉ 4275 Rte. 6A, Cummaquid, ☎ 508/362–2593) has works by currently active Cape Cod and New England artists, plus decorative antiques, beautifully displayed in an old home.

**Parnassus Book Service** (✉ Rte. 6A, Yarmouth Port, ☎ 508/362–6420), occupying a three-story 1840 former general store, has a huge selection of old and new books—Cape Cod, maritime, Americana, antiquarian, and others—and is a great place to browse. (The book stall on the shop's side is open 24 hours a day and works on the honor system; tally up your purchases and leave the money in the mail slot.) Parnassus also carries Robert Bateman's nature prints.

**Peach Tree Designs** (✉ 173 Rte. 6A, Yarmouth Port, ☎ 508/362–8317) carries home furnishings and decorative accessories, some from local craftspeople, all beautifully made.

**Pewter Crafters of Cape Cod** (✉ 933 Rte. 6A, Yarmouth Port, ☎ 508/362–3407) handcrafts traditional and contemporary pewter objects, from baby cups to tea services.

## West and South Yarmouth

*3 mi east of Hyannis.*

There's no getting around it: The part of the Cape people love to hate is Route 28. Passing through Yarmouth, it's one motel, strip mall, nightclub, and miniature golf course after another. In 1989, as *Cape Cod Life* magazine put it, "the town [began] to plant 350 trees in hopes that eventually the trees' leaves, like the fig leaf of Biblical lore, will cover the shame of unkempt overdevelopment." Regardless of the glut of tacky tourist traps, there are some interesting sights in the little villages along the way. So take it if you want to intersperse amusements with your sightseeing, because you could amuse yourself to no end on this section of Route 28. A sensible option if you want to avoid 28 entirely: Take speedy Route 6 to the exit nearest what you want to visit, then cut south across the interior.

㉑ The village of **West Yarmouth** was settled in 1643 by Yelverton Crowe, who acquired his land from an Indian sachem who told him he could have as much land as he could walk on in an hour in exchange for an "ox-chain, a copper kettle . . . and a few trinkets." The first settlers were farmers; in the 1830s, when Central Wharf near Mill Creek was built, the town turned to more commercial ventures, as headquarters for the growing packet service that ferried passengers from the Cape to Boston.

Listed on the National Register of Historic Places, the 1710 **Baxter Grist Mill,** by the shore of Mill Pond, is the only mill on Cape Cod that is powered by an inside water turbine; the others use either wind or paddle wheels. The mill was converted to the indoor metal turbine in 1860 because of the pond's low water level and the damage done to the wooden paddle wheel by winter freezes. The original metal turbine is displayed on the grounds, and a replica powers the restored mill. A videotape tells the mill's history. ⊠ *Rte. 28, W. Yarmouth,* ☎ *508/398–2231.* ☞ *Free.* ☉ *Mid-June–Labor Day, Thurs.–Sun. 4–7; mid-Sept.–mid-Oct., weekends 4–7.*

㉒ A unique and lovely walking trail, the **Yarmouth Boardwalk** stretches through swamp and marsh, and leads to the edge of Swan Pond, a pretty pond ringed with woods. ⊠ *Take Winslow Gray Rd. northeast from Rte. 28, turn right on Meadowbrook La., take to end.*

☾ An entertaining and educational stop for kids, **ZooQuarium** has sea lion shows, a petting zoo with native wildlife, wandering peacocks, pony rides in summer, aquariums, educational programs, and the new Children's Discovery Center with changing exhibits—like "Bone Up on Bones" (all about skeletons), "Zoo Nutrition," (in which kids prepare meals), and the self-explanatory "Scoop on Poop." ⊠ *674 Rte. 28, W. Yarmouth,* ☎ *508/775–8883.* ☞ *$7.50.* ☉ *Mid-Feb.–June and Sept.– late Nov., daily 9:30–5; July–Aug., daily 9:30–8.*

NEED A
BREAK?

**Jerry's Seafood and Dairy Freeze** (⊠ 654 Main St., W. Yarmouth, ☎ 508/775-9752) offers fried clams and onion rings—along with thick frappés, frozen yogurt, and soft ice cream—at good prices. It stays open until midnight in season.

㉓ **South Yarmouth** was once called Quaker Village for its large Quaker population, which settled the area in the 1770s after a smallpox epidemic wiped out the local Indian population.

The 1809 **Quaker Meeting House** is still open for meetings. Two separate entrance doors and the partition down the center were meant to divide the sexes. The adjacent cemetery has simple markers with no epitaphs, an expression of the Friends' belief that all are equal in God's eyes. Behind the cemetery is a circa 1830 one-room Quaker schoolhouse. ⊠ *Spring Hill Rd., off N. Main St.,* ☎ *508/398–3773.* ☉ *Services on Sundays, 11–noon.*

☾ **Pirate's Cove** is the most elaborate of the Cape's many miniature golf setups, with a hill, a waterfall, a stream, and an 18-hole "Blackbeard's

Challenge" course. ⊠ *728 Main St., Rte. 28, S. Yarmouth,* ☎ *508/394–6200.* ⊠ *$6.* ☉ *July–Aug., daily 9 AM–11 PM; Apr.–June and Sept.–Oct., most days 10–8.*

☾ For rainy day fun, the **Ryan Family Amusement Center** offers video-game rooms, Skee-ball, and bowling. ⊠ *1067 Rte. 28, S. Yarmouth,* ☎ *508/394–5644.* ☉ *Daily; hrs vary.*

★ The **19th Century Mercantile** is a wonderful trip to the past. Once a "ropewalk," a long narrow building used for making rope, the site was restored in 1993 and is now an old-fashioned store that sells new merchandise with an old-time flair, such as Grandpa's Pine Tar Shampoo, vintage Nottingham lace curtains, sarsaparilla, and Vinolia, the brand of soap that was used on the Titanic. ⊠ *2 N. Main St., S. Yarmouth,* ☎ *508/398–1888.*

The **Bass River** divides the southern portions of the towns of Yarmouth and Dennis. People also generally refer to the area of South Yarmouth as Bass River. Here you'll find charter boats and boat rentals, as well as a river cruise, plus seafood restaurants and markets.

## Lodging

$$$ 🏨 **Ocean Mist.** This three-story U-shaped resort sits on its own private sandy beach on Nantucket Sound. The rooms are designed to pamper: They all include wet bars, fully stocked kitchenettes, and cable TV. The duplex loft suites are a step above, with cathedral ceilings, sitting area with pull-out sofa, skylights, and one or two private balconies. There's a coffee shop, indoor heated pool, and hot tub on the property. ⊠ *97 S. Shore Dr.,* ☎ *508/398–2633 or 800/248–6478,* ℻ *508/760–3151. 31 rooms, 32 suites. Coffee shop, air-conditioning, indoor pool, hot tub, beach, coin laundry. AE, MC, V. Closed Jan.–mid-Feb.*

$$ 🏨 **Seaside.** Right on a warm, Nantucket Sound beach, with a view of scalloped beaches in both directions, this 5-acre village of Cape-style cottages (efficiencies and one- or two-bedroom units) was built in the 1940s. The decor varies from cottage to cottage, as each is individually owned. All have kitchens or kitchenettes, and many have wood-burning fireplaces. The oceanfront cottages, right off a strip of green grass set with lounge and Adirondack chairs, have the best view and decor. Cottages in the adjacent pine grove are generally very pleasant, though the least expensive units, with knotty pine walls and ceilings, have an outdated '50s look. Shoulder-season rates are very attractive. ⊠ *135 S. Shore Dr., 02664,* ☎ *508/398–2533. 39 cottages. Picnic area, beach. MC, V. Closed mid-Oct.–Apr.*

$ 🏨 **Capt. Farris House.** Steps from the Bass River Bridge dividing South Yarmouth and West Dennis and a short spin away from congested Route 28 sits this imposing 1845 Greek Revival home, built by the sea captain for whom it's named. In 1996 innkeepers Stephen and Patty Bronstein took over the newly refurbished and restored home. The eight guest rooms—large and comfortable, despite their borderline excessiveness—have either antique or canopied queen- or king-size beds, extra pillows, plush comforters, fancy drapes, TVs, phones, and tiled baths with Jacuzzis and imported bath products. All have fireplaces and sundecks. (The Honeymoon Suite has a private sundeck and a two-person Jacuzzi.) Breakfast is served in the open interior brick courtyard—a slice of Tuscany right on the Mid Cape. ⊠ *308 Old Main St., 02664,* ☎ *508/760–2818 or 800/350–9477,* ℻ *508/398–1262. 7 rooms, 1 suite. No smoking. AC, MC, V. BP.*

$ 🏨 **Manor House.** This lovely 1920s Dutch Colonial overlooks Lewis Bay and is literally steps away from the beach. Enjoy, but be back by 4 PM, when innkeeper Liz Latshaw serves tea and home-baked cookies. Each room (there are two on the first floor, four on the second) is furnished in a country antiques–style, with Laura Ashley linens, fresh flowers, claw-foot tubs and antique or canopy beds. They also each have their own individual, quaint charm—there's the "Secret Garden" (ivy print linens and wallpaper); "Howling Coyote," with a Southwestern theme; and "Whalewatch," which, with its distinctive navys and ma-roons, is the one decidedly masculine room, and one of two with water views. Bountiful breakfasts are served in the candlelit dining room, where the hand-stenciled strawberries on the ceiling match the wallpaper. ⊠ *57 Maine Ave., W. Yarmouth 02673,* ☎ *508/771–3433 or 800/962–6679. 6 rooms. Beach. No smoking. AE, MC, V. BP.*

## Nightlife and the Arts

**Betsy's Ballroom** (⊠ Yarmouth Senior Center, 528 Forest Rd., S. Yarmouth, ☎ 508/362–9538) has Saturday-night dancing year-round to bands on the Cape's largest dance floor.

**Cape Cod Irish Village** (⊠ 512 Main St., W. Yarmouth, ☎ 508/771–0100) has dancing to two- or three-piece bands performing traditional and popular Irish music year-round. The crowd is mostly couples and over-35ers.

The 90-member **Cape Cod Symphony Orchestra** (☎ 508/362–1111), under former D'Oyly Carte Opera conductor Royston Nash, gives reg-ular classical and children's concerts, with guest artists, October through May. Performances are held at Mattacheese Middle School (⊠ Off Higgins Crowell Rd.).

**Clancy's** (⊠ 175 Rte. 28, W. Yarmouth, ☎ 508/775–3332) offers live entertainment on weekends year-round.

West Yarmouth's summertime **town band concerts** are held on Mon-day in July and August at 7:30 at Mattacheese Middle School (⊠ Off Higgins Crowell Rd., ☎ 508/778–1008).

## Outdoor Activities and Sports

### BEACHES

**Flax Pond** recreation area in South Yarmouth offers freshwater swim-ming, a lifeguard, and ducks, but no sand beach, just pine-needle-covered ground. There's a piney picnic area with grills, as well as tennis and basketball courts, and plenty of parking. ⊠ *N. Main St.*

**Parker's River Beach,** a flat stretch of sand on warm Nantucket Sound, is perfect for families. It has a lifeguard, a concession stand, a gazebo and picnic area, a playground, outdoor showers, and rest rooms. ⊠ *S. Shore Dr.*

### BIKE RENTAL

**All Right Bike & Mower** (⊠ 627 Main St., W. Yarmouth, ☎ 508/790–3191) rents bikes and mopeds, and does on-site repairs.

**Outdoor Shop** (⊠ 50 Long Pond Dr., S. Yarmouth, ☎ 508/394–3819) rents bikes and scooters and makes repairs.

### FISHING

**Truman's** (⊠ Rte. 28, W. Yarmouth, ☎ 508/771–3470) can supply you with a required freshwater license and rental gear.

### GOLF

**Blue Rock Golf Course** (⊠ Off Highbank Rd., S. Yarmouth, ☎ 508/398–9295) is a highly regarded, easy to walk, 18-hole, par-3, 3,000-yard public course crossed by a pond. The pro shop rents clubs; reser-vations are mandatory in season.

**Mid Cape Racquet Club** (⊠ 193 White's Path, S. Yarmouth, ☎ 508/394–3511) has one all-weather tennis, nine indoor tennis, two racquetball, and two squash courts; indoor basketball; sauna, steam, and whirlpools; massage services; and a free weight and cardiovascular room—plus day care. Daily rates are available.

## Shopping
FACTORY OUTLETS
**The Cranberry Bog Outlet Stores** (⊠ Rte. 28, W. Yarmouth, no phone) are on the edge of a working cranberry bog. Bass, Van Heusen, and Izod are a few of the shops here.

---

# Dennis

**㉔** *7 mi west of Brewster, 5 mi north of Dennisport.*

The backstreets of Dennis still retain the Colonial charm of seafaring days. The town was named for the Reverend Josiah Dennis and incorporated in 1793. There were 379 sea captains living in Dennis when fishing, salt making, and shipbuilding were the main industries, and the elegant houses that they constructed—now museums and B&Bs—still line the streets. In 1816 Dennis resident Henry Hall discovered that adding sand to his cranberry fields' soil improved the quality and quantity of the fruit. The decades soon after that saw cranberry farming and tourism become the Cape's main commercial enterprises. The town has a number of conservation areas and nature trails (take a look at the Dennis Chamber of Commerce guide), and numerous ponds for swimming.

**West Dennis** and **Dennisport,** off Route 28 on the south shore of the Cape, are covered below.

The **Josiah Dennis Manse,** a saltbox house with add-ons, was built in 1736 for Reverend Josiah Dennis. Inside, the rooms reflect life in Reverend Dennis's day. One room is set up as a child's room, with antique furniture and toys. The keeping room has a fireplace and cooking utensils, and the attic exhibits spinning and weaving equipment. Throughout you'll see china, pewter, and portraits of sea captains. The Maritime Wing has ship models, paintings, nautical artifacts, and more. On the grounds is a 1770 one-room schoolhouse—furnished with wood-and-wrought-iron desks and chairs. ⊠ *77 Nobscussett Rd., corner of Whig St.,* ☎ *508/385–2232.* ☞ *Donations accepted.* ☉ *July–Aug., Tues. 10–noon, Thurs. 2–4.*

**㉕** On a clear day, from the top of **Scargo Tower** you'll have unbeatable panoramic views of Scargo Lake, the village's scattered houses below, Cape Cod Bay, and distant Provincetown. Winding stairs bring you to the top of the all-stone, 30-ft tower, which was rebuilt in 1890 after a fire destroyed the original wooden structure. (Don't forget to read the unsightly, but amusing, graffiti on the way up.) Expect crowds at sunrise and sunset. ⊠ *Off Rte. 6A.* ☞ *Free.* ☉ *Daily sunrise–sunset.*

NEED A
BREAK?
Stop for a bite at **Captain Frosty's,** a favorite local clam shack, with fried seafood, shellfish, and onion rings, as well as very good lobster rolls with more lobster than celery. ⊠ *219 Main St., Rte. 6A, near the Yarmouth line,* ☎ *508/385–8548.*

If you're in the mood for something cold and sweet, sample the homemade ice cream and frozen yogurt at the **Ice Cream Smuggler.** ⊠ *716 Main St., Rte. 6A, near the Dennis Public Market,* ☎ *508/385–5307.*

For Broadway-style shows and children's plays, attend a production at the **Cape Playhouse,** the most renowned summer theater in the country. In 1927 Raymond Moore, who had been working with a theatrical troupe in Provincetown, bought an 1838 former Unitarian Meeting House and converted it into a theater. (The original pews still serve as seats.) The opening performance was *The Guardsman,* starring Basil Rathbone; other stars who performed here in the early days, some in their professional stage debuts, include Bette Davis (who first worked here as an usher), Gregory Peck, Lana Turner, Ginger Rogers, Humphrey Bogart, Tallulah Bankhead, and Henry Fonda, who appeared with his then-unknown 20-year-old daughter, Jane. Cape resident Shirley Booth was such an admirer of the theater she donated her Oscar (for *Come Back Little Sheba*) and her Emmy (for *Hazel*) to the theater; both are on display in the lobby during the season. Behind-the-scene tours are also given in season; call for a schedule. The Playhouse also offers children's theater on Friday mornings during July and August. Also on the property are a restaurant, the ☞ **Cape Museum of Fine Arts,** and the **Cape Cinema,** whose exterior was designed in the style of the Congregational Church in Centerville. Inside, a 6,400 square-ft heavenly skies mural, designed by Massachusetts artist Rockwell Kent, who also designed the gold sunburst curtain, covers the ceiling. ⊠ *Main St., Rte. 6A,* ☎ *508/385–3911.*

The **Cape Museum of Fine Arts** exhibits a permanent collection of more than 850 works by Cape-associated artists. Important pieces include a portrait of a fisherman's wife by Charles Hawthorne, the father of the Provincetown art colony; a 1924 portrait of a Portuguese fisherman's daughter by William Paxton, one of the first artists to summer in Provincetown; a collection of wood-block prints by Varujan Boghosian, a member of Provincetown's Long Point Gallery cooperative; an oil sketch by Karl Knaths, who painted in Provincetown from 1919 until his death in 1971 and works by abstract expressionist Hans Hoffman and many of his students. Offerings include film festivals, lectures, art classes, and trips. ⊠ *60 Hope La., Dennis,* ☎ *508/ 385–4477.* ☞ *$3.* ☉ *Late-May–Nov., Mon.–Sat. 10–5, Sun. 1–5; Dec.–mid-May, Tues.–Sat. 10–5, Sun. 1–5.*

## Dining and Lodging

**$$$** ✕ **Red Pheasant Inn.** The Red Pheasant is one of the Cape's best cozy country inns with a consistently good kitchen. The hearty American food is prepared with elaborate sauces and herb combinations, and the atmosphere is authentically antique. Rack of lamb is served with an intense port and rosemary reduction, and exquisitely grilled veal chops come with a dense red wine and portobello mushroom sauce. Deep-fried goat cheese ravioli is another winner. Try to reserve a table in the more intimate Garden Room. The excellent, expansive wine list is a *Wine Spectator* award winner. Men may want to wear a jacket. ⊠ *905 Main St., Rte. 6A,* ☎ *508/385–2133. D, MC, V. No lunch.*

**$$** ✕ **Gina's by the Sea.** Some places are less than the sum of their parts;
**★** Gina's is more. A funky old building tucked into a sand dune, the aroma of fine northern Italian cooking blends with a fresh breeze off the bay. The dining room is tasteful, cozy, and wonderful in the fall when the fireplace is blazing. Look for lots of fresh pasta and seafood. If you don't want a long wait, come early or late, because Gina's takes no reservations. The bar gets crowded and loud, and is often fun. ⊠ *134 Taunton Ave.,* ☎ *508/385–3213. Reservations not accepted. AE, MC, V. Closed Dec.–Mar. and Mon–Wed. Oct.–Nov. No lunch Apr.–June and Sept.–Nov.*

**$$**   ✕ **Green Room Restaurant.** Like the Scargo Café, its sister restaurant down the road a piece, the Green Room relies on fresh ingredients to create fresh flavors. Salmon salsa verde is still the favorite, but a more moderately priced menu carries on the cross cultural theme—Creole shrimp is also excellent. Beneath its high cathedral ceiling, the front dining room is light and open; the bar and the back dining room are smaller and sometimes feel crowded. This is one of the few Cape restaurants that stays open until midnight in July and August. Lunch is served Wednesday and Thursday in July and August. ✉ *36 Hope La., off Rte. 6A,* ☎ *508/385–8000. AE, MC, V. Closed Nov.–May.*

**$$**   ✕ **Scargo Café.** Because the Cape Playhouse is right across the street, this café is a favorite before- and after-show haunt. Recently expanded, the dining rooms have a neo-colonial feel. Excellent early-bird specials for the show crowd include scallops cooked in Harpoon beer (a Boston brew), and fettuccine alfredo. The menu focuses on lighter fare for summer, but there's still plenty of richness, and a rather good wine cellar to complement it. Mussels Ferdinand features farm-raised mussels with a buttery Pernod sauce over pasta. An added plus: The kitchen stays open until midnight in summer. ✉ *799 Rte. 6A,* ☎ *508/385–8200. AE, D, MC, V.*

**$**   ⌂ **Four Chimneys Inn.** Russell and Kathy Tomasetti have transformed
★ their three-story, four-chimney 1881 Queen Anne Victorian gem, once home to the town doctor, into a relaxing getaway, smack dab in the heart of the Mid Cape. The eight rooms vary in size and decor, but all are tastefully furnished with cherry four-poster, wicker, or antique pine or oak beds, chenille comforters, and hand-stenciled trim. Three rooms have fireplaces, and all have views of either Scargo Lake (across the street) or of the surrounding woods and flowering gardens. You can enjoy Kathy's hearty breakfasts in the high-ceilinged dining room (home to her cranberry glass collection); the screened-in summer porch, with its lovely garden views, is an even more soothing spot to start the day. After a day of recreation, nothing could be finer than sitting under the wisteria-draped arbor, sipping afternoon tea. ✉ *946 Main St., Rte. 6A, 02638. 7 rooms, 1 suite.* ☎ *508/385–6317 or 800/874–5502,* 𝔽𝔸𝕏 *508/385–6285. 7 rooms, 1 suite. Restricted smoking. AE, D, MC. V. Closed late Oct.–late Apr. CP.*

**$**   ⌂ **Isaiah Hall B&B Inn.** Lilacs and pink roses trail the white picket fence
★ outside this 1857 Greek Revival farmhouse on a residential road on the bay side. Inside, guest rooms are decorated with country antiques, floral-print wallpapers, and homey touches such as quilts and priscilla curtains, and some have canopy beds and Oriental carpets. In the attached carriage house, rooms have stenciled white walls and knotty pine, and some have small balconies overlooking a wooded lawn with gardens, grape arbors, and berry bushes. Make-it-yourself popcorn, tea, coffee, and soft drinks are always available. ✉ *152 Whig St., 02638,* ☎ *508/385–9928 or 800/736–0160,* 𝔽𝔸𝕏 *508/385–5879. 11 rooms. Picnic area, air-conditioning, badminton, croquet. No smoking. AE, MC, V. Closed mid-Oct.–Mar. CP.*

## Nightlife and the Arts

The most renowned summer stock theater in the country is the **Cape Playhouse** (✉ Main St., off Rte. 6A, Dennis, ☎ 508/385–3911), a former 1838 former Unitarian Meeting House, where top stars appear each summer (☞ Exploring, *above*). The Playhouse also mounts children's shows on Friday mornings during July and August.

The **Reel Art Cinema** at the Cape Museum of Fine Arts (✉ 60 Hope La., Dennis, ☎ 508/385–4477) shows avant garde, classic, art, and independent films on weekends September through April.

The **Nau-Sets** (✉ Dennis Senior Center, Rte. 134, Dennis, ☎ 508/394–0257 or 508/255–5079) hold weekly square dances on Tuesday.

## Outdoor Activities and Sports

### BEACHES

**Chapin Beach** in Dennis is a lovely dune-backed bay beach with long tidal flats that at low tide allow walking far out. It has no lifeguards or services. ✉ *Chapin Beach Rd.*

**Corporation Beach** in Dennis has lifeguards, showers, rest rooms, and a food stand. At one time a packet landing owned by a corporation of townsfolk, the beautiful crescent of white sand backed by low dunes now serves a decidedly noncorporate use as a public beach. ✉ *Corporation Rd.*

For freshwater swimming, **Scargo Lake** in Dennis has two beaches that offer rest rooms and a picnic area. The sandy-bottom lake is shallow along the shore, which is good for kids. It is surrounded by woods and stocked for fishing. ✉ *Access off Rte. 6A or Scargo Hill Rd.*

### BIKING

★ The Cape's premier bike path, the **Cape Cod Rail Trail,** offers a scenic ride through the area. Following the paved right-of-way of the old Penn Central Railroad, it is about 30 mi long, stretching from South Dennis to South Wellfleet, with an extension underway. The Rail Trail passes salt marshes, cranberry bogs, ponds, and Nickerson State Park, which has its own path. Along the way you can veer off to spend an hour or two on the beach, or stop for lunch. The terrain is easy to moderate. The trail starts at the parking lot off Route 134 south of Route 6, near Theophilus Smith Road in South Dennis, and it ends at the post office in South Wellfleet. There are parking lots along the route if you want to cover only a segment: in Harwich (across from Pleasant Lake Store on Pleasant Lake Avenue) and in Brewster (at Nickerson State Park). The **Butterworth Company** (✉ 261 Stevens St., Hyannis 02601, ☎ 508/790–1111 or 800/696–2762) sells a guide to the trail for $2.95.

## Shopping

**Cape Cod Braided Rug Co.** (✉ 259 Great Western Rd., S. Dennis, ☎ 508/398–0089) specializes in braided rugs made on the premises in a variety of colors, styles, and sizes.

**Robert C. Eldred Co.** (✉ 1483 Rte. 6A, E. Dennis 02641, ☎ 508/385–3116) holds nearly two dozen auctions per year, dealing in Americana, estate jewelry, top-quality antiques, marine, Asian, American, and European art, tools, and dolls. Its "general antiques and accessories" auctions put less expensive wares on the block.

**Scargo Pottery** (✉ Dr. Lord's Rd. S, off Rte. 6A, Dennis, ☎ 508/385–3894) is set in a pine forest, where potter Harry Holl's unusual wares—such as his signature castle birdhouses—sit on tree stumps and hang from branches. Inside are the workshop and kiln, plus work by Holl's four daughters. With luck you'll catch a potter at the wheel; viewing is, in fact, encouraged.

# West Dennis and Dennisport

*3 mi east of West Yarmouth, 10 mi west of Chatham.*

㉗ Another one of those tricks of Cape geography, the village of **West Dennis** is actually south of South Dennis on the east side of the Bass River. Dennisport is farther east, near the Harwich town line.

The **Jericho House Museum** was built in 1801 for Captain Theophilus Baker. A subsequent owner named it, noticing that the walls seemed to be tumbling down. The classic Cape, with a bow roof and large central chimney, has been fully restored. Antique furnishings include 1850s portraits and Chinese and other items brought home from overseas by sea captains. Antique cranberry-harvesting and woodworking equipment, a model saltworks, marine antiques, and 19th-century sleighs and wagons are in the barn museum. Also here is a 150-piece driftwood zoo: A local man collected wood on a beach, then added eyes and beaks to bring out animal shapes. ⊠ *Trotting Park Rd. at Old Main St., W. Dennis,* ☎ *508/398–6736.* ⊡ *Donations accepted.* ⊙ *July–Aug., Wed. and Fri. 2–4 and by appointment.*

The oldest working organ in the United States and a priceless chandelier made out of Sandwich glass can be found at the 1835 **South Parish Congregational Church** (⊠ 234 Main St., ☎ 508/394–5992). The cemetery beside the church has markers dating from 1795, many of which were raised for early sea captains and read "Lost at sea." One stone in the graveyard reads, even more simply, "The Chinese Woman."

**28** The Mid Cape's last southern village is **Dennisport,** a prime summer resort area, with gray-shingled cottages, summer houses, and condominiums, and lots of white picket fences covered with rambling roses. The Union Wharf Packing Company was here in the 1850s, and the shore was lined with sail makers and ship chandlers. The beach where sea clams were once packed in the sands is now packed with sunbathers.

NEED A BREAK?   Have a slice of pizza on the patio of **Upper Crust Pizza** (⊠ 329 Old Santa Fe Trail, ☎ 505/982–0000), next to the San Miguel MiSet in a rustic mid-19th-century barn decorated with such memorabilia as a working nickelodeon, the **Sundae School Ice Cream Parlor** (⊠ 387 Lower County Rd., Dennisport, ☎ 508/394–9122; ⊙ Mid-Apr.–mid-Oct.) serves great homemade ice cream and frozen yogurt with lots of toppings, sugar-free ice cream, real whipped cream, and even old-time Moxie (a 1930s soft drink) from an antique marble soda fountain.

## Dining and Lodging

**$$**   ✕ **Swan River Restaurant and Fish Market.** From the right table, you can have a beautiful view of Swan River marsh and Nantucket sound beyond at this informal little spot that turns out great fresh fish, both traditional and creative. Besides the fried and the broiled, try Mako shark au poivre or scrod San Sebastian, simmered in garlic broth with littleneck clams. ⊠ *5 Lower County Rd., Dennisport,* ☎ *508/394–4466. AE, MC, V. Closed mid-Sept.–late May. No lunch until mid-June.*

**$–$$**   ✕ **Christine's.** A change of pace from the usual Cape fare, Christine's has a traditionally Italian menu with a number of Lebanese items and Middle Eastern touches that liven it up. Besides excellent pasta and homemade meatballs and sausage, try a Lebanese combination platter, an assortment of treats like hummus, *kafta* (finely ground beef seasoned with fresh mint, parsley, and Lebanese spices), stuffed grape leaves, and spinach or meat pies. There's a Sunday buffet brunch and a nightclub with entertainment and dancing year-round (☞ Nightlife and the Arts, *below*). ⊠ *581 Main St., Rte. 28, W. Dennis,* ☎ *508/394–7333. AE, D, MC, V.*

**$**   ✕ **Bob Briggs' Wee Packet.** Cute as can be—maybe a little too cute—this tiny dinette is decorated in classic Cape kitsch with screaming-yellow tables, seascapes on the walls, and driftwood, seashells, and little bits of moss everywhere. The food is very good, very traditional, and very plain: plenty of local seafood (broiled swordfish and fish-and-chips are very reliable) as well as sandwiches and salads. Try bread pudding with lemon sauce for dessert. You can pick up homemade freshly baked goods of all kinds—especially doughnuts, as well as made-to-order cakes—

at the adjoining bakery and doughnut shop. ✉ *79 Depot St., Dennisport,* ☎ *508/398–2181. Reservations not accepted. MC, V. Closed Oct.–Apr.*

**$** ✕ **The Breakfast Room.** Proprietor Walt McGourty presides over this classic American breakfast and lunch spot, where you'll find fast and friendly service and a regular crowd that can't seem to live without the place. Blueberry pancakes made with fresh wild blueberries from Maine and eggs Benedict are two highlights on the breakfast menu, which also offers omelets, hash browns, muffins, and bagels. Chowder, sandwiches, and salads are the main lunch offerings. ✉ *675 Rte. 28, W. Dennis,* ☎ *508/398–0581. Closed Dec.–mid-Jan. and weekdays mid-Jan.–mid-Apr.*

**$$** ⌂ **The Garlands.** There are innumerable strip motels and cottage colonies lining Old Wharf Road in Dennisport, but few places offer luxury and views to match this bi-level, motel-style complex. There are 20 units in all—16 are two-bedroom suites. Each unit has a fully equipped kitchen, phone, cable TV, bath, private sun deck or patio, and daily maid service. The oceanfront VIP suites, simply named A and B (two bedrooms) and C and D (one bedroom) are the best picks here— the floor-to-ceiling windows offer unobstructed water views. ✉ *117 Old Wharf Rd., Box 506, 02639,* ☎ *508/398–6987. 20 suites. Beach. No credit cards. Closed mid-Oct.–mid Apr.*

**$** ⌂ **Beach House Inn.** This is the kind of house you'd expect when Nantucket Sound is your backyard: shingles weathered gray from the salt air; white wicker and natural oak furniture that's practical, yet comfortable; walls of glass that frame the beauty—and sometimes ferocity—of Mother Nature. The seven rooms, on two floors, are plainly decorated with maple and wicker. Some rooms have brass or four-poster beds, and all have TV, ceiling fans, and decks, some of which overlook the front yard. The best room is undoubtedly Room 4, with its second-story waterfront deck and a private staircase leading to the inn's private beach. The common room has a TV and wide assortment of hit movies on video. You can use the barbecue grills or the fully equipped kitchen, which has a microwave, to cook meals. ✉ *61 Uncle Stephen's Way, Box 494, W. Dennis 02670,* ☎ *508/398–4575 or 617/489–4144 off-season. 7 rooms. Beach, playground. No smoking. No credit cards. CP.*

**$** ⌂ **Lighthouse Inn.** On its small private beach adjacent to West Dennis Beach, this old-style Cape resort has been in family hands since 1938. The main inn was built around a still-operational 1855 lighthouse and has five guest rooms. Scattered along a landscaped lawn are 23 individual weathered, shingled, Cape cottages (one-room to three-bedroom, no kitchens) and five multiroom buildings. Cottages have decks, knotty-pine and some painted walls, and generally nice, cabiny bedrooms. The oceanfront Guest House has a common room with a fireplace. In the main inn are a living room, a library, and a glassed-in waterfront porch with a TV. The restaurant in three waterfront rooms serves New England cuisine, seafood, and weekly shore dinners. In summer, supervised daily and evening activities and dinners for children give parents some private time. ✉ *1 Lighthouse Rd., Box 128, W. Dennis 02670,* ☎ *508/398–2244,* 🅵🅰🆇 *508/398–5658. 40 rooms, 23 cottages. Restaurant, bar, room service, pool, miniature golf, tennis court, Ping-Pong, shuffleboard, beach, fishing, billiards, nightclub, recreation room, children's programs (July–Aug.), playground. MC, V. Closed mid-Oct.–mid-May. BP.*

## Nightlife and the Arts

**Christine's** (✉ Rte. 28, W. Dennis, ☎ 508/394–7333; ☞ Dining, *above*) has entertainment nightly in season in its 300-seat showroom. Concerts, sometimes with dancing, feature name bands from the 1950s to 1970s, Top 40 bands, or jazz. There are also stand-up comedy nights and Sunday-night cabarets. Off-season, there's live entertainment and dancing to a DJ on weekends, as well as special events.

**Clancy's** (⊠ 8 Upper County Rd., Dennisport, ☎ 508/394–6661) has jazz piano year-round.

**The Sand Bar** (⊠ Lighthouse Rd., W. Dennis, ☎ 508/398–7586) features the boogie-woogie piano playing of local legend Rock King, who's been tickling the ivories—and people's funny bones—here every summer for nearly 40 years. ☺ *Closed mid-Oct.–mid-May.*

**Sundancer's** (⊠ 116 Rte. 28, W. Dennis, ☎ 508/394–1600) has dancing to a DJ, live bands and Sunday-afternoon reggae bands in season. It's closed November and December.

## Outdoor Activities and Sports

### BEACHES

The **West Dennis Beach** is one of the best on the south shore. A breakwater was started here in 1837 in an effort to protect the mouth of Bass River, but abandoned when a sandbar formed on the shore side. It is a long, wide, and popular sandy beach, stretching for 1½ mi, with marshland and the Bass River across from it. A popular spot for windsurfers, the beach also has bathhouses, lifeguards, a playground, food, and parking for 1,000 cars. ⊠ *Davis Beach Rd.*

### BOATING

**Cape Cod Boats** (⊠ Rte. 28 at Bass River Bridge, W. Dennis, ☎ 508/394–9268) rents powerboats and canoes to experienced boaters.

**Cape Cod Waterways** (⊠ 16 Rte. 28, Dennisport, ☎ 508/398–0080) rents canoes, kayaks, and manual and electric paddleboats for leisurely travel on the Swan River, as well as sailboats (delivered on the Mid Cape).

### ICE-SKATING

Fall through spring, ice-skating is available at the **Tony Kent Arena** (⊠ 8 Gages Way, S. Dennis, ☎ 508/760–2400). Keep your eyes peeled: this is where Nancy Kerrigan and Paul Wylie train.

### RUNNING

**Lifecourse** (⊠ Access and Old Bass River Rds., S. Dennis) is a 1½-mi jogging trail through woods, with 20 exercise stations along the way. It is part of a recreation area that includes basketball and handball courts, ball fields, a playground, and a picnic area.

## Shopping

**Factory Shoe Mart** (⊠ Rte. 28, Dennisport, ☎ 508/398–6000) has such brand names as Capezio, Dexter, Clark, Esprit, Nike, Reebok, L.A. Gear, and Rockport for men, women, and children.

# THE LOWER CAPE

The Lower Cape is the least developed part of the Cape, and the place that many most treasure as a result. **Brewster** and **Harwich**, rich in history and Cape flavor, are located before **Chatham,** out at the elbow, a traditional town with good shopping and strolling. South of Chatham, the Monomoy National Wildlife Refuge is a twin-island Audubon bird sanctuary. Nickerson State Park in Brewster offers plenty of recreation and the Cape's prime camping in a forest setting. The beaches, woods, swamps, walking trails, historic sights, and visitor centers of the Cape Cod National Seashore are next on the east side, with the small and pretty fishing town of **Wellfleet** opposite on the bay side. Continuing north, the sweeping dunes of **Truro** and the Province Lands follow. Last but not least, **Provincetown** is a quiet fishing village in winter. In summer it dons a motley gown and draws crowds to its galleries, crafts shops, whale-watch boats, restaurants, nightlife, and crowds— yes, crowds of people watching crowds of people.

## Brewster

**㉙** *6 mi north of Chatham, 5 mi west of Orleans.*

Named for Plymouth leader William Brewster, the area was settled in 1659, but was not incorporated as a separate town until 1803. In the early 1800s, Brewster was the terminus of a packet cargo service from Boston, and home to many seafaring families. In 1849 Thoreau wrote that "this town has more mates and masters of vessels than any other town in the country." A large number of mansions built for sea captains remain today, and quite a few have been turned into B&Bs. In the 18th and 19th centuries, the bay side of Brewster was the site of a major salt-making industry. Of the 450 saltworks operating on the Cape in the 1830s, more than 60 of them were located here. Brewster is the perfect place to learn about the natural history of the Cape. The Museum of Natural History is here, and the area is rich in conservation lands, state parks, forests, freshwater ponds, and brackish marshes.

Windmills used to be prominent in Cape Cod towns; the Brewster area once had four. The 1795 **Old Mill**—an octagonal, smock-type mill shingled in weathered pine with a roof like an upturned boat—was moved here in 1974 and has been restored. The millstones are original. At night the mill is often spotlighted and makes quite a sight.

Also on the grounds is a one-room house from 1795, the **Harris-Black House.** Once, amazingly enough, home to a family of 13, the restored 16-ft-square building is today partially furnished and dominated by a brick hearth and original woodwork. ⊠ *Off Rte. 6A,* ☎ *508/896– 9521.* ☎ *Free.* ☉ *May–June and Sept.–Oct., weekends 1–4; July– Aug., Tues.–Fri. 1–4.*

**㉚** The **Stony Brook Grist Mill,** a restored 19th-century fulling mill, is now a museum and grain mill. The scene is wonderfully picturesque, with the old mill's waterwheel slowly turning in the water of a small, tree-lined brook. Inside, exhibits include old mill equipment and looms, and you can watch cornmeal being stone-ground and get a lesson in weaving on a 100-year-old loom. Out back, across wooden bridges, a bench has a pleasant view of the pond and of the sluices leading into the mill area.

Early each spring, Stony Brook's **Herring Run** boils with alewives making their way to spawning waters. They swim in from Cape Cod Bay up Paine's Creek to Stony Brook and the ponds beyond it. The rushing stream is across the street from the mill. Farther down the path to the stream, there is an ivy-covered stone wishing well and a wooden bridge with a bench. ⊠ *Setucket Rd.* ☎ *Donations accepted.* ☉ *May– June, Thurs.–Sat. 2–5; July–Aug., Fri. 2–5.*

**㉛** For nature enthusiasts, a visit to the **Cape Cod Museum of Natural History** is a must. The museum and grounds offer guided field walks, a shop, a natural history library, lectures, classes, nature and marine exhibits, such as a working beehive and a pond- and sea-life room with live specimens, and self-guided trails through 80 acres of forest, marshland, and ponds, all rich in birds and other wildlife. The exhibit hall upstairs has a wall display of aerial photographs documenting the process by which the famous Chatham sandbar was split in two. The museum also has cruises through Nauset Marsh aboard the *Nauset Explorer,* a custom-built catamaran with an onboard spotting scope for

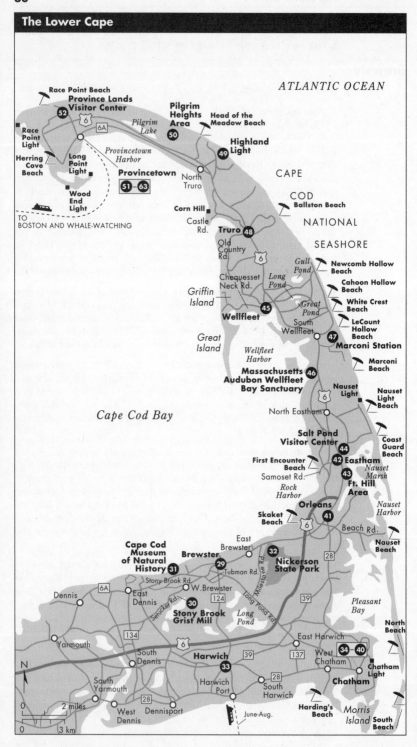

ATLANTIC OCEAN

Race Point Beach
**Province Lands Visitor Center**
52
6  6A

Race Point Light

**Pilgrim Heights Area**
50

Head of the Meadow Beach

*Pilgrim Lake*

**Highland Light**
49

Herring Cove Beach

Long Point Light

*Provincetown Harbor*

Wood End Light

**Provincetown**
51 – 63

North Truro

CAPE

COD

Ballston Beach

NATIONAL

Corn Hill

Castle Rd.

**Truro**
48

SEASHORE

Old Country Rd.
6

Newcomb Hollow Beach

*Gull Pond*

Cahoon Hollow Beach

Chequesset Neck Rd.

*Long Pond*

White Crest Beach

*Griffin Island*

*Great Pond*

LeCount Hollow Beach

**Wellfleet**
45

South Wellfleet

**Marconi Station**
47

*Great Island*

*Wellfleet Harbor*

Marconi Beach

**Massachusetts Audubon Wellfleet Bay Sanctuary**
46

Nauset Light

*Cape Cod Bay*

6

North Eastham

Nauset Light Beach

**Salt Pond Visitor Center**
44

Coast Guard Beach

**First Encounter Beach**

Samoset Rd.

42 **Eastham**

*Nauset Marsh*

43
**Ft. Hill Area**

*Rock Harbor*

**Orleans**
41

*Nauset Harbor*

Skaket Beach

6

Beach Rd.

**Nauset Beach**

**Cape Cod Museum of Natural History**
31

East Brewster

**Brewster**
29

32

**Nickerson State Park**

Tubman Rd.

28

Stony Brook Rd.

W. Brewster

*Pleasant Bay*

Dennis

6A

East Dennis

Serucker Rd.

30

124

**Stony Brook Grist Mill**

*Long Pond*

39

North Beach

Yarmouth

134

East Harwich

34 – 40

South Dennis

6

**Harwich**
33

39

137

West Chatham

Chatham Light

South Yarmouth

Harwich Port

28

South Harwich

**Chatham**

N

West Dennis

28

Dennisport

June-Aug.

**Harding's Beach**

*Morris Island*

South Beach

0    2 miles

0    3 km

TO BOSTON AND WHALE-WATCHING

close-up views of shore birds. Trips last two hours and are guided by a naturalist. ⊠ *Main St., Rte. 6A,* ☎ *508/896–3867 or 800/479–3867 in MA.* ⌨ *$5.* ☉ *Mon.–Sat. 9:30–4:30, Sun. noon–4:30.*

Set on a re-created 19th-century common with a picnic area, the **New England Fire & History Museum** exhibits 35 antique vehicles, including the only surviving 1929 Mercedes-Benz fire engine, the late Boston Pops conductor Arthur Fiedler's private collection of fire-fighting memorabilia, 14 mannequins in historical firefighter uniforms depicting firefighters through the centuries, a Victorian apothecary shop, an animated diorama of the Chicago Fire of 1871 complete with smoke and fire, a historic working forge, and medicinal herb gardens. Guided tours are given. ⊠ *1439 Main St., Rte. 6A,* ☎ *508/896–5711.* ⌨ *$5.* ☉ *Mid-May–Labor Day, weekdays 10–4, weekends noon–4; Labor Day–Columbus Day, weekends noon–4.*

At the junction of Route 124, the **Brewster Store** is a local landmark. Built in 1852, it is a typical New England general store—God love 'em!—providing such essentials as the daily papers, penny candy, and benches out front for conversation. It's a good stop for quick grocery-type refreshments, as the bicycles piled up out front in summer attest. Upstairs, the old front of the store has been re-created, and memorabilia from antique toys to World War II bond posters are displayed. Downstairs is an antique nickelodeon that you can play. ⊠ *1935 Main St., Rte. 6A,* ☎ *508/896–3744.*

Known as the Church of the Sea Captains, the handsome **First Parish Church,** with Gothic windows and a capped bell tower, is full of pews marked with the names of famous Brewster seamen. Out back is an old graveyard where militiamen, clergy, farmers, and captains rest side by side.

☾ Mesmerizing entertainment is provided at the church by **Mimsy Puppets.** Morning performances of fairy tales and folktales for children are given on Thursdays through July and August. ⊠ *Main St., Rte. 6A,* ☎ *508/896–5577.*

☾ For a look at some non-native species, visit the 20-acre **Bassett Wild Animal Farm,** home to domestic and exotic birds, a lion, a tiger, monkeys, llamas, and the animals at a petting zoo. Hayrides, pony rides, a snack bar, and a picnic area are available. ⊠ *Tubman Rd. between Rtes. 124 and 137,* ☎ *508/896–3224.* ⌨ *$5.75.* ☉ *Mid-May–mid-Sept., daily 10–5.*

In an 1830s house, the **Brewster Historical Society Museum** is made up of a sea captains' room with paintings and artifacts, an 1890 barbershop, a child's room with antique toys and clothing, a room of women's period gowns and accessories, and other exhibits on local history and architecture. Out back there is a ¼-mi nature trail over dunes leading to the bay. ⊠ *Main St., Rte. 6A,* ☎ *508/896–7593.* ⌨ *Free.* ☉ *May–June and Sept.–Oct., weekends 1–4; July–Aug., Tues.–Fri. 1–4.*

For a lovely hike or run through the local wilds, **Punkhorn Parklands,** studded with freshwater kettle-hole ponds, has 45 mi of scenic trails meandering through meadows, marshes, and pine forests. ⊠ *Run Hill Rd.*

**③②** The 1,961 acres encompassed by **Nickerson State Park** were once part of a vast estate belonging to Roland C. Nickerson, son of Samuel Nickerson, a Chatham native who became a multimillionaire and founder of the First National Bank of Chicago. At their long, private beach or their hunting lodge, Roland and his wife, Addie, lavishly entertained such visitors as President Grover Cleveland in English country-house style, with coachmen dressed in tails and top hats and a bugler announcing carriages entering the front gates.

The estate was like a village unto itself. Its gardens provided much of the household's food, supplemented by game from its woods and fish from its ponds. It also had its own electric plant and a nine-hole golf course by the water. The enormous mansion Samuel built for his son in 1886 burned to the ground 20 years later, and Roland died two weeks after the event. The even grander stone mansion built to replace it in 1908 is now part of the Ocean Edge resort (☞ Dining and Lodging, *below*). In 1934 Addie donated the land for the state park in memory of Roland and their son, who died during the 1918 flu epidemic.

The Park itself consists of acres of oak, pitch pine, hemlock, and spruce forest dotted with seven freshwater kettle ponds that were formed by glacial action. Some ponds are stocked with trout for fishing. Other recreational opportunities include canoeing, sailing, motorboating, biking along 8 mi of paved trails that have access to the Rail Trail, picnicking, cross-country skiing in winter and bird-watching—thrushes, wrens, warblers, woodpeckers, finches, larks, Canada geese, cormorants, great blue herons, hawks, owls, osprey and others frequent the park. Occasionally, red foxes and white-tailed deer are spotted in the woods. Tent and RV camping is extremely popular here (☞ Dining and Lodging, *below*), and visitor programs are offered in season. A map of the park is available on-site. ⊠ *Rte. 6A,* ☎ *508/896–3491.*

## Dining and Lodging

**$$$$**  ✕ **Chillingsworth.** This crown jewel of Cape restaurants is extremely
★  formal, terribly pricey, and outstanding in every upscale way. The classic French menu and wine cellar continue to win award after award. Every night, the seven-course table d'hôte menu ($40–$56) rotates through an assortment of appetizers, entrées and "amusements." Recent favorites have been a super-rich risotto, roast lobster, and grilled venison. At dinner, a more modest bistro menu is served in the Garden Room, a sort of patio in the front of the restaurant. The whole experience is an extravagance. ⊠ *2449 Main St., Rte. 6A,* ☎ *508/896–3640. AE, DC, MC, V. Closed Mon. mid-June–Thanksgiving; closed some weekdays Memorial Day–mid-June and mid-Oct.–Thanksgiving; closed entirely Thanksgiving–Memorial Day.*

**$$$$**  ✕ **High Brewster.** Famously authentic, High Brewster is the real thing,
★  a working country inn that serves up some of the finest food on the Cape. The restored Colonial farmhouse has low ceilings and exposed beams overhead, and it overlooks a picture-perfect New England landscape. The five-course prix fixe menu ($35–$55) changes frequently as it explores classic interpretations of American regional cooking. Longtime favorites include squash soup, grilled duck breast with black currant reduction, and apple rum ice cream. Rack of lamb is always on the menu, and perfectly prepared. Summertime sees the arrival of swordfish and striped bass. A six-course tasting dinner ($40–$42) is offered once a week in the summer (Tuesday in 1997, but it may change to Monday for 1998; call to confirm). Accommodations at the inn are also available. ⊠ *964 Setucket Rd.,* ☎ *508/896–3636 and 800/203-2634. Reservations essential. AE, MC, V. Closed 1st 2 wks in Jan.; call for weekday hrs off-season. No lunch.*

**$$$**  ✕ **Brewster Fish House.** Long thought to be entirely immune to innovation or experimentation, the Fish House is catching up with the tastes of its newish clientele, and the traditional dishes that made this spot one of the Old Cape's standards are starting to take a back seat to new ideas and an expanded menu. Management doesn't seem entirely convinced that a wholesale shift to an upscale menu will carry the day, so classic scrod and boiled dinners are still playing backup to this season's experimentation with duck, rack of lamb, Cornish game hen, and tenderloin of beef in the classic style. The wine list offers a half-dozen whites and a half-dozen reds that you don't often see by the glass. ✉ *2208 Rte. 6A,* ☎ *508/896–7867. Reservations not accepted. MC, V. Call for off-season schedule.*

**$$$$** 🏨 **Ocean Edge.** This huge, self-contained resort is almost like a town. Sports facilities are superior, and activities such as concerts, tournaments, and clambakes are scheduled throughout the summer. Accommodations range from oversize hotel rooms in the conference center, with sitting areas and direct access to the health club and tennis courts, to luxurious one- to three-bedroom condominiums in the woods. All are decorated in modern style. Condominiums have washer-dryers and some fireplaces, and many units have ocean views. The resort is so spread out that you may need a car to get from your condo to the pool. Weekend MAP packages are available. ✉ *2907 Main St., Rte. 6A, 02631,* ☎ *508/896–9000 or 800/343–6074,* FAX *508/896–9123. 170 condominium units, 90 hotel rooms. 3 restaurants, pub, room service, ponds, 2 indoor and 4 outdoor pools, saunas, driving range, golf privileges, putting greens, 11 tennis courts, basketball, exercise room, beach, bicycles, children's programs, concierge. AE, D, DC, MC, V. MAP.*

**$$–$$$$** 🏨 **Captain Freeman Inn.** A splendid 1866 Victorian built for a packet ★ schooner fleet owner continues to show off its original opulence with a marble fireplace, herringbone-inlay flooring, ornate Italian plaster ceiling medallions, tall windows, and 12-ft ceilings. Guest rooms have hardwood floors, antiques and Victorian reproductions, and eyelet spreads, and some have fishnet or lace canopies. Three luxury suites are in a 1989 addition—spacious bedrooms with queen-size canopy beds, sofas, fireplaces, phones, TV with VCR, mini-refrigerators, and French doors leading to small enclosed porches with private whirlpool spas. Winter weekend cooking schools share the innkeeper's skills, as do dinners (on request for an extra charge) in the off-season. ✉ *15 Breakwater Rd., 02631,* ☎ *508/896–7481 or 800/843–4664,* FAX *508/ 896–5618. 12 rooms, 9 with private bath. Pool, badminton, croquet, bicycles. No smoking. AE, MC, V. BP.*

**$$–$$$** 🏨 **Brewster Farmhouse.** Inside a restored 1846 farmhouse on historic Route 6A, these sophisticated guest rooms offer modern baths, Lane reproduction and antique furnishings, goose-down pillows, thick white towels, and turndown service with bedside sherry and chocolates. One room has a fireplace, another has a king-size canopy bed and sliders out to a private deck. In good weather, a gourmet breakfast prepared by the inn's owners, Carol and Gary Concors, is served on a patio looking onto the 2-acre backyard with apple trees and gardens on its edges and a heated pool and whirlpool in its center. ✉ *716 Main St., Rte. 6A, 02631,* ☎ *508/896–3910 or 800/892–3910,* FAX *508/896–4232. 5 rooms (including 1 2-bedroom suite), 3 with private bath. Bicycles. No smoking. AE, D, DC, MC, V. BP.*

**$–$$** 🏨 **Old Sea Pines Inn.** Fronted by a white-columned portico and ★ wraparound veranda overlooking a broad lawn, the Old Sea Pines evokes the atmosphere of a summer estate from an earlier time. The living room is spacious, with an appealing sitting area before the fireplace. A sweeping staircase leads to guest rooms decorated with reproduction wallpaper, framed old photographs, and well-matched antique furnishings. Many rooms are very large; some have fireplaces, including the inn's best room, which also has a sitting area in an enclosed sunporch. Rooms in a newer building are sparsely but well decorated, with bright white modern baths, and have cast-iron queen-size beds. The shared-bath rooms are *very* small but sweetly done, and a steal in summer. Sunday dinner and musical revue in season. ⊠ *2553 Main St., Rte. 6A, Box 1070, 02631,* ☎ *508/896–6114,* FAX *508/896–7387. 22 rooms (5 share 2½ baths), 3 suites. Restaurant. No smoking. AE, D, DC, MC, V. Closed Jan.–Mar. BP.*

**$–$$**  ☞ **The Ruddy Turnstone.** Built in the early 1800s, this beautifully preserved Cape homestead is set on 3 acres that gently slope to a marsh at the edge of the property. Pine-board floors with braided rugs in the main house, the original barn-board walls of the Carriage House, quilts, and country antiques take you back in time, while modern amenities tend to comfort. Rooms have queen-size beds and large baths, most with shower/tub combos. An upstairs common sitting room offers an incredible view of the marsh and the bay beyond, another common room has a fireplace and small library. A delicious country breakfast is served at little tables on the porch or in the dining room. ⊠ *463 Rte. 6A, Brewster 02631,* ☎ *508/385–9871 or 800/654–1995. 5 rooms. Air-conditioning. No smoking. MC, V. Closed Jan. and Feb. BP.*

**$**  ⚠ **Nickerson State Park.** The Cape's largest and most popular camping site is on 2,000 acres teeming with wildlife, white pine, hemlock, and spruce forest, and jammed with opportunities for trout fishing, walking, or biking along 8 mi of paved trails; canoeing; sailing; motorboating; and bird-watching. Basic sites cost $6; 23 premium sites, situated on the edges of ponds, are $7. RVs must be self-contained. Facilities include showers, bathrooms, picnic tables, barbecue areas, and a store. Maps and schedules of park programs are available at the park entrance. ⊠ *3488 Main St., Rte. 6A, E. Brewster 02631,* ☎ *508/896–3491, reservations* ☎ *508/896–4615,* FAX *508/896–3103. 418 sites. No credit cards. Open Oct.15–Apr. 15 to self-contained campers only.*

## Nightlife and the Arts

The **Cape Cod Museum of Natural History** (⊠ Rte. 6A, Brewster, ☎ 508/896–3867 or 800/479–3867 in MA) has stargazing sessions and "Live at the Museum" world music concerts held Wednesdays at 7:30, early July through August in the auditorium.

The **Woodshed** (⊠ Rte. 6A, Brewster, ☎ 508/896–7771), the rustic bar at the Brewster Inn, is a good place to soak up local color and listen to pop duos or bands that perform most nights.

## Outdoor Activities and Sports

### BEACHES

**Flax Pond** in Nickerson State Park in Brewster, surrounded by pines, offers picnic areas, a bathhouse, and water sports rentals.

**Breakwater Landing, Linnell Landing, Paine's Creek, Point of Rocks,** and **Robin's Hill** bay beaches all have access to the flats that at low tide make for very interesting exploration.

### BICYCLING

The **Cape Cod Rail Trail Bikeway** (☞ Dennis, *above*) cuts through Brewster with many access points: Long Pond Road, Underpass Road, and Millstone Road to name a few.

**The Rail Trail Bike Shop** (⊠ 302 Underpass Rd., ☎ 508/896–8200) rents bikes, including children's bikes, and offers free parking and a picnic area, with easy access to the Rail Trail.

### BOATING

**Jack's Boat Rentals** (⊠ Flax Pond, Nickerson State Park, ☎ 508/896–8556) rents canoes, kayaks, Seacycles, Sunfish, pedal boats, and sailboards; guide-led kayak and canoe tours are also offered.

FISHING

Many of Brewster's freshwater ponds offer good fishing for perch, pickerel, and more; five ponds are well stocked with trout. Especially good for fishing is Cliff Pond in Nickerson State Park.

GOLF

**Captain's Golf Course** (⊠ 1000 Freeman's Way, Brewster, ☎ 508/896–5100), with 18 holes, was voted among the top 75 public courses in the country by *Golf Digest* in 1996.

**Ocean Edge Golf Course** (⊠ Villagers Rd., Rte. 6A, ☎ 508/896–5911), an 18-hole, par-72 course winding around five ponds, features Scottish-style pot bunkers and challenging terrain. Three-day residential or commuter golf schools are offered in spring and early summer.

HORSEBACK RIDING

**Moby Dick Farm** (⊠ 179 Great Fields Rd., ☎ 508/896–3544) offers instruction and trail rides to qualified riders.

**Woodsong Farm** (⊠ 121 Lund Farm Way, ☎ 508/896–5555) offers instruction and day programs but no trail rides; they also have a horsemanship program for children 5–18.

TENNIS

The **Ocean Edge** resort (⊠ Rte. 6A, Brewster, ☎ 508/896–9000) has five clay and six Plexipave courts. It offers lessons and round-robins and hosts Tim Gullikson's World-Class Tennis School, with weekend packages and video analysis.

## Shopping

**B. D. Hutchinson** (⊠ 1274 Long Pond Rd., Brewster, ☎ 508/896–6395), a watch- and clock maker, sells antique and collectible watches, clocks, and music boxes.

**Kemp Pottery** (⊠ 258 Rte. 6A, W. Brewster, ☎ 508/385–5782) has functional and decorative stoneware and porcelain, fountains, garden sculpture, pottery sinks, and stained glass.

**Kingsland Manor** (⊠ 440 Main St., Rte. 6A, W. Brewster, ☎ 508/385–9741) has ivy covering the facade, fountains in the courtyard, and everything "from tin to Tiffany"—including English hunting horns, full-size antique street lamps, garden and house furniture, weather vanes, jewelry, and chandeliers.

**Kings Way Books and Antiques** (⊠ 774 Main St., Rte. 6A, Brewster, ☎ 508/896–3639) sells out-of-print and rare books, including a large medieval section, plus small antiques, china, glass, silver, coins, and linens.

**Punkhorn Bookshop** (⊠ 672 Rte. 6A, Brewster, ☎ 508/896–2114), an antiquarian and out-of-print book seller, specializes in natural history, the Cape and region, fine arts, and biography, and sells antique prints and maps.

**The Spectrum** (⊠ 369 Rte. 6A, Brewster, ☎ 508/385–3322) purveys imaginative American arts and crafts, including pottery, stained glass, art glass, and more.

**Sydenstricker Galleries** (⊠ Rte. 6A, Brewster, ☎ 508/385–3272) features glassware handcrafted by a unique process, which you can watch in progress.

# Harwich

**㉝** *3 mi east of Dennisport, 1 mi south of Brewster.*

Originally known as Setucket, Harwich separated from Brewster in 1694, and was re-named after the famous seaport in England. Like other townships on the Cape, Harwich is actually a cluster of seven small villages, including Harwich Port. Three naturally sheltered harbors make the town, like its English namesake, a popular spot for boaters. Wychmere Harbor is particularly beautiful.

The Cape's famous cranberry industry took off in Harwich in 1844, and Alvin Cahoon was its principle grower at the time. Each September Harwich holds a great **Cranberry Festival** to celebrate the importance of this indigenous berry; festival dates for 1998 are September 11–20. There are cranberry bogs throughout Harwich.

Once a private school, the pillared 1844 Greek Revival building of **Brooks Academy** now houses the museum of the **Harwich Historical Society.** In addition to a large photo-history collection and exhibits on artist Charles Cahoon (grandson of Alvin), the socio-technological history of the cranberry culture, and shoemaking, the museum displays antique clothing and textiles, china and glass, fans, toys, and much more. There is also an extensive genealogical collection for researchers. On the grounds is a powder house that was used to store gunpowder during the Revolutionary War, as well as a restored 1872 outhouse. ⊠ *80 Parallel St.,* ☎ *508/432–8089.* ☞ *Donations accepted.* ☉ *June–mid-Oct., Thurs.–Sun. 1–4.*

☙ **Brooks Park** on Main Street is a good place for kids to stretch their legs, with a playground, picnic tables, a ballpark, tennis courts, and a bandstand where summer concerts are held.

☙ **Batter's Box** has softball and baseball batting cages and pitching machines, including one with fastballs up to 90 mph, a bumper-boat pool, and a video-arcade room. ⊠ *322 Main St., Rte. 28, Harwich Port,* ☎ *508/430–1155.* ☞ *$1.50 for 10 pitches, $5 for 40 pitches, or $10 for 100 pitches. $5 per bumper-pool ride.* ☉ *Apr.–May and Sept.–mid-Oct., Mon.–Sat. 11–7, Sun. 11–9; June–Aug., daily 9 AM–11 PM.*

For kids who have spent too much time in the car watching you drive,
☙ a spin behind the wheel of one of 20 top-of-the-line go-carts at **Bud's Go-Karts** may be just the thing. ⊠ *9 Sisson Rd., off Rte. 28, Harwich Port,* ☎ *508/432–4964.* ☞ *$5 for 6 mins.* ☉ *June–Labor Day, Mon.–Sat. 9 AM–11 PM, Sun. 1–11.*

☙ **The Trampoline Center** has 12 trampolines—a perfect activity to release some of that boundless, bouncing energy. They are set up at ground level over pits for safety. ⊠ *296 Rte. 28, W. Harwich,* ☎ *508/432–8717.* ☞ *$3.75 for 10 mins.* ☉ *Apr.–mid-June, weekends (hrs vary widely, call ahead); mid-June–Labor Day, daily 9 AM–11 PM.*

## Dining and Lodging

$$$  ✕ **Cape Sea Grill.** One person's fancy can be another person's stuffy, and Cape Sea Grill flirts with that line. There are times when the dining room feels like you're visiting someone's prissy aunt, but the food is first-rate and tends to put you at ease. Appetizers like Wellfleet oysters and pecan-smoked, free-range chicken set a nice stage. The best main dishes are Chatham cod, and a swordfish dish wrapped in applewood smoked bacon. Spanish-style paella takes its cue from a version that originates in the mountains of Spain, where toasted pasta thickened with bread, parsley, and almonds takes the place of the traditional rice. The Twin Brûlée dessert sampler changes now and then—pray for the espresso and vanilla combo. The wine list is excellent, but decidedly upscale. ⊠ *31 Sea St., Harwich Port,* ☎ *508/432-4745. AE, MC, V. Closed Nov.–Mar., and most evenings after Columbus Day. No lunch.*

$$$–$$$$  ▦ **Beach House Inn.** Right on the beach, and a private, warm, sandy Nantucket Sound beach at that, this inn is reminiscent of an earlier era. Renovated in 1995, construction has transformed the inn into a

turn-of-the-century seacoast resort, with a turret, a wraparound porch, and elegant suites. From umbrella tables on the deck you can watch the sweep of coast and surrounding grasslands while enjoying breakfast. Off-season, there's a fire burning in the common room. Guest rooms, off a central hallway, are beautifully maintained and simply decorated, with country-antique furnishings; some baths have whirlpool tubs. One suite has three walls of glass overlooking the ocean, with a Jacuzzi in the bath and a king-size bed. ⊠ *4 Braddock La., Harwich Port 02646,* ☎ *508/432–4444 or 800/870–4405. 12 rooms. MC, V. Closed Mar. CP.*

**$$–$$$**   🖬 **Augustus Snow House.** This grand old Victorian is the epitome of
**★**    elegance. Common areas include the oak-paneled front room, with tall windows and a fireplace, and a wicker-filled screened porch. The stately dining room is the setting for the three-course breakfast, with dishes like baked pears in raspberry and cream sauce. Up a wide staircase, the guest rooms are individually decorated with Victorian-print wallpapers, luxurious carpets, and fine antiques and reproduction furnishings. All also have quilts, TVs, phones, fans, and antique brass bathroom and lighting fixtures; three also have whirlpool tubs. Downstairs, a lovely garden-like tearoom is open on Thursday, Friday, and Saturday for afternoon tea. ⊠ *528 Main St., Harwich Port 02646,* ☎ *508/430–0528 or 800/320–0528. 5 rooms. Air-conditioning. AE, MC, V. BP.*

**$**   🖬 **Handkerchief Shoals Motel.** About 2 mi from Harwich Port and 3 mi from downtown Chatham, this single-story property is set back from the highway and surrounded by well-maintained lawn and trees. Rooms are large and sparkling clean, with sitting and desk areas, tiled baths, cable TV, phones, fridge, and microwave—a very good value. ⊠ *Rte. 28, Box 306, S. Harwich 02661,* ☎ *508/432–2200. 26 units. Pool, Ping-Pong. D, MC, V. Closed mid-Oct.–mid-Apr.*

## Nightlife and the Arts

**Bishop's Terrace** (⊠ Rte. 28, W. Harwich, ☎ 508/432–0253) restaurant has dancing to jazz most Saturday nights in its small lounge, which is set in a converted barn with tools hanging from barn-board walls. Several nights in season, a pop-classical pianist performs.

**Jake Rooney's Pub** (⊠ Rte. 28, Harwich Port, ☎ 508/430–1100) has a comfortable pub atmosphere, live entertainment five nights a week, and good light and full menus.

☾ **Harwich Junior Theatre** (⊠ Division St., ☎ 508/432–2002) gives theater classes for children year-round and presents four summer productions.

**Irish Pub** (⊠ 126 Main St., Rte. 28, W. Harwich, ☎ 508/432–8808) has dancing to bands—playing Irish, American, and dance music—as well as sing-alongs, pool, darts, sports TV, and pub food in the bar.

**Town band concerts** in Harwich take place in summer on Tuesday at 7:30 in Brooks Park (☎ 508/432–1600).

## Outdoor Activities and Sports

### BEACHES

Harwich has 22 beaches, more than any other Cape town. Most of the ocean beaches are on Nantucket Sound, where the water is a bit calmer and warmer. Freshwater pond beaches are also abundant.

### BOATING

**Cape Water Sports** (⊠ Rte. 28, Harwich Port, ☎ 508/432–7079) rents Sunfish, Hobie Cats, Lasers, powerboats, sailboards, day sailers, and canoes and gives instructions.

1998 marks the eighth annual **Sails Around Cape Cod** regatta (☎ 508/430–1165). The late-August race circumnavigates the Cape, a distance of 140 nautical mi, beginning and ending in Harwich Port. Dates for 1998 are August 14–16/21–23.

Fishing trips are operated on a walk-on basis from spring through fall on the *Golden Eagle* (⊠ Wychmere Harbor, Harwich Port, ☎ 508/432–5611).

**Cranberry Valley Golf Course** (⊠ 183 Oak St., Harwich, ☎ 508/430–7560) has a championship layout of 18 well-groomed holes surrounded by cranberry bogs.

**Deer Meadow Riding Stables** (⊠ Rte. 137, E. Harwich, ☎ 508/432–6580) offers trail rides by reservation.

Water temperatures vary from 50°F to 70°F in Nantucket Sound and from 40°F to 65°F on the ocean side and in Cape Cod Bay. Check with a dive shop about local conditions and sites, as well as about dive boats. Wrecks in area waters include a steamship and schooners. Rentals, instruction, group dives, and information are available through **Cape Cod Divers** (⊠ 815 Main St., Rte. 28, Harwich Port, ☎ 508/432–9035 or 800/348–4641).

**Cape Cod Divers** (☞ Scuba Diving, *above*) has a heated, indoor pool and swimming lessons year-round.

The **Melrose Tennis Center** (⊠ 792 Main St., Harwich Port, ☎ 508/430–7012) offers three Omni courts (synthetic grass and a sand layer, so it plays soft) and six Har-Tru courts, a pro shop, lessons, and daily round-robins and mixed doubles.

## Shopping

**Merlyn Auctions** (⊠ 204 Main St., N. Harwich, ☎ 508/432–5863) are held every Saturday night, with moderate to inexpensive prices for antique and used furniture.

# Chatham

**34** *5 mi east of Harwich, 1 mi west of Orleans.*

In 1656 William Nickerson traded a boat for the 17 square mi of land that make up Chatham. In 1712 the area separated from Eastham and was incorporated as a town—the surnames of the Pilgrims who first settled here still dominate the census list. Although Chatham was originally a farming community, the sea finally lured townspeople to turn to fishing for their livelihood, an industry that has held strong to this day.

Situated at the bent elbow of the Cape, with water nearly surrounding it, Chatham has all the charm of a quiet seaside resort, with relatively little commercialism. And it *is* charming: gray-shingled houses with tidy awnings and cheerful flower gardens, an attractive Main Street with crafts and antiques stores alongside homey coffee shops, and a five-and-ten. It's a traditional town, where elegant summer cottages share the view with stately houses of purebred Yankee architecture, including some of the finest bow-roof houses in the country. There is none of Provincetown's flash, yet it's not overly quaint—well-to-do without being ostentatious, casual and fun but refined, and never tacky.

**35** Authentic all the way, the **Railroad Museum** is set in a restored 1887 depot. Exhibits include a walk-through 1910 New York Central ca-

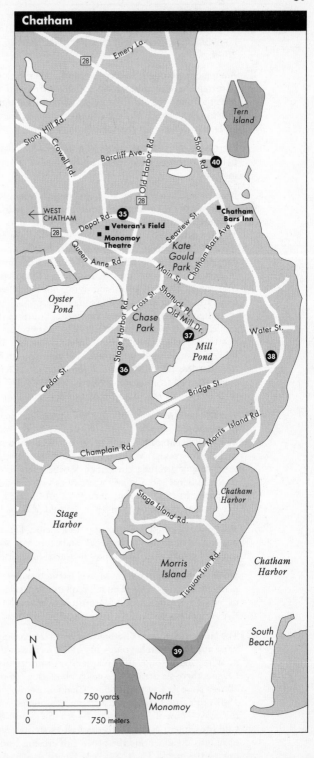

**Chatham**

Emery La.

28

Tern
Island

Stony Hill Rd.

Crowell Rd.

Barcliff Ave.

Old Harbor Rd.

Shore Rd.

40

28

← WEST
CHATHAM

Depot Rd.

35

Veteran's Field

Seaview St.

Chatham
Bars Inn

28

Monomoy
Theatre

Chatham Bars Ave.

Kate
Gould
Park

Queen Anne Rd.

Main St.

Oyster
Pond

Stage Harbor Rd.

Chase
Park

Cross St.

Shattuck Pl.

Old Mill Dr.

37

Water St.

Mill
Pond

38

Cedar St.

36

Bridge St.

Morris Island Rd.

Champlain Rd.

Stage Island Rd.

Chatham
Harbor

Stage
Harbor

Morris
Island

Tisquan-Tum Rd.

Chatham
Harbor

South
Beach

N

39

0        750 yards

0        750 meters

North
Monomoy

boose, old photographs, equipment, thousands of train models, and a new diorama of the 1915 Chatham rail yards. ⊠ *Depot Rd., Chatham, no phone.* ✉ *Donations accepted.* ☉ *Mid-June–mid-Sept., Tues.–Sat. 10–4.*

☺ A unique playground, the **Play-a-round,** a multilevel wooden playground of turrets, twisting tubular slides, jungle gyms, and more, was designed with the input of local children and built by volunteers. There's a section for people with disabilities and a fenced-in area for small children. ⊠ *Depot Rd., across from Railroad Museum.*

On **Queen Anne Road** around Oyster Pond, half-Cape houses, open fields, and rolling pastures reveal the area's Colonial and agricultural history.

**㊱** Built by a sea captain in 1752 and occupied by his descendants until it was sold to the Chatham Historical Society in 1926, the **Atwood House Museum** has a gambrel roof, variable-width floor planks, fireplaces, an old kitchen with a wide hearth and a beehive oven, and some antique dolls and toys. The New Gallery displays portraits of local sea captains. The Joseph C. Lincoln Room has the manuscripts, first editions, and mementos of the Chatham writer. In the basement is an antique tool room. The 1974 Durand Wing houses collections of seashells from around the world and threaded Sandwich glass, as well as Parian ware figures, unglazed porcelain vases, figurines, and busts. In a remodeled freight shed are murals (1932–45) by Alice Stallknecht Wight portraying religious scenes in Chatham settings. They have been exhibited at major galleries around the country. On the grounds are an herb garden, the old turret and lens from the Chatham Light, and a camp rescued from eroding North Beach. ⊠ *347 Stage Harbor Rd.,* ☎ *508/945–2493.* ✉ *$3.* ☉ *Mid-June–Sept., Tues.–Fri. 1–4.*

**㊲** The **Old Grist Mill** was built in 1797 by Colonel Benjamin Godfrey for the purpose of grinding corn. How practical the mill actually was is a matter of debate: For it to work properly, a wind speed of at least 20 mph was necessary. Winds over 25 mph required the miller to reef the sails, or quit grinding altogether. Moved to its present location from Miller Hill in 1956, the mill was extensively renovated and opened to the public in 1957. Closed since 1992 for budgetary reasons, the mill may or may not be open in 1998. ⊠ *Old Mill Dr.*

**Mill Pond** is a lovely place to stop for a picnic. There's fishing from the bridge, and often the "bullrakers" can be seen at work, plying the pond's muddy bottom with 20-ft rakes in search of shellfish.

**Stage Harbor,** sheltered by Morris Island, is where Samuel de Champlain anchored in 1606. The street on its north side is, not surprisingly, called Champlain Road. A skirmish here between Europeans and Native Americans marked the first bloodshed in New England.

**㊳** The famous view from **Chatham Light**—of the harbor, the offshore sandbars, and the ocean beyond—justifies the crowds that gather to share it. When fog shrouds the area, pierced with darting beams from the beacon, there's a dreamlike quality about it. Coin-operated telescopes allow a close look at the famous "Chatham Break," the result of a fierce 1987 nor'easter that blasted a channel through a barrier beach just off the coast—now it is known as North and South beaches. The **Cape Cod Museum of Natural History** in Brewster (☞ *above*) has a display of photos documenting the process of erosion leading up to and following the break. The U.S. Coast Guard Auxiliary, which supervises the lighthouse, offers tours on the first and third Wednesdays, April–September. The lighthouse is also open by appointment, and on three special occasions during the year: Seafest, an annual tribute to the maritime industry held in mid-October; mid-May's Cape Cod Maritime Week; and June's Cape Heritage Week. Note that the phone number provided is for an auxiliary member, Peggy Fisher, who may be succeeded in 1998 by another auxiliary member. ⊠ *Main St.,* ☎ *508/945–0719.*

**Monomoy National Wildlife Refuge** is a 2,750-acre preserve including the Monomoy Islands, a fragile, 9-mi-long barrier-beach area south of Chatham. Once a fishing village, Monomoy's North and South islands were created when a storm divided the former Monomoy Island in 1978. Monomoy was itself separated from the mainland in a 1958 storm. A paradise for bird-watchers, the island is an important stop along the North Atlantic Flyway for migratory waterfowl and shore birds—peak migration times are May and late July. It also provides nesting and resting grounds for 285 species, including gulls—great black-backed, herring, and laughing—and several tern species. White-tailed deer also live on the islands, and harbor and gray seals frequent the shores in winter. The only structure on the islands is the **South Monomoy Lighthouse.** Built in 1849, the shiny red-orange structure, along with the keeper's house, was refurbished in 1988.

Monomoy is a quiet, peaceful place of sand and beach grass, of tidal flats, dunes, marshes, freshwater ponds, thickets of bayberry and beach plum, and a few pines. Because the refuge harbors several endangered species, visitors' activities are limited. Certain areas are fenced off to protect nesting areas of terns and the endangered piping plover. The Massachusetts Audubon Society in South Wellfleet and other groups conduct tours of the island. The **Rip Ryder** (☎ 508/587–4540 or 508/945–5450) or the **Water Taxi** (☎ 508/430–2346) will take you over in season for some lone bird-watching. Rates vary greatly: If you catch a ride out with the seal-watching cruise, it's $14 per person; if you charter a private ride, rates are $45 per person round trip or $10 per person for three or more passengers.

**㊴** The **Monomoy National Wildlife Refuge headquarters,** on the misleadingly named Morris Island, has a visitor center and bookstore (☎ 508/945–0594; ⏱ daily 8–5, with gaps) where you can pick up pamphlets on Monomoy and the birds, wildlife, flora and fauna found there. A ¾-mi interpretive walking trail, closed at high tide, around Morris Island gives a good view of the refuge and the surrounding waters.

**㊵** The **Fish Pier** bustles with activity when Chatham's fishing fleet returns, sometime between noon and 2 PM, depending on the tide. The unloading of the boats is a big local event, drawing crowds who watch it all from an observation deck. From their fishing grounds 3 to 100 mi offshore, fishermen bring in haddock, cod, flounder, lobster, halibut, and pollack, which is packed in ice and shipped to New York and Boston or sold at the fish market here. You might also see sand sharks being unloaded. They typically go fresh and frozen to French and German markets. Also here is *The Provider,* a monument to the town's fishing industry, featuring a hand pulling a fish-filled net from the sea. ⊠ *Shore Rd. and Barcliff Ave., Chatham.*

The **Chatham Winery** is open for tastings of the fruit and flower wines produced here. ⊠ *1291 Rte. 28, Chatham,* ☎ *508/945–0300.*

The process of glassblowing is fascinating to watch. Demonstrations are performed daily at the **Chatham Glass Company.** ⊠ *17 Balfour La., Chatham,* ☎ *508/945–5547.*

## Dining and Lodging

$$$ ✕ **Christian's.** New owners have taken over an old restaurant, always an insecure moment for the local crowd. The general hope is that Christian's won't change too much. The food here has been consistently good, and what can't change is the pleasure of people-watching from the upstairs deck at lunch. Downstairs, the dining room is now done in an Old Cape/country-French decor: Pilgrim blue with lace. Lunch and dinner are also both French and New England: boneless roast duck with raspberry sauce, and flaky sautéed sole with lobster and lemon-butter sauce. The mahogany-paneled piano bar and deck upstairs serve a light, seafood-based, movie-theme menu. At press time (fall 1997), renovations are planned for the winter, so the appearance may change somewhat. ✉ *443 Main St.,* ☎ *508/945–3362. AE, D, DC, MC, V. Closed weekdays Jan.–Mar.*

$$–$$$ ✕ **The Impudent Oyster.** Just off the beaten path in downtown Chatham, what once was an offbeat scene has now become popular with an older crowd. In summer, bouillabaisse and seafood *fra diablo* (a spicy tomato sauce) are on the menu. The barbecued tuna with orange juice, soy sauce, vinegar and cumin, is a consistent, year-round standout. ✉ *15 Chatham Bars Ave.,* ☎ *508/945–3545. AE, MC, V.*

$$ ✕ **Chatham Squire.** What had been a bar scene and not much more
★ has now evolved into an excellent dining experience. This is the kind of place where if you order anything local, you can't go wrong. The fish is as fresh and good as you get on Cape Cod, and the kitchen continues to innovate while not forgetting its Cape roots. The calamari is always tender, the oysters a lovely mouthful. The Squire is not as inexpensive as it was (or as its exterior implies), but it is much finer. ✉ *487 Main St.,* ☎ *508/945–0945. Reservations not accepted. AE, D. MC, V.*

$$ ✕ **The Sou'wester.** You're more likely to see someone in a yellow slicker and rubber boots than a sports jacket and loafers in this small spot, the most unpretentious bar/restaurant in Chatham, if not the Outer Cape. The decor could hardly even be called that: Eight booths, low light, and what look like fake Oriental rugs for wallhangings. The handwritten menu is heavy on steak tips, sirloin, and pork chops, with fish cakes and beans for the fishermen patrons who haven't had their fill of fish for the day. A long-neck Bud seems the appropriate accompaniment. On weekends there's good, loud, get-down live music in the bar, known as the Lincoln Lodge. ✉ *1528 Route 29,* ☎ *508/945-4424. AE, MC, V.*

$$ ✕ **Vining's Bistro.** Chatham's restaurants have tended to the conser-
★ vative end of the culinary spectrum, but this relatively new bistro is convincing everyone that it deserves mention among the most inventive in the area. The wood grill is the center of attention here, where the chef employs zesty rubs and spices from all over the globe. The wonderful, exotic Bangkok fisherman's stew has almost too much seafood crammed in; spit-roasted Jamaican chicken competes with a Portobello mushroom sandwich as the restaurant's signature dish. Go for the "Mahogany Fire Noodles," an awe-inspiring shrimp and chicken dish. The restaurant is upstairs at the Gallery building, and many of the windows look out on the art below. ✉ *595 Main St.,* ☎ *508/945–5033. Reservations not accepted. AE, D, MC, V. Closed mid-Jan.–Apr.*

$$$ ✕🖼 **Chatham Bars Inn.** This 1914 landmark inn is an oceanfront resort in the old style. High above the beach on a wind-swept bluff, the inn enjoys a fine view of the ocean through massive floor-to-ceiling windows. In matters culinary, the hotel has diversified its formal dining room menu, which now leans on current American hits like rich, seared foie gras, and steamed swordfish with clams. Grand Sunday dinner

buffets occasion ice carvings and a raw bar (including caviar and half lobsters). Lavish buffet breakfasts are served daily year-round. The North Beach Tavern is open for more casual lunch and dinner, and the Beach House Bar opens in season for breakfast and lunch down by the water. The inn consists of a main building flanked with wings—on the ground floor the grand lobby gives way to the formal restaurant on one side and a porch-fronted lounge on the other—and 26 one- to eight-bedroom cottages on 20 landscaped acres overlooking the ocean. The inn has been renovated, but decor is not particularly striking or unique. Some rooms have ocean-view porches or decks and traditional Cape-style furnishings. Many families return because of summertime children's programs. In spite of its history, however, the value for the money is not exceptional. ⊠ *Shore Rd., Chatham 02633,* ☎ *508/945–0096 or 800/527–4884,* FAX *508/945–5491. 136 rooms, 20 suites. 3 restaurants, bar, lobby lounge, pool, putting green, 4 tennis courts, exercise room, volleyball, beach, baby-sitting, children's programs. AE, DC, MC, V.*

$$$$ ★ 🏨 **Wequassett Inn.** This exquisite, traditional resort offers first-rate accommodations in 20 Cape-style cottages along a little bay and on 22 acres of woods, along with luxurious dining, attentive service, evening entertainment, and plenty of sunning and sporting opportunities. Guest rooms have received design awards. Decor is Early American, with country pine furniture and such homey touches as handmade quilts and duck decoys. ⊠ *Pleasant Bay, Chatham 02633,* ☎ *508/432–5400 or 800/ 225–7125,* FAX *508/432–5032. 102 rooms, 2 suites. Restaurant, grill, piano bar, room service, pool, 4 tennis courts, exercise room, windsurfing, boating. AE, D, DC, MC, V. Closed Nov.–Apr. MAP.*

$$$–$$$$ ★ 🏨 **Queen Anne Inn.** Built in 1840 as a wedding present for the daughter of a famous clipper ship captain, the building first opened as an inn in 1874, at which time a north wing was added; a south wing was added much later, in the 1960s. The quite grand resulting structure has large guest rooms furnished in a casual yet elegant style. Some have working fireplaces, private balconies, and hot tubs. All have phones and TVs. Lingering and lounging are encouraged—around the large heated outdoor pool, at the tables on the veranda, in front of the fireplace in the cozy sitting room, and in the plush parlor. The tastefully designed restaurant, which serves a delicious breakfast, is also open for dinner. ⊠ *70 Queen Anne Rd., Chatham 02633,* ☎ *508/945–0394 or 800/545–4667,* FAX *508/945–4884. 31 rooms. Restaurant, bar, airconditioning, heated outdoor pool, spa, 3 tennis courts. AE, D, MC, V.*

$$–$$$$ ★ 🏨 **Captain's House Inn.** Finely preserved architectural details, superb taste in decorating, opulent home-baked goods, and an overall feeling of warmth and quiet comfort are just part of what makes Jan and Dave McMaster's inn one of the Cape's finest. Each room in the four inn buildings has its own personality. Some are quite large, most have fireplaces; some are lacy and feminine, others refined and elegant. The gorgeous Hiram Harding Room in the bow-roofed Captain's Cottage has 200-year-old hand-hewn ceiling beams, a wall of raised walnut paneling, and a large, central working fireplace. The new luxury suites in the stables are particularly spacious and have every amenity imaginable, including whirlpools, fireplaces, TV/VCR, mini-refrigerators, coffeemakers, and private balconies. ⊠ *371 Old Harbor Rd., Chatham 02633,* ☎ *508/945–0127,* FAX *508/945–0866. 16 rooms, 3 suites. Croquet, bicycles. No smoking. AE, MC, V. BP.*

$$–$$$ ★ 🏨 **Moses Nickerson House.** Warm, thoughtful service and a love of fine antiques and decorating characterize this B&B. Guest rooms in the 1839 house feature queen beds and wide-board pine floors, and some have gas-log fireplaces. Each room has an individual look: one with dark woods, leathers, Ralph Lauren fabrics, and English hunting antiques;

another with a high canopy bed and a Nantucket hand-hooked rug. Bathrooms, however, are small. Breakfast is served in a pretty glassed-in sunroom with views of the garden and fishpond. ⊠ *364 Old Harbor Rd., Chatham 02633,* ☎ *508/945–5859 or 800/628–6972. 7 rooms. No smoking. AE, MC, V. BP.*

## Nightlife and the Arts

**Chatham Squire** (⊠ 487 Main St., Chatham, ☎ 508/945–0945), with four separate bars, including a raw bar, is a rollicking year-round local hangout, drawing a young crowd to the bar side and a mixed crowd of locals to the restaurant (☞ Dining and Lodging, *above*).

**Monomoy Theatre** (⊠ 776 Main St., Chatham, ☎ 508/945–1589) presents summer productions—thrillers, musicals, classics, modern drama—by the Ohio University Players.

Chatham's summer **town band concerts** (⊠ Kate Gould Park, Main St., ☎ 508/945–5199) begin at 8 on Fridays and draw up to 6,000 people. As many as 500 fox-trot on the roped-off dance floor, and there are special dances for children and sing-alongs for all.

**Wequassett Inn** (⊠ Pleasant Bay, Chatham, ☎ 508/432–5400) has a jazz duo or piano music nightly in its lounge (jacket requested).

## Outdoor Activities and Sports

### BASEBALL

The **Cape Cod Baseball League** (☎ 508/432–6909), begun in 1885, is an invitational league of college players that boasts Carlton Fisk, Ron Darling, and the late Thurman Munson as alumni. Considered the country's best summer league, it is scouted by all the major-league teams. Ten teams play a 44-game season from mid-June to mid-August; games held at Veterans Field are free (☞ Close-Up, *below*). ⊠ *Main and Depot Sts. by the rotary.*

**Baseball clinics** (☎ 508/432–6909) for children 6–8, 9–12, and 13–17 are offered in one-week sessions by the Chatham A's in summer. ⊠ *Veterans Field, Main and Depot Sts. by the rotary.*

### BEACHES

If you're looking for a crowd-free sandy beach, boats at Chatham Harbor—such as the **Water Taxi** (☎ 508/430–2346)—will ferry you across to North Beach ($12 per person), which is a sand spit adjoining Orleans's Nauset Beach, to South Beach ($15), or to Monomoy ($45, or $10 per person for three or more passengers).

**Outermost Harbor Marine** runs shuttles to South Beach. ⊠ *Morris Island Rd.,* ☎ *508/945–2030.* ⊠ *$9.*

**Harding's Beach,** west of Chatham center, is open to the public and charges daily parking fees to nonresidents in season.

### BIKING

**Bert & Carol's Lawnmower & Bicycle Shop** (⊠ 347 Orleans Rd., Rte. 28, N. Chatham, ☎ 508/945–0137) rents a variety of bikes.

### BOATING

**Monomoy Sail & Cycle** (⊠ 275 Orleans Rd., N. Chatham, ☎ 508/945–0811) rents sailboards and Sunfish.

### GOLF

**Chatham Seaside Links** (⊠ Seaview St., ☎ 508/945–4774), a nine-hole course, is good for beginners.

### SURFING

The Outer Cape beaches, including North Beach in Chatham, are the best spots for surfing, which is tops when there's a storm offshore.

TENNIS

**Chatham Bars Inn** (✉ Shore Rd., Chatham, ☎ 508/945–0096, ext. 1155) offers four waterfront all-weather tennis courts, lessons, and a pro shop.

## Shopping

**Main Street** in Chatham is a busy shopping area with upscale and conservative merchandise. Here you'll find galleries, crafts, clothing stores, bookstores, and a few good antiques shops.

**Cape Cod Cooperage** (✉ 1150 Queen Anne Rd., at Rte. 137, Chatham, ☎ 508/432–0788), set in an old barn, sells woodenware made in a century-old tradition by an on-site cooper (barrel maker), as well as hand-decorated furniture, craft supplies, and more.

**Chatham Glass Co.** (✉ 17 Balfour La., W. Chatham, ☎ 508/945–5547) is a glassworks where you can watch glass being blown and buy it, too—objects including marbles, Christmas ornaments, jewelry, and art glass.

**Chatham Jam and Jelly Shop** (✉ 10 Vineyard Ave., at Rte. 28, W. Chatham, ☎ 508/945–3052) sells preserves—cranberry with strawberries, Maine wild blueberry—nutty conserves, and ice-cream toppings, all made on-site in small batches.

**Fancy's Farmhouse in the Cornfield** (✉ Rte. 28, W. Chatham, ☎ 508/945–1949) sells local and exotic produce, fresh-baked breads, pies, and pastries, dried flowers, baskets, frozen prepared gourmet foods, spices, and potpourri.

**Marion's Pie Shop** (✉ 2022 Main St., Rte. 28, W. Chatham, ☎ 508/432–9439) sells homemade and home-style fruit breads, pastries, prepared foods such as lasagna, Boston baked beans, chowder base, and, of course, pies, both savory and sweet.

**The Spyglass** (✉ 618 Main St., Chatham, ☎ 508/945–9686) carries telescopes, barometers, writing boxes, maps, and other nautical antiques.

**Yellow Umbrella Books** (✉ 501 Main St., Chatham, ☎ 508/945–0144) has an excellent selection of new books, many about Cape Cod, as well as used books.

*En Route*  North of Chatham, Route 28 winds through wooded upland toward **Pleasant Bay,** from which a number of country roads will take you to a nice view of Nauset spit and the islands in the bay.

# Orleans

**41** *4 mi southwest of Eastham, 35 mi east of the Sagamore Bridge.*

Named for Louis-Philippe de Bourbon, duc d'Orléans, who reputedly visited the area during his exile from France in the 1790s, Orleans was incorporated as a town in 1797. Historically, it has the distinction of being the only spot in the continental United States to have received enemy fire during either world war. In July 1918 a German submarine fired on commercial barges off the coast. Four were sunk, and one shell is reported to have fallen on American soil. Orleans is a mix of quiet seaside village and bustling commercial center—today's commercial hub of the Lower Cape.

A walk along Rock Harbor Road, a pleasant winding street lined with gray-shingled Cape houses, white picket fences, and neat gardens leads to the bay-side **Rock Harbor,** a former packet landing and the site of a War of 1812 skirmish in which the Orleans militia kept a British warship from docking. In the 19th century Orleans had an active saltworks, and a flourishing packet service between Rock Harbor and Boston developed. Today the former packet landing is the base of charter-fishing and party boats in season, as well as of a small commercial fishing fleet whose catch hits the counters at the fish market and small restaurant here. Sunsets over the harbor are memorable.

# THE CAPE COD BASEBALL LEAGUE

**A**T THE BASEBALL HALL OF **FAME** in Cooperstown, NY, you'll find a poster announcing a showdown between arch rivals Sandwich and Barnstable. The date? July 4, 1885. In the hundred-plus years since that day, the Cape's ball-playing tradition has continued unabated, and if you're a sports fan, a visit to a Cape Cod Baseball League summer game is a must. Seeing a game on the Cape is to come into contact with baseball's roots. You'll remember why you love the sport, and you'll have a newfound sense of why it became *the* national pastime.

As they have since the 1950s, top-ranked college baseball players from around the country descend on the Cape when school lets out, in time to begin the season in mid-June. This is no sandlot, catch-as-catch-can scene. Each player joins one of the league's 10 teams, which are based in Bourne, Wareham, Falmouth, Cotuit, Hyannis, Dennis-Yarmouth, Harwich, Brewster, Chatham, and Orleans. Players lodge with local families and work day jobs cutting lawns, painting houses, or giving baseball clinics in town parks. In the evenings, though, their lives are given over to baseball, as they don uniforms and head for the field.

To judge by the past, Cape League veterans are tomorrow's major league stars. By latest count, one of every eight active major league ballplayers spent a summer in the Cape League on the way up. The Boston Red Sox's recent starting infield—Mo Vaughn, John Valentin, Tim Naehring, and Nomar Garciaparra—all played in the Cape League in different years, for different teams. Add names like Frank Thomas, Jeff Bagwell, Albert Belle (before he lost his manners), Will Clark, and Walt Weiss, and you begin to get a sense of the quality on display.

As good as the ball is—and you'll often see bunches of major league scouts at a game—another great reason to come out to the ballpark is . . . the ballpark. Chatham's **Veterans Field** is the Cape's 3Com Park at Candlestick Point, because, much like the San Francisco version, fog tends to engulf the games. Orleans's **Eldredge Park** is a local favorite—immaculate, cozy, and comfortable. Some have bleachers, while in others it's up to you to bring your own chair or blanket and stretch out behind a dugout or baseline. Kids are free to roam and can even try for foul balls, which they are, however, asked to return because, after all, balls don't grow on trees. When hunger hits, the ice-cream truck and hot-dog stand are never far away.

Games start at either 5 or 7 PM, depending on whether the field has lights (there are occasionally afternoon games). Each team plays 44 games in a season, so finding one is rarely a problem. And best of all, they're always free. The Cape's baseball scene is so American, the ambience so relaxed and refreshing, that it's tempting to invoke the old *Field of Dreams* analogy. But the league needs no Hollywood comparison. This is the real thing. It was built a long time ago, and they are still coming.

*–by Seth Rolbein*

Another French connection, the 1890 **French Cable Station Museum** was the stateside landing point for the 3,000-mi-long transatlantic cable that originated in Brittany. Another cable laid between Orleans and New York City completed the France–New York link, and many important messages were communicated through the station. In World War I it was an essential connection between Army headquarters in Washington and the American Expeditionary Force in France, and it was under guard by the marines. By 1959 telephone service had rendered the station obsolete, and it was closed. Its equipment is still in place. ⊠ *41 S. Orleans Rd.,* ☎ *508/240–1735.* 🖭 *Donations accepted.* ⊙ *June–Labor Day, Mon.–Sat. 10–4.*

The **Jonathan Young Windmill** is a landmark from the days of salt-making in Orleans, when it would pump saltwater into shallow vaults for evaporation. A program explaining the history and operation of the mill demonstrates the old millstone and grinding process. ⊠ *Rte. 6A and Town Cove, Orleans,* ⊙ *July–Aug., daily 11–4; June and Sept., weekends 11–4.*

Ċ The **Academy of Performing Arts** offers two-week sessions of theater, music, and dance classes, with a show at the end of each session, to children 8 to 12 years old. There are also year-round classes for ages 4 to adult in dance, music, and drama. ⊠ *120 Main St., Orleans,* ☎ *508/255–5510.*

## Dining and Lodging

$$$ ✕ **Captain Linnell House.** Grecian columns grace the front of the house, giving you the uncomfortable feeling that Tara has been relocated to the Cape. Inside, lots of antiques and lace grace the small, cozy dining rooms. The traditional menu relies on local seafood as well as a number of very good meat dishes, like an always-perfect rack of lamb. The Sunday buffet brunch is much too filling—unless you can control yourself. The large, newly-reconstructed Carriage Room is popular for weddings and banquets. ⊠ *137 Skaket Beach Rd.,* ☎ *508/255–3400. Reservations essential. AE, MC, V.*

$$$ ✕ **Nauset Beach Club.** What once was the unsung local hero has now
★ become widely known for its fine dining. Thanks to its location in a former residence, you might feel like you're eating in someone's living room, but the sophisticated food, rooted in contemporary Italian cuisine, is beyond even that of the finest home cooks. *Zuppa di pesce* (Italian seafood stew) features shrimp, scallops, calamari, and lobster in a sauce Provençale on pasta. The penne *puttanesca* (with olives, garlic, peppers, and anchovies) delivers a whole medley of tastes. There are only two drawbacks: The bar is too small and uncomfortable for the long waits (reservations are accepted only for parties of six or more), and the attitude can be off-putting—even in big, lavish parties, for instance, if one diner wants to order only an appetizer, it will be charged as a dinner. ⊠ *222 Main St, E. Orleans,* ☎ *508/255–8547. AE, D, DC, M, V. No lunch. No dinner Sun. and Mon. mid-October–Memorial Day.*

$$ ✕ **Kadee's Lobster & Clam Bar.** A summer landmark that harkens back to an era when drive-ins were the norm and not the exception, Kadee's serves good clams and fish-and-chips that you can grab on the way to the beach from the take-out window. Or sit down in the smoke-free indoor dining room for steamers and mussels, pasta and seafood stews, or their famous Portuguese kale soup. There's a gift shop as well, and a miniature golf course out back. The only serious problem here is that, for what they serve, the prices seem to have gone through the roof. ⊠ *Main St.,* ☎ *508/255–6184. Reservations not accepted. MC,*

*V. Closed day after Labor Day–week before Memorial Day and week-days in early June.*

**$–$$** ✕ **The Yardarm.** Orleans' version of a roadhouse, the Yardarm is al-ways smoky and full of locals. The tube over the bar is likely to be tuned to a sports game, and the only pool table in town is always busy (though not in use when dinner is being served). The hearty, well-cooked food gives great value, especially the baked sole, barbecued ribs or chicken, and the pot roast (thanks to the portions, expect to take some of your dinner home). It's also a great place to stop in for a burger-and-beer lunch. ✉ *Rte. 28,* ☎ *508/255–4840. Reservations not ac-cepted. AE, DC, MC, V.*

**$** ✕ **Land Ho!** Orleans's flagship local scene is definitely fun and bois-terous even if the food doesn't always live up to the atmosphere (and is almost always overpriced). Hanging from the rafters are dozens of homemade wooden signs; anybody who's anybody in town has one made up. Under them is the blackboard menu, which offers typical tav-ern fare: clams, burgers, chowder, and fish-and-chips. The conversa-tion is generally tastier (and saltier) than the chowder. It's a good place for a rainy day lunch. ✉ *Rte. 6A and Cove Rd.,* ☎ *508/255–5165. Reservations not accepted. AE, MC, V.*

**$** ✕ **Lobster Claw.** If you're over 6 ft tall, keep an eye out for the fish-ing nets hanging from the ceiling inside this goofy little spot. Tables are lacquered turquoise, portions are huge, and the lobster roll is one of the best. This is a fun, authentic place to drop in for lunch, and your kids are likely to love it. ✉ *Rte. 6A,* ☎ *508/255–1800. Reservations not accepted. AE, D, DC, MC, V.*

**$–$$** ⊞ **Kadee's Gray Elephant.** A mile from Nauset Beach, next to shops,
★ a grocery store, a farm stand, a post office, and **Kadee's Lobster & Clam Bar** (☞ *above*), this 200-year-old house offers small vacation studio and one-bedroom apartments. Unique on the Cape, they are a cheer-ful riot of color, from wicker painted lavender or green to bright pink bows and flowers painted on furniture to beds layered in quilts and comforters mixing plaids and florals to colorfully framed art. Kitchens are fully equipped with microwaves, attractive glassware, irons and boards—and even lobster crackers. ✉ *216 Main St., Box 86, E. Or-leans 02643,* ☎ *508/255–7608,* ℻ *508/240–2976. 6–8 apartments. Restaurant, miniature golf, gift shop. MC, V.*

## Nightlife and the Arts

THE ARTS

The **Addison Holmes Gallery** (✉ 43 Rte. 28, Orleans, ☎ 508/255–6200), housed in four rooms of a quaint brick-red Cape, represents over two dozen area artists and offers a broad range of contemporary works, many of which are inspired by life on Cape Cod. A new sculp-ture garden opened in 1997, a perfect complement to the tasteful gallery.

*Cape Cod Antiques & Arts* (✉ Box 2824, Orleans 02653–0039, ☎ 508/255–2121 or 800/660–8999), a monthly supplement of the *Register* and the *Cape Codder* available at local newsstands, is chock-full of in-formation on galleries, upcoming shows, Cape artists, antiques shops, auctions, and so forth.

**Tree's Place** (✉ Rte. 6A, at Rte. 28, Orleans, ☎ 508/255–1330) dis-plays the work of New England artists, including Robert Vickery, Don Stone, and Elizabeth Mumford (whose popular folk art is bordered in mottoes and poetic phrases). Champagne openings are held on Satur-day nights in summer.

The **Academy Playhouse** (✉ 120 Main St., Orleans, ☎ 508/255–1963), one of the oldest community theaters on the Cape, presents 12 or 13 productions year-round, including original works.

The **Cape & Islands Chamber Music Festival** (✉ Box 2721, Orleans 02653, ☎ 508/255–9509) is three weeks of top-caliber performances, including a jazz night, at various locations in August.

The 90-member **Cape Cod Symphony Orchestra** (☎ 508/362–1111) sails into Orleans on Saturday, August 22, 1998, for one of its three Sounds of Summer Pops Concert (the other two are in Mashpee and Hyannis). The performance takes place at the Eldridge Field Park.

## Outdoor Activities and Sports

BEACHES

The town-owned **Nauset Beach**—not to be confused with Nauset Light Beach up a ways at the National Seashore—is a 10-mi-long, wide sweep of sandy ocean beach with low dunes and large waves good for bodysurfing or board surfing. There are lifeguards, rest rooms, showers, and a food concession. It's open to off-road vehicles. ✉ *Beach Rd.*

**Skaket Beach** on Cape Cod Bay is a sandy stretch with calm warm water good for children. There are rest rooms, lifeguards, and a snack bar. ✉ *Skaket Beach Rd.*

BOATING

**Arey's Pond Boat Yard** (✉ Off Rte. 28, S. Orleans, ☎ 508/255–0994) has a sailing school with individual and group lessons.

FISHING

Many of Orleans's freshwater ponds offer good fishing for perch, pickerel, trout, and more. The required license, along with rental gear, is available at the **Goose Hummock Shop** (✉ Rte. 6A, Orleans, ☎ 508/255–0455).

**Rock Harbor Charter Boat Fleet** (✉ Rock Harbor, Orleans, ☎ 508/255–9757 or 800/287–1771 in MA) goes for bass and blues in the bay from spring through fall. Walk-ons and charters are both available.

The *Flying Mist* (☎ 508/255–0880) leaves Rock Harbor for both half- and full-day sport fishing trips. Cleaning, smoking, freezing, and stuffing—as in taxidermy—services are available.

ICE-SKATING

Ice-skating is available at the **Charles Moore Arena** (✉ O'Connor Way, Orleans, ☎ 508/255–2971). Kids 9–14 ice-skate to DJ-spun rock and flashing lights at **Rock Night**, which takes place Fridays from 8 to 10 PM.

SPORTS EQUIPMENT RENTALS

**Nauset Sports** (✉ Rte. 6A, Orleans, ☎ 508/255–4742; ✉ Rte. 6, N. Eastham, ☎ 508/255–2219) rents surf, body, skim, and sail boards, kayaks, wet suits, in-line skates, and tennis racquets.

SURFING

The Outer Cape beaches, including **Nauset Beach** in Orleans, are the best spots for surfing, especially when there's a storm offshore. For a surf report—water temperature, weather, surf, tanning factor—call 508/240–2229.

**Pump House Surf Co.** (✉ 9 Rte. 6A, Orleans, ☎ 508/240–2226) rents wet suits and sells other surf gear.

## Shopping

**Baseball Shop** (⊠ 26 Main St., Orleans, ☎ 508/240–1063) sells licensed products relating to baseball and other sports—new and collectible cards (and non-sports cards) as well as hats, clothing, and videos.
**Bird Watcher's General Store** (⊠ 36 Rte. 6A, Orleans, ☎ 508/255–6974 or 800/562–1512) sells nearly everything avian but the birds themselves: feeders, paintings, houses, books, binoculars, fountains, calls—ad infinitum.
**Clambake Celebration** (⊠ 9 West Rd., Orleans, ☎ 508/255–3289 or 800/423–4038) prepares full clambakes, including lobster, clams, mussels, corn, potatoes, onions, and sausage, for you to take away—or it'll deliver or air-ship year-round. The food is layered in seaweed in a pot and ready to steam.
**Fancy's Farm Stand** (⊠ 199 Main St., E. Orleans, ☎ 508/255–1949) sells local and exotic produce, fresh-baked breads, pies, and pastries, dried flowers, baskets, frozen prepared gourmet foods, spices, and potpourri.
**Hannah** (⊠ 47 Main St., Orleans, ☎ 508/255–8234) has women's fashions with flair, by such labels as Hannah, No Saint, and Angel Heart.
**Heaven Scent You** (⊠ 13 Cove Rd., Orleans, ☎ 508/240–2508) offers massage, spa services, and beauty treatments—everything you need for some relaxation and rejuvenation.
**Kemp Pottery** (⊠ Rte. 6A, Orleans, ☎ 508/255–5853) has functional and decorative stoneware and porcelain, fountains, garden sculpture, sinks, and stained glass.
**Tree's Place** (⊠ Rte. 6A, at Rte. 28, Orleans, ☎ 508/255–1330), one of the Cape's best and most original shops, has a collection of hand-crafted kaleidoscopes, as well as art glass, hand-painted porcelain and pottery, handblown stemware, Russian lacquer boxes, jewelry, and imported ceramic tiles as well as fine art.

# Eastham

**㊷** *6 mi south of Wellfleet.*

It was here in 1620 that Myles Standish and company landed on First Encounter Beach and met the Nauset Indians. The meeting was peaceful, but the Pilgrims moved on to Plymouth anyway. Nearly a quarter century later, they returned to settle the area, which they originally called by its Indian name, Nawsett. Eastham was incorporated as a town on June 7, 1651.

Like many of the towns on the Cape, Eastham started as a farming community, later turning to the sea and to salt-making for its livelihood—at one time there were more than 50 saltworks in town. A more atypical industry here was asparagus-growing; from the late 1800s through the 1920s, Eastham was known as the "Asparagus Capital."

**㊸** The road to the National Seashore's **Ft. Hill Area** winds past the **Captain Edward Penniman House** (☞ *below*) ending at a parking area with a lovely view of old farmland traced with stone fences that rolls gently down to **Nauset Marsh** (☞ *below*) and a red-maple swamp. Appreciated by bird-watchers and nature photographers, the 1-mi Red Maple Swamp Trail begins outside the Penniman house and winds through the area, branching into two separate paths, one of which eventually turns into a boardwalk that meanders through wetlands. The other path leads directly to Skiff Hill, an overlook with benches and informative plaques that quote Samuel de Champlain's account of the area when he moored off Nauset Marsh in 1605. Also on Skiff Hill is Indian Rock, a large boulder that was moved to the hill from the marsh below. Once used by Indians as a sharpening stone, the rock is cut with deep grooves and smoothed in circles where ax-heads were whetted.

The French Second Empire–style **Captain Edward Penniman House** was built in 1868 for a whaling captain. The impressive exterior is noted for its mansard roof, its cupola, which once commanded a dramatic view of bay and sea, and the whale-jawbone entrance gate. Call the National Seashore at 508/255–3421 to find out when the interior, currently being renovated section by section, is open for guided tours by reservation or for browsing through changing exhibits. ⊠ *Ft. Hill Area, Eastham.*

The park on Route 6 at Samoset Road has as its centerpiece the **Eastham Windmill,** the oldest windmill on Cape Cod. A smock mill built in Plymouth in the early 1680s, it was moved to this site in 1793 and is the only Cape windmill still on the site where it was used commercially. The mill was restored by local shipwreck historian William Quinn and friends. ⊠ *Rte. 6.* 🎫 *Free.* ☉ *Late June–Labor Day, Mon.–Sat. 10–5, Sun. 1–5.*

Frozen in time, the 1741 **Swift-Daley House** was once the home of Gustavus Swift, founder of the Swift meat-packing company. Inside the full Cape with bow roof you'll find beautiful pumpkin-pine woodwork and wide-board floors; a ship's-cabin staircase, which, like the bow roof, was built by ships' carpenters; and fireplaces in every downstairs room. The Colonial-era furnishings include an old cannonball rope bed, tools, a melodeon, and a ceremonial quilt decorated with beads and coins. Antique clothing includes a stunning 1850 wedding dress. Out back is a tool museum. ⊠ *Rte. 6, no phone.* 🎫 *Free.* ☉ *July–Aug., weekdays 1:30–4.*

A great spot for watching sunsets over the bay, **First Encounter Beach** is drenched in history. Near the parking lot, a bronze marker commemorates the first encounter between local Indians and passengers from the *Mayflower,* led by Captain Myles Standish, who explored the entire area for five weeks in November and December 1620 before moving on to Plymouth. The remains of a Navy target ship retired after 25 years of battering now rest on a sandbar about 1 mi out.

★ The Cape's most expansive national treasure, the **Cape Cod National Seashore** was established in 1961 under the administration of President John F. Kennedy, for whom Cape Cod was home and haven. The 44,000-acre seashore encompasses and protects 30 mi of superb ocean beaches, great rolling dunes, swamps, marshes, and wetlands, pitch pine and scrub oak forest, all kinds of wildlife, and a number of historic structures. Self-guided nature trails, as well as biking and horse trails, lace through these landscapes. From here, hiking trails lead to a red-maple swamp, **Nauset Marsh,** and **Salt Pond,** in which breeding shellfish are suspended from floating "nurseries"—their offspring will later be used to seed the flats. Also here, the Buttonbush Trail is a nature path for people with vision impairments. A hike or bike ride to Coast Guard Beach leads to a turnout looking out over marsh and sea. A section of the cliff here was washed away in 1990, revealing remains of a prehistoric dwelling.

④④ **Salt Pond Visitor Center** is the first visitor center of the Cape Cod National Seashore; the other, the **Province Lands Visitor Center,** is in Provincetown (☞ Provincetown, *below*). The Salt Pond center offers guided walks, tours, boat trips, demonstrations, and lectures from mid-April through Thanksgiving, as well as evening beach walks, campfire talks, and other programs in summer.

The Salt Pond Visitor Center includes a museum with several displays: on whaling and the old saltworks, early Cape Cod artifacts including scrimshaw, the journal that Mrs. Penniman kept while on a whaling voyage with her husband, and some of the Pennimans' possessions, such as their tea service and the captain's top hat. A bookstore and an air-conditioned auditorium showing films on geology, sea rescues, whaling, Henry David Thoreau, and Marconi are also in the visitor center. Something's up most summer evenings at the outdoor amphitheater, from slide-show talks to military-band concerts. ⊠ *Off Rte. 6, Eastham,* ☎ *508/255–3421.* ⊡ *Free.* ☉ *Mar.–June and Sept.–Dec., daily 9–4:30; July–Aug., daily 9–5; Jan.–Feb., weekends 9–4:30.*

Roads and bicycle trails lead to **Coast Guard and Nauset Light beaches,** which begin an unbroken 30-mi stretch of barrier beach extending to Provincetown—the "Cape Cod Beach" of Thoreau's 1865 classic, *Cape Cod.* You can still walk its length, as Thoreau did, though the Atlantic continues to claim more of the Cape's eastern shore every year. The site of the famous beach cottage of Henry Beston's 1928 book, *The Outermost House,* is to the south, near the end of Nauset spit. Designated as a literary landmark in 1964, the cottage was, alas, completely destroyed in the Great Blizzard of February 1978.

Moved 350 ft back from its perch at cliff's edge in 1996, the much-photographed red and white **Nauset Light** still tops the bluff where the "Three Sisters" lighthouses once stood; the Sisters themselves can be seen in a little landlocked park surrounded by trees, reached by paved walkways off Nauset Light Beach's parking lot. How the lighthouses got there is a long story. In brief—in 1838 three brick lighthouses were built 150 ft apart on the bluffs in Eastham, overlooking a particularly dangerous area of shoals (shifting underwater sandbars). In 1892, after the eroding cliff dropped the towers into the ocean, they were replaced with three wooden towers. In 1918 two were moved away, as was the third in 1923. Eventually the National Park Service acquired the Three Sisters and brought them together here, where they would be safe, rather than returning them to the eroding coast. The Fresnel lens from the last working lighthouse is on display at the **Salt Pond Visitor Center** (☞ *above*). Lectures on, and guided walks to, the lighthouses are conducted throughout the season. In 1997 tours of Nauset Light were given on Sunday between 4:30 and 7:30 PM in July and August, on Sunday from 1–4 PM after Labor Day, and by appointment; the plan was to repeat this schedule in 1998, but call in advance to confirm. ☎ *508/240–2612.*

## Dining and Lodging

$   ✕ **Lori's Family Restaurant.** Omelettes, pancakes, French toast, waffles—admit it, they're a part of being on vacation. Fresh mushrooms and sprouts to accompany the eggs are typical of the perfect, extra touches that make every plate special. Lori's serves only breakfasts (it's open daily from 6 AM to 1 PM), although the restaurant transforms into Mitchel's, a casual bistro, in the evening. ⊠ *Main St. Mercantile, Rte. 6, N. Eastham,* ☎ *508/255–4803. Reservations not accepted. MC, V.*

$$$–$$$$   ⊡ **Four Points by Sheraton.** At the entrance to the National Seashore, this Sheraton offers standard modern-decor rooms with two double beds or one king-size bed, and tiled baths. Rooms have views of the pool, the parking lot, or the woods. Outside rooms are a little bigger and brighter and have mini-refrigerators. Children under 17 stay free. ⊠ *Rte. 6, Eastham 02642,* ☎ *508/255–5000 or 800/533–3986,* ℻ *508/240–1870. 107 rooms, 2 suites. Restaurant, lobby lounge, no-smoking rooms, room service, 1 indoor and 1 outdoor pool, saunas, 2 tennis courts, exercise room, video games. AE, D, DC, MC, V.*

# In case you want to see the world.

**At American Express, we're here to make your journey a smooth one. So we have over 1,700 travel service locations in over 120 countries ready to help. What else would you expect from the world's largest travel agency?**

do more

**Travel**

# In case you want to be welcomed there.

We're here to see that you're always welcomed at establishments everywhere. That's why millions of people carry the American Express® Card – for peace of mind, confidence, and security, around the world or just around the corner.

do more®

Cards

# And just in case.

We're here with American Express® Travelers Cheques and Cheques *for Two.*® They're the safest way to carry money on your vacation and the surest way to get a refund, practically anywhere, anytime.

Another way we help you...

do more

**Travelers Cheques**

**$$–$$$** ⊞ **Penny House Inn.** Tucked behind a wave of privet hedge, this rambling gray-shingle inn offers mostly spacious rooms, furnished with a combination of antiques, collectibles, and wicker. Deluxe rooms have air-conditioning, mini-refrigerators, and phones. Captain's Quarters, the grandest room, has a sitting area with a wood-burning fireplace, a king-size brass bed, miniature library, and a lovely view of the garden. Common areas include the Great Room, with a fireplace and lots of windows, a combination sunroom/library, and a garden patio set with umbrella tables. Afternoon tea is available. ⊠ *4885 County Rd., Rte. 6, N. Eastham 02651,* ☎ *508/255–6632 or 800/554–1751,* FAX *508/255–4893. 11 rooms. No smoking. AE, D, MC, V. BP.*

**$$–$$$** ⊞ **Whalewalk Inn.** Named for the widow's walk atop the building, this 1830 whaling master's home is situated on 3 acres of rolling lawns, gardens, and meadows. Lots of windows give the place an open, airy feeling, while wide-board pine floors, fireplaces, and 19th-century country antiques provide historical appeal. Guest rooms are spacious, with wooden or brass twin, double, or queen-size beds, floral fabrics, and country-cottage furniture. Several accommodations are available, including rooms in the main inn and suites with fully equipped kitchens in the converted barn and guest house. A secluded saltbox cottage has a fireplace, kitchen, and private patio. A carriage house was added in 1997; some of its rooms have fireplaces and whirlpool tubs; all have air-conditioning. Breakfast is served each morning in the cheerful sunroom. ⊠ *220 Bridge Rd., Eastham 02642,* ☎ *508/255–0617. 11 rooms, 5 suites. Bicycles. No smoking. MC, V. BP.*

**$** ⊞ **Hostelling International–Mid Cape.** On 3 wooded acres near the Cape Cod Rail Trail and a 15-minute walk to the bay, this hostel has eight cabins that sleep six to eight each; two of them can be used as family cabins. There is a common area and kitchen, volleyball nets, a Ping-Pong table, and a number of programs are available for guests. ⊠ *75 Goody Hallet Dr., Eastham 02642,* ☎ *508/255–2785.* ☾ *Mid-May– mid-Sept.*

**$** ⚘ **Atlantic Oaks Campground.** Next to the Four Points Sheraton, this campground is less than a mile north of the Salt Pond Visitor Center. Primarily a 100-site RV camp, it offers limited tenting as well. The setting is a pine and oak forest, and you're minutes from the Seashore. There are bikes for rent and there's direct access to the Cape Cod Rail Trail. RV hookups, including cable TV, cost $32 for two people; tent sites are $24 for two people. Showers are free. ⊠ *Rte. 6, Eastham 02642,* ☎ *508/255–1437 or 800/332–2267. 100 RV sites, 30 tent sites. Bicycles, playground, coin laundry. Closed Nov.–Apr.*

## Nightlife and the Arts

The **Cape Cod National Seashore** offers summer evening programs, such as slide shows, sunset beach walks, concerts by local groups or military bands, and campfire sing-alongs, at its **Salt Pond Amphitheater** (⊠ Eastham, ☎ 508/255–3421).

**First Encounter Coffee House** (⊠ Chapel in the Pines, Samoset Rd., Eastham, ☎ 508/255–5438), presents a mixture of professional and local folk and blues in a no-smoking, no-alcohol environment, with refreshments available during intermission, and books national acts the first and third Saturday of each month. It's closed May and September.

**Four Points by Sheraton** (⊠ Rte. 6, ☎ 508/255–5000; ☞ Lodging, *above*) often features a folk guitarist–comedian–sing-along leader in its lounge. In summer there's some kind of entertainment most nights.

## Outdoor Activities and Sports

BEACHES

On the bay side of the Outer Cape, **First Encounter Beach** is open to the public and charges daily parking fees to nonresidents in season.

**Coast Guard Beach** in Eastham is a long beach backed by low grass and heathland. It has no parking lot of its own, so park at the Salt Pond Visitor Center and take the free shuttle to the beach.

**Nauset Light Beach,** adjacent to Coast Guard Beach, continues the landscape of long sandy beach backed by grass and heathland, with the lighthouse for a little extra Cape atmosphere.

BICYCLING

**Nauset Trail** is maintained by the National Seashore and stretches 1⅗ mi, from Salt Pond Visitor Center in Eastham through groves of apple and locust trees to Coast Guard Beach.

The **Little Capistrano Bike Shop** (✉ Rte. 6, Eastham, ☎ 508/255–6515) rents a variety of different bikes and trailers, and is conveniently located between the Cape Cod Rail Trail and the National Seashore Bike Trail.

HEALTH AND FITNESS CLUBS

The **Norseman Athletic Club** (✉ Rte. 6, N. Eastham, ☎ 508/255–6370 or 508/255–6371) has four racquetball, two squash, six indoor tennis, and two indoor basketball courts, an Olympic-size heated pool indoors, swimming lessons, aerobics and children's self-defense classes, along with a pro shop and a restaurant.

# Wellfleet and South Wellfleet

*10 mi north of Eastham, 13 mi southeast of Provincetown.*

㊺ Once the center of a large oyster industry and, with Truro, a Colonial whaling and codfishing port, **Wellfleet** is now a center for artists and writers. Less than 2 mi wide, it is one of the more tastefully developed Cape resort towns, with a number of fine restaurants, historic houses, art galleries, and a good old main street in the village proper. South Wellfleet borders North Eastham and is home to a wonderful Audubon sanctuary and a drive-in theater that doubles on weekends as a flea market.

★ ㊻ A trip to the Cape isn't complete without a visit to the delightful and informative **Massachusetts Audubon Wellfleet Bay Sanctuary.** A 1,000-acre haven for more than 250 species of birds attracted by the varied habitats found here, the sanctuary is the jewel of the Massachusetts Audubon Society. It is a superb place for walking, birding, and looking west over the salt marsh and bay at wondrous sunsets.

The Audubon Society hosts a great variety of naturalist-led wildlife tours year-round. This part of the Cape, including the Monomoy Islands, has a number of fascinating natural spots. There are birding, canoe, and kayak trips, bay cruises, bird and insect walks, hikes, snorkeling, and winter seal cruises. There are also camps for children in July and August and weeklong field schools for adults. Phone reservations are required ahead of time, as some of the programs are very popular. ✉ *Off Rte. 6, Box 236, S. Wellfleet 02663, ☎ 508/349–2615. ⚏ $3. ☉ Daily 8 AM–dusk.*

㊼ **Marconi Station** on the Atlantic side of the Cape's forearm is the site of the first transatlantic wireless station erected on the U.S. mainland. From here, Italian radio and wireless-telegraphy pioneer Guglielmo Marconi sent the first American wireless message to Europe—"most cordial greetings and good wishes" from President Theodore Roosevelt to Edward VII of England—on January 18, 1903. The station broadcast news for 15 years. An outdoor shelter contains a model of the original station, of which only fragments remain as a result of cliff erosion—parts of the tower bases are sometimes visible on the beach below, where they fell. The Seashore's administrative headquarters is here, and though it is not an official visitor center, it can provide information at times when the centers are closed. Inside there is a mock-up of the spark-gap transmitter used by Marconi. Off the parking lot a 1¼-mi trail and boardwalk lead through the **Atlantic White Cedar Swamp**—it's one of the most beautiful trails on the Seashore. Free maps and guides are available at the trailhead. **Marconi Beach,** south of the station, is another of the National Seashore's ocean beaches (☞ Beaches, *below*). ✉ *Marconi Site Rd., S. Wellfleet,* ☎ *508/349–3785.* ☉ *Daily 8–4:30.*

For a **scenic loop** through a classic Cape landscape near Wellfleet's Atlantic beaches—with scrub and pines on the left, heathland meeting cliffs and ocean below on the right—take LeCount Hollow Road just north of the Marconi Station turnoff. All of the beaches on this strip rest at the bottom of a tall, grass-covered dune, which lend dramatic character to this outermost shore. The first of the four, **LeCount Hollow,** is restricted in season, as is the last, **Newcomb Hollow,** with a scalloped shoreline of golden sand. In between, **White Crest** and **Cahoon Hollow** are town-managed public beaches. On breezy days, hang gliders fly from the cliffs. Cahoon has a hot restaurant and dancing spot, the Beachcomber (☞ Nightlife and the Arts, *below*). Backtrack to Cahoon Hollow Road and turn west for the southernmost entrance to the town of Wellfleet proper, across Route 6.

The **First Congregational Church,** a handsome 1850 Greek Revival building, is said to have the only town clock in the world to strike on ship's bells. The church's interior is lovely, with pale blue walls, a brass chandelier hanging from an enormous gilt ceiling rosette, subtly colored stained-glass windows, and pews curved to form an amphitheater facing the altar and the 1873, 738-pipe Hook and Hastings tracker-action organ. To the right is a Tiffany-style window depicting a clipper ship, with a dedication to the memory of a sea captain. Concerts are given in July and August on Sundays at 8 PM. ✉ *Main St., Wellfleet.*

For a glimpse into Wellfleet's past, the **Wellfleet Historical Society Museum** exhibits furniture, paintings, shipwreck salvage, needlework, navigation equipment, early photographs, Indian artifacts, clothing, and more. The society's Samuel Rider House is no longer open to the public. ✉ *Main St., Wellfleet,* ☎ *508/349–9157.* ⚑ *$1.* ☉ *Late June–mid-Sept., Tues.–Sat. 2–5.*

The comfortable **Wellfleet Public Library** is a great place to spend a rainy afternoon. There is an extensive selection of periodicals and plenty of books to cozy up with. The library is a reflection of the literary life of Wellfleet, as are the two wonderful bookstores in town (☞ Shopping, *below*). Among the writers who have spent time here are Mary McCarthy, Edmund Wilson, Annie Dillard, and Marge Piercy. ✉ *55 W. Main St.,* ☎ *508/349–0310.*

**Main Street** is a good place to start if you're in the mood for shopping (☞ Shopping, *below*). **Commercial Street** has all the flavor of the fishing town that Wellfleet remains. Galleries and shops occupy small weathered-shingle houses, some of which look like fishing shacks. A good stroll around town would take in Commercial and Main streets, ending perhaps at **Uncle Tim's Bridge.** The short walk across this arcing landmark—

with its beautiful, much-photographed view over marshland and a tidal creek—leads to a small wooded island. It's a great waterside spot. Follow Commercial Street to the **Wellfleet Pier,** busy with fishing boats, sailboats, yachts, charters, and party boats. At the twice-daily low tides you can shellfish on the tidal flats for oysters, clams, and quahogs (for a permit, call 508/349–9818).

Chequesset Neck Road makes for a pretty 2½-mi drive from the harbor to the bay past Sunset Hill—a great place to catch one. At the end, on the left, is a parking lot and wooded picnic area, from which nature trails lead off to **Great Island,** perfect for the beachcomber and solitude seeker. The "island" is actually a peninsula connected by a sand spit that tidal action built up. The more than 7 mi of trails that wind along the inner marshes and the water are the most difficult on the Seashore, since they're mostly in soft sand. In the 17th century there were lookout towers here for shore whaling, as well as a tavern. Animals were pastured, and oystering and cranberry harvesting were undertaken. By 1800 the hardwood forest that had covered the island had been cut down for use in ship- and home-building. The pitch pines and other growth you see today were introduced in the 1830s to keep the soil from washing into the sea. The National Seashore offers occasional guided hikes on Great Island, and, from February through April, seal walks. To the right of the Great Island lot, a road leads to **Griffin Island,** with its own walking trail. ⊠ *Off Chequesset Neck Rd., about 3 mi from Wellfleet center.*

## Dining and Lodging

**$$$**  ✕ **Aesop's Tables.** A worthy choice for a special dinner, Aesop's spe-
★     cializes in seafood entrées and appetizers that often take the local Wellfleet oyster to new heights. Inside this 1805 captain's house are five dining rooms; aim for a table on the porch overlooking the center of town. The signature dish is an exotic bouillabaisse, with mounds of fresh-off-the-boat seafood in a hot saffron broth (the homemade sourdough bread is the perfect accompaniment). The marinated duck breast is another favorite. When it's time for dessert, you're likely to see lots of diners tempting fate with Death by Chocolate, a heavy mousse cake. Some nights in summer, there's live jazz in the tavern, where the mood and the menu are more casual. ⊠ *316 Main St.,* ☎ *508/349–6450. AE, DC, MC, V. Closed Columbus Day–Mother's Day. No lunch.*

**$$**   ✕ **Finely JP's.** This unassuming little roadside spot right on Route 6
★     gives no hint that chef John Pontius consistently turns out wonderful food full of the best Italian influences and ingredients. The dining room is small and noisy, but the fish and pasta dishes (which emphasize good olive oil and plenty of lemon) silence all. Appetizers are especially good, among them a warm spinach and scallop salad or blackened beef with charred pepper relish. The Wellfleet paella draws raves, but the chef is not afraid to cook a Delmonico steak either. Don't confuse this place with PJ's, a fast-fried-food-and-ice-cream hangout just a few miles to the north. ⊠ *Rte. 6, S. Wellfleet,* ☎ *508/349–7500. Reservations not accepted. D, MC, V. Closed Mon.–Wed. Thanksgiving–Memorial Day; Mon.–Tues. Memorial Day–mid-June and Oct.–Thanksgiving; Tues. Labor Day–Oct.*

**$$**   ✕ **Painter's.** Named for chef-owner Kate Painter (not for all the local
★     artists), this fresh, new restaurant combines sophisticated, fashionable food with service of the same caliber. Ten to 15 seafood dishes run from traditional Portuguese fare like clam *Cataplana* (hearty clam stew with a spicy red wine sauce and a white-wine meat sauce) to a new American version of pan-seared tuna, encrusted with sesame and mustard seeds. All the desserts are homemade as well; the flourless chocolate Armagnac fig cake begs to be devoured. A revamped upstairs studio bar

features a lighter sandwich menu, plus live jazz Monday through Wednesday in season (July 4–Labor Day). If you don't make it home, breakfast is also served. Call for more information about Kate's special theme-dinners on major holidays in season. ⊠ *50 Main St.,* ☎ *508/ 349–3003. AE, MC, V. Closed Dec.–Apr.*

$$ ✕ **Serena's.** The name is Italian for "mermaid," and images of the Sirens grace the restaurant everywhere. A family place that caters to kids, Serena's offers an assortment of "blackboard specials" every night. The Italian menu is short on the contemporary northern cuisine you'll find at many other places; red sauces still prevail here. Try the seafood *fra diavolo,* a bouillabaisselike stew that can be ordered hot and extremely hot. The bar side of the restaurant is reserved for smokers. ⊠ *Rte. 6,* ☎ *508/349–9370. Reservations not accepted. AE, D, DC, MC, V. No lunch.*

$–$$ ✕ **Bayside Lobster Hutt.** There is no hutt , though there are plenty of lobsters, and fishing nets draped on the side of the building. The name and nets are reminders of an eccentric owner and former fisherman, well loved in Wellfleet, recently deceased, whose family carries on the business. Diners share meals with strangers at long tables and often as not wind up neighbors on the beach the next day. The food is nothing fancy, but the emphasis is always on freshness and quality; sea clam pie (in fact a clam potpie) typifies the authenticity of the fare. You're encouraged to bring your own beer or wine. ⊠ *91 Commercial St.,* ☎ *508/349–6333. Reservations not accepted. No credit cards. BYOB. Closed mid-Sept. through Memorial Day.*

$ ✕ **Flying Fish Café.** Anywhere, anytime, on any vacation, a breakfast spot like this is a welcome discovery. Great omelettes, "green eggs and ham" (green veggies, ham, and cheese), or something called "risky home fries" (home fries with cheese and veggies)—you'll probably end up at the Flying Fish more than once. Lunch and dinner are also served. ⊠ *Briar La., east of Main St.,* ☎ *508/349–3100. Reservations not accepted. MC, V. Closed mid-Oct.–mid-Apr.*

$$–$$$ ⊞ **Surf Side Colony Cottages.** There's very little in the way of accom-
★ modations on the Atlantic shore of the Outer Cape, so these one- to three-bedroom cottages are an especially good find. Scattered on either side of Ocean View Drive, they range from units in a piney grove to well-equipped ocean-side cottages—the two closest to the water have the best views—a one-minute walk from Maguire's Landing town beach, a beautiful, wide strand of sand, dunes, and surf. Though the exteriors are retro-Florida, with pastel shingles and flat roofs, cottage interiors are tastefully decorated in Cape style, including knotty-pine paneling. All units have phones, wood-burning fireplaces, screened porches, kitchens, (mostly) tiled baths, rattan furniture, carpeting, and grills. Some have roof decks with an ocean view and outdoor showers. Ocean-side cottages have dishwashers. ⊠ *Box 937, S. Wellfleet 02663,* ☎ *508/349–3959,* 🖷 *508/349–3959. 18 cottages. Picnic area, coin laundry. MC, V. 1-wk minimum in summer. Closed Nov.–Mar.*

$–$$ ⊞ **Wellfleet Motel & Lodge.** A mile from Marconi Beach, opposite the Audubon sanctuary is this well-maintained and tasteful highway-side complex, which sits on 12 wooded acres. In the single-story motel, built in 1964, the rooms, which are renovated each winter, are decorated in pale blues, yellows, greens, and peach; rattan and wicker pieces add to their summery feel. Rooms in the two-story lodge are bright and spacious, with oak furniture, king- or queen-size beds, and balconies or patios. Rooms are well-equipped with appliances. The property offers direct access to the Cape Cod Rail Trail. ⊠ *Rte. 6, Box 606, 02663,* ☎ *508/ 349–3535 or 800/852–2900,* 🖷 *508/349–1192. 57 rooms, 8 suites.*

*Bar, coffee shop, air-conditioning, 2 pools, hot tub, basketball, conference room. AE, DC, MC, V. Lodge and facilities closed Dec.–Mar.*

$ 🏨 **Holden Inn.** If you're watching your budget and can deal with a few fashion don'ts, like the occasional shag carpet, try this old-timey place just out of the town center. Rooms have hardwood floors and are clean and simply decorated with grandma's-house wallpapers, ruffled sheers or country-style curtains, and antiques like brass-and-white iron or spindle beds or a marble-top table. Private baths with old porcelain sinks are available in adjacent 1840 and 1890 buildings. The lodge has shared baths, an outdoor shower, and a large, screened-in porch with a lovely view of the bay and Great Island far below. The main house has a common room and a screened front porch with rockers and a bay view through trees. There are picnic tables and gardens in the backyard. ✉ *140 Commercial St., Box 816, 02667,* ☎ *508/349–3450. 27 rooms, 14 with shared bath. No credit cards. Closed mid-Oct.–mid-Apr.*

$ 🏨 **Inn at Duck Creek.** Close to Route 6, set on 5 wooded acres by a duck pond, creek, and salt marsh, this old inn consists of the circa-1815 main building and two other old houses. Rooms in the main inn (except rustic third-floor rooms) and in Saltworks have a simple charm. Typical furnishings include claw-foot tubs, country antiques, lace curtains, chenille spreads, and rag rugs on hardwood floors. The two-room Carriage House is cabiny, with rough barn-board and plaster walls. The inn's parlors, screened porches, and marsh-view deck invite relaxation. There's fine dining at Sweet Seasons or pub dining with entertainment at the Tavern Room (☞ Nightlife and the Arts, *below*). ✉ *70 E. Main St., Box 364, 02667,* ☎ *508/349–9333,* FAX *508/349–0234. 25 rooms, 17 with private bath. 2 restaurants. AE, MC, V. Closed mid-Oct.–mid-May. 2-night minimum weekends in season. CP.*

## Nightlife and the Arts

**Beachcomber** (✉ Cahoon Hollow Beach, off Rte. 6, Wellfleet, ☎ 508/349–6055)—right on the beach, with national touring acts most nights, weekend happy hours with live reggae, and dancing nightly in summer—is big with the college crowd at night and on breaks from the sun. Indoors or at tables by the beachfront bar, order from a menu featuring fun appetizers, salads, burgers, seafood, barbecue, frozen drinks, and a raw bar.

The **Tavern Room** (✉ Main St., Wellfleet, ☎ 508/349–7369), set in an 1800s building, with beamed ceiling, fireplace, and a bar covered in nautical charts, features live entertainment from jazz to pop to country to Latin ensembles. Munchies are served alongside the menu of traditional and Latin/Caribbean-inspired dishes.

A number of organizations sponsor outdoor activities at night, including the **Massachusetts Audubon Wellfleet Bay Sanctuary**'s bat walks, night hikes, and lecture series (☞ *above*).

The drive-in movie is alive and well on Cape Cod at the **Wellfleet Drive-In Theater** (✉ Rte. 6, ☎ 508/349–7176 or 508/255–9619); films start at dusk nightly in season, and there's a miniature golf course.

On Saturday evenings in July and August during the **Wellfleet Gallery Crawl,** Wellfleet's art galleries are open for cocktail receptions to celebrate show openings. Walk from gallery to gallery meeting the featured artists and checking out their works.

The **Wellfleet Harbor Actors Theater** (✉ WHAT; box office on Main St., ☎ 508/349–6835) presents world premieres of American plays, satires, farces, and black comedies in its mid-May–mid-October season.

## Outdoor Activities and Sports

BEACHES

**Public Beaches.** The spectacular dune-drop Atlantic beaches **White Crest** and **Cahoon Hollow** charge daily parking fees of $10 to non-residents in season only. Cahoon Hollow has lifeguards, rest rooms, and a restaurant and music club on the sand. **Marconi Beach** charges $5, or $3 for bikes and walk-ins. And **Mayo Beach** just west of Wellfleet Harbor is free. You can also park free at the small lot by **Great Island** on the bay.

**Restricted Beaches.** Resident or temporary resident stickers are required for access to Wellfleet beaches in season only. For the rest of the year, anyone can visit them for free. **LeCount Hollow** and **Newcomb Hollow** are the restricted ocean beaches; **Indian Neck** harbor beach and the **Duck Hollow** bay beach past Great Island are also restricted. For more information call the Wellfleet Chamber of Commerce at 508/349–2510.

The **Wellfleet Ponds,** nestled in the woods between Route 6 and the ocean, were formed by glaciers and are fed by underground springs. Mild temperatures and clear, clean water make swimming pleasant for the whole family, a refreshing change from the bracing, salty surf of the Atlantic. They are also perfect for canoeing, sailing, or kayaking (boats are available from Jack's Boat Rentals, ☞ *below*). These fragile ecosystems have called for restricted use, and a Wellfleet beach sticker is required in season. Motorized boats are not allowed.

BIKING

**Black Duck Sports Shop** (⊠ Rte. 6, S. Wellfleet, ☎ 508/349–9801; off-season, ☎ 508/349–2335) rents a variety of bikes.

BOATING

**Jack's Boat Rentals** (⊠ Gull Pond, Wellfleet, ☎ 508/349–7553; ⊠ Rte. 6, Wellfleet, 508/349–9808 for long-term rentals) provides canoes, kayaks, sailboards, Sunfish, and sailboards. Guided tours are also available.

FISHING

Fishing trips are operated on a walk-on basis from spring through fall on the *Navigator* (⊠ Town Pier, Wellfleet, ☎ 508/349–6003). Rods, reels, and bait are included.

GOLF AND TENNIS

**Chequessett Yacht & Country Club** (⊠ Chequessett Neck Rd., Wellfleet, ☎ 508/349–3704) is a semiprivate club (public use on space-available basis) with a nine-hole golf course and five hard-surface tennis courts by the bay. Lessons are available.

**Oliver's** (⊠ Rte. 6, Wellfleet, ☎ 508/349–3330) has one Truflex and seven clay courts, offers lessons, and arranges matches.

SURFING

The Atlantic coast Marconi and White Crest beaches are the best spots for surfing.

## Shopping

**Blue Heron Gallery** (⊠ Bank St., Wellfleet, ☎ 508/349–6724) is one of the Cape's best galleries, with contemporary works—including Cape scenes, jewelry, sculpture, and pottery—by regional and nationally recognized artists, including Steve Allrich and Del Filardi.

**Eccentricity** (⊠ 361 Main St., Wellfleet, ☎ 508/349–7554) keeps the corner of Main and Briar offbeat. One of the most interesting stores on the Cape, it sells gorgeous kimonos, ethnic-inspired cotton clothing, carved African wooden sculpture and barbershop paintings, odd Mexican items, and trinkets.

**Hannah** (⊠ Main St., Wellfleet, ☎ 508/349–9884) has tasteful and interesting women's clothing. There is another Hannah in Orleans.

**Herridge Books** (⊠ Main St. between Rte. 6 and town center, Wellfleet, ☎ 508/349–1323) is a perfect store for a town that has hosted so many writers. Its dignified literary fiction, art and architecture, literary biography and letters, mystery, Americana, sports—et cetera—sections are full of used books in very nice condition. Herridge also carries new books on the Cape and its history. Bibliophiles will want to spend a whole day here, rain or shine.

**Karol Richardson** (⊠ 3 W. Main St., Wellfleet, ☎ 508/349–6378) fashions women's wear in luxurious fabrics and sells interesting shoes, hats, and jewelry. The store may look and sound a little New York for Wellfleet, but the clothing is original.

**Kendall Art Gallery** (⊠ 40 E. Main St., Wellfleet, ☎ 508/349–2482) carries contemporary art, including Harry Marinsky's bronzes in the sculpture garden, John French's brightly colored ceramic renderings of real and whimsical building facades, and watercolors by Walter Dorrell.

**Off Center** (⊠ Off Main St., Wellfleet, diagonally across from Eccentricity, ☎ 508/349–3634), another of the Eccentricity (☞ *above*) owners' enterprises, sells mainstream yet stylish women's clothing.

**Wellfleet Drive-In Theater** (⊠ Rte. 6, Eastham–Wellfleet line, ☎ 508/349–2520) is the site of a giant flea market (mid-April–June and September–October, weekends and Monday holidays 8–4; July and August, Monday holidays, Wednesday, Thursday, and weekends 8–4). There's a snack bar and playground.

**West Main Books** (⊠ 41 W. Main St., ☎ 508/349–2095) anchors the west end of Wellfleet with another outstanding store, this one selling new books. Thoughtful and obviously well-read buyers cater to an equally intelligent clientele, making this one of the Cape's finest bookstores.

*En Route*  If you're in the mood for a quiet, lovely ride winding through what the Cape might have looked like before Europeans arrived, follow **Old County Road** from Wellfleet to Truro on the bay side. It's bumpy and beautiful—let's hope it stays that way—with a stream or two to pass. So cycle on it, or drive slowly, or just stop and walk to take in the nature around you.

## Truro

 *7 mi southeast of Provincetown.*

Settled in 1697, Truro has had several names. Originally called Pamet after the local Indians, in 1705 the name was changed to Dangerfield in response to all of the sailing mishaps off its shores. "Truroe" was the final choice, and Truro became the namesake of a Cornish town that homesick settlers thought it to resemble. The town relied on the sea for its income—whaling, shipbuilding, and codfishing were the main industries.

Today a town of high dunes, estuaries, and rivers fringed by grasses, rolling moors, and houses sheltered in tiny valleys, Truro is a popular retreat of artists, writers, politicos, and numerous vacationing psychoanalysts. Edward Hopper summered here from 1930 to 1967, finding the Cape light ideal for his austere brand of realism. In 1997 Vice President Al Gore and his family vacationed here for two weeks in August.

One of the largest towns in terms of area—almost 43 square mi—it is the smallest in population, with about 1,400 year-round residents. If you thought Wellfleet's downtown was small, wait until you see—or don't see—Truro's. It's a post office, a town hall, a shop or two. You'll know it by the sign that says DOWNTOWN TRURO at a little plaza entrance. There's also a library, a firehouse, and a police station, but that's about all. It is the Cape's narrowest town, and from a high perch you can see the Atlantic Ocean on one side and Cape Cod Bay on the other.

NEED A BREAK?

**Jams** has the answer for whoever is looking for a great picnic lunch, sandwiches, fresh produce, a bottle of wine or Evian, a sweet treat— you get the picture. You won't be able to *stay* for lunch, unless you're content getting your knees knocked on the bench outside, but with the best part of the Cape just outside the door, who would want to? ⊠ *Off Rte. 6 in Truro center,* ☎ *508/349–1616. Phone orders accepted.*

For a few hours of exploring, take the kids to **Pamet Harbor.** At low tide you can walk out on the flats and discover the creatures of the salt marsh. A nearby plaque identifies the plants and animals and tells of the ecological importance of the area. ⊠ *Depot Rd.*

On **Corn Hill,** a tablet commemorates the finding of a buried cache of corn by Standish and the *Mayflower* crew. They took it to Plymouth and used it as seed, returning later to pay the Indians for the corn they'd taken.

**Truro Center for the Arts at Castle Hill,** housed in a converted 19th-century barn, offers summer arts-and-crafts workshops for children, as well as courses and single classes in art, crafts, photography, and writing for adults. Teachers have included notable New York– and Provincetown-based artists. ⊠ *10 Meeting House Rd., Box 756, Truro 02666,* ☎ *508/349–7511.*

Built at the turn of the century as a summer hotel, the **Truro Historical Museum** is now a repository of 17th-century firearms, mementos of shipwrecks, early fishing and whaling gear, ship models, a pirate's chest, scrimshaw, and more. One room exhibits wood carvings, paintings, blown glass, and ship models by Courtney Allen, artist and founder of Truro's historical society. An excellent self-guided historical tour of town is available at the gift shop. ⊠ *Lighthouse Rd., N. Truro,* ☎ *508/487–3397.* ▣ *$3.* ☉ *Memorial Day–Sept., daily 10–5.*

❹❾ Truly a breathtaking sight, **Highland Light,** also called Cape Cod Light, is the Cape's oldest lighthouse. It was the last to have become automated, in 1986. The first light on this site, powered by 24 whale-oil lamps, began warning ships of Truro's treacherous sandbars in 1798— the dreaded Peaked Hills Bars alone, to the north, have claimed hundreds of ships. The current light, a 66-ft tower built in 1857, is powered by two 1,000-watt bulbs that are reflected by a huge Fresnel lens. Its beacon is visible for 20 mi.

One of four active lighthouses on the Outer Cape, Highland Light has the distinction of being listed in the National Register of Historic Places. Thoreau used it as a stopover in his travels across the Cape's backside, as the Atlantic side of the Outer Cape is called. Erosion threatened to cut this lighthouse from the 117-ft cliff on which it stood and drop it into the sea. Thanks to a concerted effort by local citizens and lighthouse lovers, the necessary funds were raised and the lighthouse was moved back 450 ft to safety in 1996. The lighthouse was opened to the public during the 1997 summer season.

The gardens of **A Touch of Heaven** are a beautiful place and—unlike most other gardens—are meant to be picked. The lawns are set with benches and birdbaths. The flowers are so abundant you don't feel guilty gathering a bunch to take home. ⊠ *Pond Village Heights, N. Truro, no phone.*

**⑤⓪** At the **Pilgrim Heights Area** of the Cape Cod National Seashore, a short trail leads to the spring where a Pilgrim exploring party stopped to refill their casks, tasting their first New England water. Walking through this still-wild area of oak, pitch pine, bayberry, blueberry, beach plum, and azalea gives you a taste of what it was like for these voyagers in search of a new home. "Being thus passed the vast ocean . . ." William Bradford wrote in *Of Plimoth Plantation*, "they had no friends to welcome them, no inns to entertain them or refresh their weather-beaten bodies; no houses, or much less towns to repair to, to seek for succour."

From an overlook you can see the bluffs of High Head, where glaciers pushed a mass of earth before melting and receding. Another path leads to a swamp, and a bike trail leads to Head of the Meadow Beach, often a less crowded alternative to others in the area.

### Dining and Lodging

$$–$$$     ✕ **Adrian's.** Adrian Cyr moved his restaurant to a high bluff overlooking Pilgrim Lake and all of Provincetown in 1993, and his cooking seems to have elevated with the locale. Cayenne-crusted grilled salmon with maple-mustard sauce is one of the well-prepared dishes. Tuscan salad is always a hit, with tomatoes, olives, fresh basil, plenty of garlic, and balsamic vinegar. Pizzas, too, are standouts, especially one made with cornmeal dough and topped with shrimp and artichokes. The outdoor deck, always a fun spot, bursts at the very popular breakfasts and brunches, served every day in season. ⊠ *In the Outer Reach Hotel, Rte. 6, N. Truro,* ☎ *508/487–4360. Reservations not accepted. AE, M, V. Closed mid-Oct.–Memorial Day.*

$$–$$$     ✕ **Whitman House.** Another of the Cape's traditional inn-restaurants, the Whitman House achieves its elegance with original, Colonial ambience. Dinner is never a surprise, which can be comforting or exasperating, depending on your mood. Items like surf-and-turf are a big hit with a largely older clientele. It's strangely a matter of pride to say that there's nothing new at the Whitman House. ⊠ *Rte. 6, at mi marker 108.5,* ☎ *508/487–1740. AE, DC, MC, V. Closed Jan.–Mar. No lunch.*

$–$$     🏠 **East Harbour.** This meticulously maintained complex outside Provincetown offers simple accommodations ranged around a manicured lawn separated from the bay beach by low grasses. The two-bedroom cottages have paneled walls, Colonial-style furnishings, and full kitchens. Motel rooms have large picture windows, mini-refrigerators and coffeemakers, light paneling, and '60s motel-style furnishings. A newer apartment is all white and bright, with skylights, Shaker-reproduction furnishings, a modern kitchen, and second-floor views of the harbor from a private deck. All units have individual heat, microwaves, phones, and cable TV. ⊠ *618 Shore Rd., Box 183, N. Truro 02652,* ☎ *508/ 487–0505,* ℻ *508/487–6693. 7 cottages, 9 rooms, 1 apartment. Picnic area, beach, coin laundry. AE, D, MC, V. In season, 1-wk minimum for cottages, 2-night minimum for rooms. Closed Nov.–Mar.*

$–$$     🏠 **South Hollow Vineyards Inn.** Set on five acres of working vineyard, this 1836 former farmstead has been transformed into an elegant inn. The decor reflects the wine motif: Antique casks and presses stand in the corners, and deep greens and burgundies predominate. The house is also decorated with country-style antiques and Oriental carpets. Guest rooms have four-poster king- and queen-size beds and modern tiled baths. Homegrown berries often garnish dishes at breakfast, which is served in the sunroom or, in nice weather, outside on the patio. A large common sundeck has sweeping views of the vineyard, and is a perfect place to relax with a book. ☎ *Rte. 6A, N. Truro 02652, ☎ 508/487–6200, FAX 508/487–4248. 5 rooms. BP. No smoking. MC, V.*

$ ☎ **Hostelling International–Truro.** In a former Coast Guard station right on the dunes, this handsome, well-placed hostel has 42 beds, kitchen facilities, a common area, and naturalist-led programs. It's also right by the ½-mi Cranberry Bog Trail, which takes you by a refurbished cranberry bog and an old bog house. ⊠ *N. Pamet Rd., Box 402, Truro 02666, ☎ 508/349–3889; ☼ Mid-June–Labor Day.*

## Outdoor Activities and Sports

### BEACH

**Head of the Meadow Beach** in Truro, part of the National Seashore, is often less crowded than others in the area. It has only temporary rest room facilities and no showers since the bathhouse burned down.

### BICYCLING

The **Head of the Meadow Trail** is 2 mi of easy cycling between dunes and salt marshes from High Head Road, off Route 6A in North Truro, to the Head of the Meadow Beach parking lot. Bird-watchers also love the area.

**Bayside Bikes** (⊠ 102 Rte. 6A, N. Truro, ☎ 508/487–5735) rents bicycles by the hour, day, or week, with easy access to the Head of the Meadow Trail.

### GOLF

The **Highland Golf Links** (⊠ Lighthouse Rd., N. Truro, ☎ 508/487–9201), a nine-hole, par-36 course on a cliff overlooking the Atlantic, is unique for its resemblance to Scottish links, instead of the well-manicured and less challenging courses that are more typical.

### SURFING

**Long Nook** in Truro is a good surfing spot.

## Shopping

**Whitman House Gift Shop** (⊠ Rte. 6, N. Truro, ☎ 508/487–3204) sells Amish quilts and other country items.

# Provincetown

**51** *27 mi north of Orleans, 62 mi from the Sagamore Bridge.*

The Cape's smallest town in area and the second-smallest in year-round population, Provincetown's 8 square mi are packed with history. The curled fist at the very tip of the Cape, Provincetown's shores curve protectively around a natural harbor, the perfect spot for visitors from any epoch to anchor. Historical records show that Thorvald, brother of Viking Leif Erikson, came ashore here in AD 1004 to repair the keel of his boat, and consequently named the area "Kjalarness," or Cape of the Keel. Bartholomew Gosnold came to Provincetown in 1602 and named the area Cape Cod after the abundant codfish he found in the local waters.

The Pilgrims remain Provincetown's most famous visitors. On Monday, November 21, 1620, the *Mayflower* dropped anchor in Provincetown Harbor after a difficult 63-day voyage; while in the Harbor they signed the Mayflower Compact, the first document to declare a democratic form of government in America. Ever practical, one of the first things the Pilgrims did was to come ashore to wash their clothes, thus start-

## Provincetown

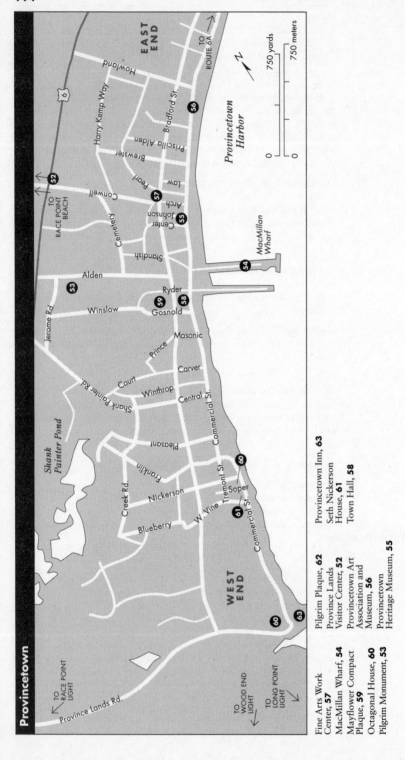

Fine Arts Work
Center, **57**
MacMillan Wharf, **54**
Mayflower Compact
Plaque, **59**
Octagonal House, **60**
Pilgrim Monument, **53**

Pilgrim Plaque, **62**
Province Lands
Visitor Center, **52**
Provincetown Art
Association and
Museum, **56**
Provincetown
Heritage Museum, **55**

Provincetown Inn, **63**
Seth Nickerson
House, **61**
Town Hall, **58**

ing the ages-old New England tradition of Monday washday. They stayed in the area for five weeks before moving on to Plymouth. Plaques and parks commemorate the landing throughout town.

During the American Revolution, Provincetown Harbor was controlled by the British, who used it as a port from which to sail to Boston and launch attacks on Colonial and French vessels. In November of 1778 the 64-gun British frigate *Somerset* ran aground and was wrecked off Provincetown's Race Point—every 60 years or so the shifting sands uncover her remains.

Incorporated as a town in 1727, Provincetown was for many decades a bustling seaport, with fishing and whaling as its major industries. Fishing is still an important source of income for many Provincetown natives, while the town ranks fourth in the world as a whale-watching, rather than hunting, mecca.

Artists began coming here in 1899 for that unique Cape Cod light. Provincetown has the distinguished reputation of being the nation's oldest continuous art colony. Poets, writers, and actors have also been part of the art scene. Eugene O'Neill's first plays were written and produced here, and the Fine Arts Work Center continues to have in its ranks some of the most important writers of our time.

During the early 1900s, Provincetown became known as Greenwich Village North. Artists from New York and Europe discovered the town's unspoiled beauty, special light, lively community, and colorful Portuguese flavor. By 1916, with five art schools flourishing here, painters' easels were nearly as common as shells on the beach. This bohemian community, along with the availability of inexpensive summer lodgings, attracted young rebels and writers as well, including John Reed (*Ten Days That Shook the World*) and Mary Heaton Vorse (*Footnote to Folly*), who in 1915 began the Cape's first significant theater group, the Provincetown Players. The young, then-unknown Eugene O'Neill joined them in 1916, when his *Bound East for Cardiff* premiered in a tiny wharf-side fish house in the East End that was minimally fitted out as a theater. After 1916 the Players moved on to New York. Their theater, at present-day 571 Commercial Street, is long since gone, but a model of it and of the old Lewis Wharf on which it stood is on display at the Pilgrim Monument museum.

Provincetown is a place of creativity, sometimes startling originality, and great diversity. In the busy downtown, Portuguese-American fishermen mix with painters, poets, writers, whale-watching families, cruise-ship passengers on brief stopovers, and many lesbian and gay residents and visitors, for whom P-town, as it's almost universally known, is one of the most popular East Coast seashore spots. In summer Commercial Street is packed with sightseers and shoppers after the treasures of the overwhelming number of galleries and crafts shops. At night, raucous music and people spill out of bars, drag shows, and sing-along lounges galore. It's a fun, crazy place, with the extra dimension of the fishing fleet unloading their catch at MacMillan Wharf, in the center of the action.

The town's main thoroughfare, Commercial Street, is 3 mi from end to end. In season, driving from one end of the main street to the other could take forever—walking is definitely the way to go. A casual stroll will allow you to see the many different architectural styles (Victorian, Second Empire, Gothic, and Greek Revival, to name a few) used to build impressive houses for wealthy sea captains and merchants. Be on the lookout for blue plaques fastened to housefronts explaining their historical significance—practically the entire town has been designated part of the Provincetown Historic District. The Historical Society puts out a series of walking-tour pamphlets, available for about $1 each at many shops in town, with maps and information on the history of many buildings and the more or less famous folk who have occupied them. You may also want to pick up a free Provincetown gallery guide.

The center of town is where the crowds and most of the touristy shops are. The quiet East End is mostly residential, with some top galleries, and the similarly quiet West End has a number of small inns with neat lawns and elaborate gardens. Narrated trolley sightseeing tours make the downtown circuit throughout the day, or the romantically inclined can hire a horse-drawn carriage (☎ 508/487–4246) from the stand in front of the Town Hall.

Near the Provincetown border, **massive dunes** actually meet the road in places, turning Route 6 into a sand-swept highway. Scattered among the dunes are primitive cottages, called dune shacks, built from flotsam and other found materials, that have provided atmospheric as well as cheap lodgings to a number of famous artists and writers over the years—among them poet Harry Kemp, Eugene O'Neill, e. e. cummings, Jack Kerouac, and Norman Mailer. The few surviving shacks are privately leased from the National Seashore, whose proposal to demolish the shacks was halted by their inclusion on the National Register of Historic Places in 1988. The dunes are fragile, but there are paths leading through them to the ocean.

The **Province Lands** begin at High Head in Truro and stretch to the tip of Provincetown. The area is scattered with ponds, cranberry bogs, and scrub. Bike and walking trails lace through forests of stunted pines, beech, and oak and across desertlike expanses of rolling dunes. Protected against development, the Province Lands are the "wilds" of the Cape.

A beautiful spot to stop for lunch before biking to the beach, the **Beech Forest picnic area** borders a small pond covered with water lilies. The adjacent bike trails lead to Herring Cove and Race Point beaches. ⊠ *Race Point Rd., Provincetown.*

**52** Inside the **Province Lands Visitor Center** you'll find literature and nature-related gifts, frequent short films—on local geology, the U.S. Life Saving Service, and more—and exhibits on the life of the dunes and the shore. You can also pick up information on guided walks, birding trips, lectures, bonfires, and other current programs throughout the Seashore, as well as on the Province Lands' own beaches, Race Point and Herring Cove, and the walking, biking, and horse trails. Don't miss the wonderful 360-degree view of the dunes and the surrounding ocean from the observation deck. ⊠ *Race Point Rd.,* ☎ *508/487–1256.* ☉ *Apr.–Nov., daily 9–5.*

Not far from the present Coast Guard Station is the **Old Harbor Station,** a U.S. Life Saving Service building towed here by barge from Chatham in 1977 to rescue it from an eroding beach. It is reached by a boardwalk across the sand, and plaques along the way tell about the lifesaving service and the whales seen offshore. Inside are displays of such equipment as Lyle guns, which shot rescue lines out to ships in distress when seas were too violent to launch a surfboat, and breeches buoys, in which passengers were hauled across those lines to safety. On Thursday nights in summer there are reenactments of the old-fashioned lifesaving procedure. ⊠ *Race Point Beach, no phone.* ☜ *Donations accepted; Thurs. night $3.* ☉ *July–Aug., daily 10–4.*

**53** The first thing you'll see in Provincetown is the **Pilgrim Monument,** stretching into the sky. The incongruous edifice commemorates the first landing of the Pilgrims in the New World and their signing of the Mayflower Compact, America's first rules of self-governance, before they set off from Provincetown Harbor to explore the mainland. Climb the 252-ft-high tower (116 steps and 60 ramps) for a panoramic view— dunes on one side, harbor on the other, and the entire bay side of Cape Cod beyond. At the base is a museum of Lower Cape and Provincetown history, with exhibits on whaling, shipwrecks, and scrimshaw, a diorama of the *Mayflower* and another of a glass factory.

The tower was erected of granite shipped from Maine, according to a design modeled on a tower in Siena, Italy. President Theodore Roosevelt laid the cornerstone in 1907, and President Taft attended the 1910 dedication. On Thanksgiving Eve, in a ceremony that includes a museum tour and open house, 5,000 white and gold lights that drape the tower are illuminated—a display that can be seen as far away as the Cape Cod Canal. They are lit nightly into the New Year. ✉ *High Pole Hill,* ☎ *508/487–1310.* ▨ *$5.* ☉ *Apr.–June and Sept.– Nov., daily 9–5; July and Aug., daily 9–7; last admission 45 min. before closing.*

**54** **MacMillan Wharf,** with its large municipal parking facility, is a sensible place to start a tour of town. It is one of the remaining five of 54 wharves that once jutted into the bay. The wharf is the base for P-town whale-watch boats, fishing charters, and party boats. The **Chamber of Commerce** is also here, with all kinds of information and events schedules. Kiosks at the wharf have rest room and parking lot locations, bus schedules, and other information for visitors.

In 1997 renovations were completed on the former **Provincetown Marina building** on MacMillan Wharf, and it opened as the **Whydah Museum,** the new home for artifacts and treasure recovered from the *Whydah,* a pirate ship that sank off the coast of Wellfleet more than 265 years ago, and the only pirate shipwreck ever authenticated in the world. The new museum is both entertaining and educational, with a display featuring the conservation and restoration processes at work on actual artifacts, and a history component that gives the untold story of 18th-century pirating life. Eventually the museum hopes to collect and house all of the recovered pieces, some of which are presently on loan to other museums across the country. ✉ *16 MacMillan Wharf,* ☎ *508/487–7955.* ▨ *$5.* ☉ *Memorial Day–Labor Day, daily 9–7; Apr.–Memorial Day and Labor Day–Oct. 15, daily 10–5; Oct. 15– Dec., weekends 10–5. Closed Jan.–Mar.*

**55** Housed in an 1860 Methodist church listed on the Register of Historical Places, the **Provincetown Heritage Museum** displays scenes of domestic life and workplaces from centuries past. Exhibits include antique fire-fighting equipment, fishing artifacts, artwork donated by Provincetown-related artists, as well as antique prints and watercolors of schooners, wax figures, and a half-scale model, 66 ft long, built by a master shipbuilder, of the fishing schooner *Rose Dorothea,* which won the Lipton Cup in 1907. ✉ *356 Commercial St.,* ☎ *508/487–7098.* ▨ *$3.* ☉ *Mid-June–Columbus Day, daily 10–6.*

---

NEED A BREAK?

The **Provincetown Portuguese Bakery** (✉ 299 Commercial St., ☎ 508/487-1803) puts out fresh Portuguese breads and pastries, and serves breakfast and lunch all day from March to October. Novices should ask the counter-person which pastries to try. It's open until 11 PM in summer.

---

Founded in 1914 to collect and show the works of Provincetown-associated artists, the **Provincetown Art Association and Museum** (PAAM) houses a 1,650-piece permanent collection, displayed in changing exhibits that combine up-and-comers with established artists of the 20th century. Some of the art hung in the four bright galleries is for sale. The museum store has books by or about local artists, authors, and topics, as well as posters, crafts, cards, and gift items. PAAM-sponsored year-round courses (one day and longer) offer the opportunity of studying under such talents as Sal Del Deo, Carol Westcott Wharf, and Tony Vevers. ⊠ *460 Commercial St.,* ☎ *508/487–1750.* ⊑ *$3.* ☉ *Nov.–Apr., weekends noon–4 and by appointment; Memorial Day–Labor Day, daily noon–5 and 8–10; May, Sept., and Oct., Fri.–Sun. noon–5. Gallery open by appointment Mon.–Thurs.*

To see or hear the work of up-and-coming artists, visit the gallery or attend a reading at the **Fine Arts Work Center.** A nonprofit organization founded in 1968, the FAWC sponsors 10 writers and 10 artists from October through May each year with a place to work, a stipend to live on, and access to artists and teachers. A new Summer Program offers open-enrollment workshops in both writing and the visual arts. The buildings in the complex around the center, which it owns, were formerly part of Day's Lumber Yard Studios, built above a lumberyard by a patron of the arts to provide poor artists with cheap accommodations. Robert Motherwell, Hans Hoffmann, and Helen Frankenthaler have been among the studios' roster of residents over the years. ⊠ *24 Pearl St.,* ☎ *508/487–9960.*

The former studios of two noted artists, Edwin W. Dickinson, at 46 Pearl Street, and Charles W. Hawthorne, at 48 Pearl Street, give an idea of the proliferation of now-famous artists that worked in the area. Hawthorne's **Cape Cod School of Art,** established here in 1899, (now managed by Lois Griffel, a former student of Henry Hensche, one of Hawthorne's disciples) put the town on its path to becoming a major art colony. ⊠ *22 Brewster St.,* ☎ *508/487–0101.*

The **Duarte Motors** parking lot now sits on the site of the former railroad depot. The train ran across Bradford and Commercial streets to the wharf, where it picked up passengers from the frequent Boston boats and crates of iced fish from the Commercial Street fish sheds. Train service ended in 1960. ⊠ *Bradford St.*

The **Town Hall** was used by the Provincetown Art Association as its first exhibit space and still exhibits paintings donated to the town over the years, including Provincetown scenes by Charles Hawthorne and WPA-era murals by Ross Moffett. ⊠ *260 Commercial St.,* ☎ *508/487–7000.*

In a little park behind the town hall, is the **Mayflower Compact plaque,** carved in bas-relief by sculptor Cyrus Dalin, depicting the historic signing. ⊠ *Bradford St.*

NEED A BREAK?   The **Provincetown Fudge Factory** (⊠ 210 Commercial St., ☎ 508/487-2850), across from the post office, makes silky peanut-butter cups, fudge, chocolates, saltwater taffy, yard-long licorice whips, and custom-flavor frozen yogurt—a must stop for those with a discriminating sweet tooth, and they'll ship their goodies to you, too.

At 74 Commercial Street in the West End, between Soper and Nickerson streets, is an interesting piece of Provincetown architecture: an **octagonal house** built in 1850.

**⑥¹** The oldest building in town, dating from 1746, is on the corner of Commercial and Soper Streets. The small Cape-style **Seth Nickerson House**—a private home—was built by a ship's carpenter, with massive pegged, hand-hewn oak beams and wide-board floors.

**⑥²** The bronze **Pilgrim Plaque,** set into a boulder at the center of a little park, commemorates the first footfall of the Pilgrims onto Cape soil—Provincetown's humble equivalent of the Plymouth Rock. ⊠ *End of Commercial St.*

**⑥³** The **Provincetown Inn** houses a series of 19 murals, painted in the 1930s from old postcards, depicting life in the 19th-century town. ⊠ *1 Commercial St.,* ☎ *508/487–9500.*

For a day of fairly private beachcombing and great views, you can walk across the stone jetty at low tide to **Long Point,** a sand spit with two lighthouses and two Civil War bunkers, named Fort Useless and Fort Ridiculous because they were hardly needed. It's a 2-mi walk across soft sand—beware of poison ivy and deer ticks if you detour from the path—or hire a boat at Flyer's (☎ 508/487–0898) to drop you off and pick you up.

## Dining and Lodging

**$$$** ✕ **Café Edwige.** Delicious contemporary American food, relaxed and
★ friendly service, an eclectic, homey upstairs setting—Café Edwige delivers all this night after night. Begin with a Maine crab cake, warm goat cheese on crostini, or a refreshing field-green-and-wild-lettuce salad, then consider lobster and Wellfleet scallops over pasta with a wild mushroom, Asiago, and tomato broth, or planked local codfish with roasted corn and shiitakes. Don't pass on the wonderful desserts such as the Cape Cod cranberry crumble served with vanilla ice cream. The tasty breakfast here is a P-town institution. ⊠ *333 Commercial St.,* ☎ *508/ 487–2008. AE, MC, V. Closed Nov.–May.*

**$$$** ✕ **Ciro's and Sal's.** From the decor to the pasta-dominated Italian cuisine, this spot would be well at home in Boston's North End. The shrimp scampi is steeped in garlic, well beyond the USDA minimum daily requirement, to be sure. The old aluminum graters now in use as lampshades add a touch of whimsy. ⊠ *4 Kiley Ct.,* ☎ *508/487–0049. MC, V. Closed Mon.–Thurs. Nov.–Memorial Day. No lunch.*

**$$$** ✕ **Front Street.** A friendly, casual bistro, Front Street is so good, so con-
★ sistent, and so romantic that many locals now consider it the best restaurant in P-town. Well versed in classic Italian cooking ("We can do that with our eyes closed"), chef-owners Donna Aliperti and Kathleen Cotter also venture into Mediterranean regions like Greece, southern France, and even north Africa. Duck smoked in Chinese black tea is dense and luscious, served with a different lusty sauce every day (current favorites are fresh tropical fruit with mixed peppercorns). An aromatic smoked salmon chowder comes and goes from the menu (there ought to be a town meeting to put it on permanently!). Littleneck clams and calamari with butternut squash in a roast garlic broth is essentially a simple dish, but its unique flavors show off these traditional ingredients in a new light. The wine list is also a winner. ⊠ *230 Commercial St.,* ☎ *508/487–9715. Closed Jan. 1–mid-May. AE, D, MC, V.*

**$$$** ✕ **Martin House.** Provincetown can be such a zoo that it is a genuine
★ relief to find an island of calm like Martin House right on crazy Commercial Street. The house's 1850s woodwork lends a historic feeling that is graced with Provincetown paintings. Changes in the kitchen have lessened the excitement around the restaurant in recent years, but the food is still ambitious and contemporary. Oysters Claudia matches native Wellfleets-on-the-half-shell with an Asian dipping sauce, wasabi, and pickled ginger. Pan-roasted lamb rib eye comes with an exotic mint

chutney and quinoa–wild rice tabbouleh. ⊠ *157 Commercial St.,* ☎ *508/487–1327. AE, D, DC, MC, V. Closed Wed. and Thu., Jan.–Mar.*

**$$$**    ✕ **The Mews.** This P-town favorite is still going strong. Downstairs,
★    the main dining room opens onto magnificent harbor views. The Mews menu focuses on seafood, but does it with a cross-cultural flair. Some favorites are rich and spicy Wellfleet scallops, shrimp, and crab mousse in a wonton on grilled filet mignon, and a decadent French lentil ragout with chipotle aioli sauce. A piano bar upstairs serves lunch (weekdays in summer) and dinner from a light café menu. Brunch is also served every day in season. ⊠ *429 Commercial St.,* ☎ *508/487–1500. Reservations not accepted. AE, D, DC, MC, V. Closed Jan. No lunch weekdays off-season or Sat. between Columbus Day and Memorial Day.*

**$$–$$$**    ✕ **Lobster Pot.** Provincetown's Lobster Pot is fit to do battle with all the Lobster Pots anywhere (and everywhere) on the Cape. As you enter you'll pass through one of the hardest-working kitchens on the Cape, which turns out classic New England cooking: lobsters galore, generous and filling seafood platters, and some of the best chowder around. The upstairs deck overlooks the harbor; there's also a take-out lobster market and bakery on the premises. ⊠ *321 Commercial St.,* ☎ *508/487–0842. Reservations not accepted. AE, D, DC, MC, V. Closed Jan.*

**$$–$$$**    ✕ **Lorraine's.** More than simply South of the Border, Lorraine's Mexican is stylish, with a nicely expanded menu of new American cuisine that draws deliciously from the flavors of the Southwest. The small, intimate dining room has a cherrywood bar (home of P-town's best margarita) and lots of local color. Lorraine's paella is excellent, and the *mujeres* clams (an appetizer with citrus juice and cilantro) are addictive. Also wonderful is a dish called simply *carnitas* (slow-cooked pork with a sauce of sour cream, guacamole, and salsa). To find this little place, you'll have to navigate a narrow alley. Lorraine's 2, a tiny outpost in Whaler's Wharf (☎ *508/487–8600*), serves breakfast, lunch, and dinner and offers takeout. ⊠ *229 Commercial St.,* ☎ *508/ 487–6074. MC, V. No lunch. Closed Jan.–Mar.*

**$$–$$$**    ✕ **The Moors.** If the Moors looks like it was just tossed up by an errant wave, that's because in a way, it was. After a fire destroyed the place years ago, locals scoured the outer beaches for driftwood, which was used to rebuild Mylan Costa's establishment. Buoys and nets now hang everywhere—an authentic testament to the fine Portuguese seafood the Moors has been serving for almost 40 years. Specialties include a delicious soup made from *chouriço* (pork sausage) and linguiça sausage, sea clam pie, and marinated swordfish steaks. Complete Portuguese dinners with a choice of three entrées are offered ($16—$20). There's entertainment in the lounge (☞ *Nightlife and the Arts, below*). ⊠ *5 Bradford St. W,* ☎ *508/487–0840. AE, D, DC, MC, V. Closed Nov.–Mar.*

**$$–$$$**    ✕ **Napi's.** No visit to Provincetown is complete without dinner at
★    Napi's, an excellent Mediterranean-style restaurant that serves zesty meals from a large and impressive menu. Most of the items are Greco-Roman, like delicious shrimp feta (shrimp flambé in ouzo and Metaxa, served with a tomato, garlic, and onion sauce). The menu is strong on vegetarian items with an international flair. The dining room is tasteful and casual, and the staff is attentive. If you arrive early and see a man at the bar drinking a glass of muscadet, hunched over the *New York Times* crossword puzzle, introduce yourself: It's Napi himself. ⊠ *7 Freeman St.,* ☎ *508/487–1145. Reservations essential. AE, D, DC, MC, V. No lunch June–mid-Sept.*

**$$–$$$**    ✕ **Sal's Place.** Sal has sold the restaurant that bears his name, so he's no longer cooking commercially (though he is still painting). Even so, this little waterfront trattoria still serves up massive portions of southern Italian specialties. Calamari and linguine, *vongole* (tiny clams) over pasta, and delicious vegetarian spinach lasagna are all longtime favorites. Determined carnivores will want to tackle the steak *pizzaiola*: a massive 25-ounce rib steak served with marinara sauce. Opinions on the decor depend on one's commitment to the authenticity of wine bottles in wicker, but the back dining room overlooks the harbor—the best decor there is. The front dining room is smoke free. ⊠ *99 Commercial St., ☎ 508/487–1279. MC, V. Closed Nov.–Apr. and Mon.–Thurs. Oct. 2–mid-June. No lunch.*

**$$** ✕ **Bubala's by the Bay.** Personality abounds at this once-was-a-run-
★ down, funky joint. When they took over in 1993, Bubala's owners pumped up the funk as well as the quality—the building is painted bright yellow, adorned on top with campy carved birds. The result is an inviting mello-delic seaside '60s feel. The kitchen rarely stops, serving breakfast, lunch, and dinner, with a strong assortment of vegetarian offerings and local seafood. Bubala's buys right off the local boats, and treats fish right when they arrive. The wine list is priced practically at retail, and the U-shape bar picks up into the evening. The whole experience is loud and lively and unique. ⊠ *183 Commercial St., ☎ 508/ 487–0773. AE, D, MC, V. Closed Halloween–April Fool's Day.*

**$$** ✕ **Dancing Lobster Café Trattoria.** Chef and owner Pepe Berg put in
★ some time at Harry's Bar in Venice and brought back to P-town rich Italian flavors and stylish whimsy. Now he's moved again, this time to one of Provincetown's flagship buildings, the former Flagship Restaurant in the East End. This should be a marriage made in culinary heaven, matching a great local chef with a wonderful waterfront ark of a building. As for the food, it's innovative and sophisticated—outstanding in every way. Appetizers are pure classics—carpaccio is a favorite—and the *zuppa di pesce* (Italian seafood stew) with couscous is wonderfully strange and filling. The lobster may not dance, but the customers will—with pleasure. ⊠ *463 Commercial St., ☎ 508/487– 0900. Closed mid-Nov.–mid-May.*

**$$** ✕ **Euro Island Grill & Café.** Jamaican specialties like conch chowder, conch fritters, and jerk chicken are staples on the menu, but traditional seafood is available as well. For what it is, it's somewhat overpriced, but the upstairs deck, overlooking Commercial Street and town hall, is one of the nicest places in town for people-watching. Drinks at the outdoor bar (especially the Red Stripe beer) are the perfect accompaniment to the raw-bar offerings. ⊠ *258 Commercial St., ☎ 508/487– 2505. AE, D, DC, MC, V. Closed Columbus Day–mid-May.*

**$** ✕ **Mojo's.** Provincetown's fast-food institution, where anybody who
★ knows anything about food and happens to be in a hurry goes to grab a bite. The tiniest of kitchens turns out one of the most varied and eclectic menus in town, everything from fresh-cut french fries to fried clams with the bellies, steak subs, tacos, tofu burgers (and the regular kind), tacos, hummus, pizzas, salads, and the best fried fish. How they crank it out, so fast and so good, is anybody's guess. It's hard to believe Mojo's has been around for 25 years, but it's even harder to believe it hasn't always been a part of Provincetown. There's some seating in the back at picnic tables. ⊠ *5 Ryder St. ext., ☎ 508/487–3140. Reservations not accepted. No credit cards. Closed at some times mid-Oct.–early May, depending on weather and crowds.*

**$$$–$$$$** ⊡ **The Brass Key.** Conveniently close to restaurants, shops, galleries,
★ and nightlife, this complex of buildings is fast becoming Provincetown's most luxurious resort. The main house, originally an 1828 sea captain's home, was beautifully restored a number of years ago. Since then, the property has expanded to include several other buildings and cottages, all renovated with the same care and attention to detail. The tastefully

painted and papered rooms are furnished with country-style antiques. All have Bose stereos, phones, mini-refrigerators, and TV/VCRs (there's a videocassette library). Deluxe rooms have gas fireplaces and whirlpools. A widow's walk sundeck has a panoramic view of Cape Cod Bay. In season, complimentary cocktails are served in the courtyard; in winter, wine is served before a roaring fire in the common room. Long favored by a largely gay clientele, the owners and staff make everyone feel welcome and pampered. ⊠ *12 Carver St., 02657,* ☎ *508/487–9005 or 800/842–9858,* ℻ *508/487–9020. 34 rooms. Air-conditioning, no-smoking rooms, in-room VCRs, pool, spa. AE, MC, V. CP.*

**\$\$–\$\$\$**   🏨 **Best Western Chateau Motor Inn.** The personal attention of owners Charlotte and Bill Gordon, whose family has run the place since it was built in 1958 (additions completed in 1972, a total remodeling in 1995), shows in the landscaped lawn and garden that surround the pool and inside in the well-maintained modern hotel rooms, with wall-to-wall carpeting and tiled baths. Atop a hill with expansive views from picture windows of marsh, dunes, and sea, the motel is a longish walk to the center of town, yet removed from it all. Children under 18 stay free. ⊠ *105 Bradford St. Ext., Box 558, 02657,* ☎ *508/487–1286 or 800/528–1234,* ℻ *508/487–3557. 54 rooms. Pool. AE, D, DC, MC, V. Closed Nov.–May 1. CP.*

**\$\$–\$\$\$**   🏨 **Hargood House.** This apartment complex on the water, ½ mi from
**★**   the town center, is a great option for longer stays. Many of the individually decorated units have decks and large water-view windows; all have full kitchens, modern baths, and phones. No. 8 is like a light, bright beach house on the water, with three glass walls, cathedral ceilings, private deck, dining table and chairs. No. 20 has a fireplace and a home-style kitchen. Rental is mostly by the week in season; two-night minimum off-season. Pets are welcome. ⊠ *493 Commercial St., 02657–2413,* ☎ ℻ *508/487–9133. 19 apartments. Grills, beach. AE, MC, V.*

**\$–\$\$\$**   🏨 **The Masthead.** Hidden away in the quiet west end of Commercial Street, the Masthead is a charming cluster of shingled houses that overlook a lush lawn, a 450-ft-long boardwalk, and a private beach. Spacious rooms, efficiencies, apartments, and fully-outfitted cottages offer a wide range lodging options. The cottages, for instance, which sleep seven, are an ideal choice for families or larger groups and for longer stays (children under 12 stay free). The deepwater mooring right on the property makes this a great choice if you're boating into town. This is classic Provincetown—friendly, unpretentious, and homey. ⊠ *31–41 Commercial St., Box 577, 02657,* ☎ *508/487–0523 or 800/395–5095,* ℻ *508/487–9251. 7 apartments, 3 cottages, 2 efficiencies, 9 rooms. Beach, dock. AE, D, DC, MC, V.*

**\$\$**   🏨 **Anchor Inn.** A short walk from town center, this 1912 shingled guest house with turret and porch offers small rooms with sliding doors that open out to a deck overlooking the water. The decor is very simple but pleasant, with light-painted wood furnishings, and all rooms have ceiling fans. Ask if you would like air-conditioning or a king-size bed. ⊠ *175 Commercial St., 02657,* ☎ *508/487–0432 or 800/858–2657 outside MA,* ℻ *508/487–6280. 25 rooms. Beach. AE, MC, V.*

**\$\$**   🏨 **Fairbanks Inn.** On the next street over from Commercial Street, this comfortable and nicely decorated inn includes the 1776 main house and auxiliary buildings. Guest rooms have four-poster or canopy beds, Oriental rugs on wide-board floors, and antique furnishings, and some have fireplaces or kitchens. The wicker-filled sunporch and the garden are good places for afternoon cocktails. ⊠ *90 Bradford St., 02657,* ☎ *508/487–0386. 13 rooms, 1 efficiency, 1 2-bedroom apartment. AE, MC, V. CP.*

**$–$$** ⊞ **The Captain and His Ship.** Built in 1887 for a sea captain, this stately, mansard-roof Victorian is only a short walk from the center of town, and across the street from the Provincetown bay. Jim Baer, who has owned the house since 1980, has elegantly decorated it with a nautical theme—model ships and ship prints grace the immaculate, spacious rooms. They have water views and are furnished with period antiques and Oriental carpets. All have color TVs, VCRs, phones, mini-refrigerators, and hair dryers. Some have private decks. A small back-yard garden patio, set with umbrella tables, is a pleasant place to relax. ⊠ *164 Commercial St., 02657,* ☎ *508/487–1850 or 800/400–2278. 8 rooms. Air-conditioning, in-room VCRs. No smoking. Limited parking. MC, V. Closed Nov.–Apr. CP.*

**$** ⊞ **The Meadows.** At the far west end of Bradford Street, between the town center and the beach, these motel-like rooms may not have the convenience of an in-town locale, but they do offer a good value, especially for families. The property is well-maintained and nicely land-scaped. Rooms are bright, clean, and comfortable, furnished in standard motel style. All have TVs with HBO and mini-refrigerators. Bicycle rentals and tennis courts are nearby. ⊞ *Bradford St. Ext., 02657,* ☎ *508/487–0880. 20 rooms. MC, V.*

## Nightlife and the Arts

THE ARTS

The **Beach Plum Music Festival,** held in August at the Provincetown Town Hall (⊠ 260 Commercial St., ☎ 508/349–6874), is a series of popular folk and jazz concerts by such performers as Wynton Marsalis, Arlo Guthrie, Queen Ida, and Holly Near.

**Provincetown Art Association and Museum** (⊠ 460 Commercial St., ☎ 508/487–1750) has along with its collection a gift shop with many books on Provincetown and its artists.

**Provincetown Fiction Workshop** (⊠ 4 Duncan La., ☎ 508/487–2855), directed by award-winning author Dean Albarelli, offers short-term writing classes and literature discussion groups year-round.

The **Provincetown Playhouse Mews Series** (⊠ Town Hall, 260 Commercial St., ☎ 508/487–0955) presents classical chamber, folk, ethnic, and jazz concerts in summer.

The **Provincetown Theatre Company** (☎ 508/487–4715) presents classics, modern drama, and new works by local authors year-round, as well as staged readings and playwriting workshops.

The **Provincetown Repertory Theatre** (⊠ Box 812, ☎ 508/487–0600) stages a summer lineup of classic and modern drama, presenting Equity actors and local talent.

NIGHTLIFE

**Atlantic House** (⊠ 6 Masonic Pl., ☎ 508/487–3821), the grandfather of the gay night scene and the only gay bar open year-round, is frequented mostly by men. There are several lounge areas and an outdoor patio.

**Boatslip Beach Club** (161 Commercial St., ☎ 508/487–1669) holds a mixed gay and lesbian tea dance daily from 3:30 to 6:30 on the out-door pool deck. The club has indoor and outdoor dance floors. There is ballroom dancing here, as well as two-stepping Thursday through Sunday nights.

The **Cape Cod National Seashore** offers summer evening programs, such as slide shows, sunset beach walks, concerts (local groups, military bands), and sing-alongs, at its Province Lands Visitor Center (⊠

Provincetown, ☎ 508/487–1256), and sunset campfire talks on the beaches in Provincetown.

**Club Euro** (✉ 258 Commercial St., ☎ 508/487–2505) has weekend concerts by big names in world music (including African music, Jamaican reggae, Chicago blues, and Cajun zydeco) in a great room—an 1843 Congregational church, later a movie theater—done in an eerie ocean dreamscape: oceanic sea-green walls with a half-submerged, three-dimensional mermaid, spouting fish, and a black ceiling high above. Pool tables and a late-night menu are available.

**Crown & Anchor Complex** (247 Commercial St., ☎ 508/487–1430), in the heart of town, incorporates a number of bars under one roof, including a disco, a leather bar, a cabaret of gay and straight comics and drag shows, a pool bar, a seafood restaurant, and a game room with pool tables and video games.

**The Moors Restaurant** (✉ Bradford St. Ext., ☎ 508/487–0840) has presented Lenny Grandchamp in its lounge for nearly 20 years. He plays the piano, sings, tells jokes, and leads sing-alongs up to six nights a week in season.

**Napi's** (✉ 7 Freeman St., ☎ 508/487–1145) offers easy-listening piano in its upstairs lounge on weekends, nightly in season.

**Pied Piper** (✉ 193A Commercial St., ☎ 508/487–1527), P-town's only women-owned and -operated club, draws hordes of gay men to its post-tea dance gathering at 6:30 every evening in July and August and weekends in the shoulder seasons. Later in the evening, though, the crowd is mostly, though not exclusively, women. The club has a deck overlooking the harbor, a small dance floor with a good sound system, and two bars. Every Tuesday night at 10 (in season only) is "Puttin' On The Hits," an amateur talent show.

## Outdoor Activities and Sports

### BEACHES

All the **National Seashore beaches** have lifeguards, showers, and rest rooms. Only Herring Cove has food. From mid-June through Labor Day, parking costs $5 per day, or $20 for a yearly pass good at all Seashore beaches; walk-ins pay $3. Senior citizens with Golden Eagle passports and persons with disabilities and Golden Age Passports are admitted free.

**Race Point Beach** has a remote feeling, with a wide swath of sand stretching far off into the distance around the point and Coast Guard Station. Behind the beach is pure duneland, and bike trails lead off the parking lot. Because of its position, on a point facing north, the beach gets sun all day long, whereas the east-coast beaches get fullest sun early in the day.

**Herring Cove Beach** is calmer than, a little warmer than, and not as pretty as Race Point Beach, since the parking lot isn't hidden behind dunes. There's a hot dog stand here, and sunsets from the parking lot to the right of the bathhouse are great.

### BIKING

The **Province Lands Trail** is a 5¼-mi loop off the Beech Forest parking lot on Race Point Road in Provincetown, with spurs to Herring Cove and Race Point beaches and to Bennett Pond. The paths wind up and down hills amid dunes, marshes, woods, and ponds, affording spectacular views. More than 8 mi of bike trails lace through the dunes, cranberry bogs, and scrub pine of the National Seashore, with many access points, including Herring Cove and Race Point. The Beech Forest trail offers an especially nice ride through a shady forest to Bennet Pond.

**Arnold's** (✉ 329 Commercial St., ☎ 508/487–0844) rents all types of bikes, children's included.

**Nelson's Riding Stables and Bike Rentals** (✉ 43 Race Point Rd., ☎ 508/487–8849), across from the Beech Forest bike trail, rents a variety of bikes, including trailers, and has a deli with picnic-ready food and free parking.

**Ptown Bikes** (✉ 42 Bradford St., ☎ 508/487–8735; ✉ 306 Commercial St., ☎ 508/487–6718) has Trek (Commercial Street) and Mongoose (Bradford Street) mountain bikes at good rates; they also provide free locks and maps.

### BOATING
**Flyer's Boat Rental** (✉ 131A Commercial St., Provincetown, ☎ 508/487–0898) has sailboards, Sunfish, Hobies, Force 5s, Lightnings, powerboats, and rowboats. Flyer's will also shuttle you to Long Point.

### FISHING
You can go for fluke, bluefish, and striped bass on a walk-on basis from spring through fall with **Cap'n Bill & Cee Jay** (✉ MacMillan Wharf, ☎ 508/487–4330 or 800/675–6723).

### HORSEBACK RIDING
The **Province Lands Horse Trails** lead to the beaches through or past dunes, cranberry bogs, forests, and ponds.

**Nelson's Riding Stable** (✉ 43 Race Point Rd., ☎ 508/487–1112), offers trail rides by reservation.

ℭ **Bayberry Hollow Farm** (✉ W. Vine St. Ext., ☎ 508/487–6584) has pony rides year-round.

### PARASAILING
**AquaVentures** (✉ MacMillan Wharf, ☎ 800/300–3787).

### TENNIS
**Bissell's Tennis Courts** (✉ Bradford St. at Herring Cove Beach Rd., ☎ 508/487–9512. ☉ Memorial Day–Sept.) has five clay courts and offers lessons.

### WHALE-WATCHING
One of the joys of Cape Cod is spotting whales while they're swimming in and around the feeding grounds at Stellwagen Bank, about 6 mi off the tip of Provincetown. On a sunny day, the boat ride out into open ocean is part of the pleasure, but the thrill, of course, is in seeing these great creatures. You might spot minke whales, humpbacks, who put on the best show when they breach, finbacks, or perhaps the most endangered great whale species, the right whale. Dolphins are a welcome sight as well, as they play in the boats' bow waves. They are, in fact, toothed whales. Many people also come aboard for birding, especially during spring and fall migration. You can see a great variety of birds at sea—gannets, shearwaters, storm petrels, among many others.

Several boats take whale-watchers out to sea—and bring them back—with morning, afternoon, or sunset trips lasting three to four hours. All boats have food service, but remember to bring sunscreen and a sweater or jacket—the breeze makes it chilly. Some boats stock seasickness pills, but if you're susceptible, come prepared.

The municipal parking area by the harbor in Provincetown fills up by noon in summer. Consider taking a morning boat to avoid crowds and the hottest sun. And although it may be cold in April, it is one of the better months for spotting whales, who at that time have just migrated north after mating, and are very hungry. Good food is, after all, what brings the whales to this part of the Atlantic.

**Dolphin Fleet** tours are accompanied by scientists from the Center for Coastal Studies in Provincetown who provide commentary while collecting data on the whale population they've been monitoring for years. They know many of the whales by name and tell you about their habits and histories. Reservations are required. ⊠ *Ticket office in Chamber of Commerce building at MacMillan Wharf,* ☎ *508/349–1900 or 800/826–9300.* ⚐ *$18.50 (seasonal variations). Tours Apr.–Oct.*

The ***Portuguese Princess*** sails with a naturalist on board to narrate. The snack bar offers Portuguese specialties. ⊠ *Tickets available at 70 Shank Painter Rd. ticket office or at Whale Watchers General Store, 309 Commercial St.,* ☎ *508/487–2651 or 800/442–3188.* ⚐ *$15–$19.50. Tours May–Oct.*

The ***Ranger V,*** the largest whale-watch boat, also has a naturalist on board. ⊠ *Ticket office on Bradford and Standish Sts.,* ☎ *508/487–3322 or 800/992–9333.* ⚐ *$18. Tours May–mid-Oct.*

## Shopping
GALLERIES

**Albert Merola Gallery** (⊠ 424 Commercial St., ☎ 508/487–4424) has 20th-century and contemporary master prints, Picasso ceramics, and features artists from in and around Provincetown, Boston, and New York, including James Balla and Richard Baker.

**Berta Walker Gallery** (⊠ 208 Bradford St., ☎ 508/487–6411) deals in Provincetown-affiliated artists, including Selina Trieff and Nancy Whorf, working in various media.

**Long Point Gallery** (⊠ 492 Commercial St., ☎ 508/487–1795) founded in 1977, is a cooperative of 17 well-established artists—including Varujan Boghosian, Robert Motherwell, Paul Resika, Judith Rothschild, and Tony Vevers.

**The William Scott Gallery** (439 Commercial St., ☎ 508/487–4040), an interesting new gallery, primarily shows contemporary works such as John Dowd's reflective, realistic Cape scapes.

GENERAL

**Giardelli Antonelli** (⊠ 417 Commercial St., ☎ 508/487–3016) specializes in handmade, quality clothing by local designers. Hand-knit sweaters are especially popular at the upscale shop.

**Impulse** (⊠ 188 Commercial St., ☎ 508/487–1154) has contemporary American crafts, including jewelry and an extraordinary kaleidoscope collection. The Autograph Gallery features framed photographs, letters, and documents signed by celebrities.

**Kidstuff** (⊠ 381 Commercial St., ☎ 508/487–0714) carries unusual, colorful children's wear.

**Northern Lights Leather** (⊠ 361 Commercial St., ☎ 508/487–9376) has high-fashion clothing, boots, shoes, and accessories of very fine, soft leather, plus silk clothing. Its shop around the corner sells all kinds of hammocks—rope, wood, Yucatán multicolor cotton—and has a waterfront deck where you can try them out.

**Remembrances of Things Past** (⊠ 376 Commercial St., ☎ 508/487–9443) deals with articles from the 1920s to the 1960s, including Bakelite and other jewelry, telephones, neon items, ephemera, and autographed celebrity photographs.

**Southwest Connection** (⊠ 241 Commercial St. in Whaler's Wharf, ☎ 508/487–9739) carries furniture, ironwork, pottery, baskets, and Zapotec rugs from the Southwest and Mexico.

**Tim's Used Books** (⊠ 242 Commercial St., ☎ 508/487–0005) has volumes of volumes, rooms of used-but-in-good-shape books, including some rare and out of print texts. It's a great place to browse for that perfect book to read on vacation.

**West End Antiques** (⊠ 146 Commercial St., ☎ 508/487–6723) specializes in variety: $4 postcards, a $3,000 model ship, handmade dolls, and better-quality glassware—Steuben, Orrefors, and Hawkes.

# CAPE COD A TO Z

## Arriving and Departing

### By Boat

Ferries ply the waters between Boston, Plymouth, and Provincetown in season. Between the Cape and Islands, year-round boats run between Woods Hole, Falmouth, Hyannis, and Martha's Vineyard, and between Hyannis and Nantucket. There is also seasonal Cape–Islands service from these ports, and from Harwich Port east of Hyannis. Martha's Vineyard A to Z and Nantucket A to Z in Chapters 3 and 4 have details on Cape–Island ferry service.

#### TO AND FROM BOSTON

**Bay State Cruise Company** makes the three-hour trip between Commonwealth Pier in Boston and MacMillan Wharf in Provincetown from Memorial Day to Columbus Day. ☎ *617/723–7800 in Boston, ☎ 508/ 487–9284 in Provincetown. ⌑ One-way/same-day round-trip: $18/$30, $5/$10 bicycles.*

#### TO AND FROM PLYMOUTH

**Capt. John Boats'** passenger ferry makes the 1½- to 2-hour trip between Plymouth's Town Wharf and Provincetown from Memorial Day to mid-June, weekends; mid-June to September, daily. Schedules allow for day excursions. ☎ *508/747–2400 or 800/242–2469 in MA. ⌑ Round-trip: $24.*

#### HARWICH PORT–NANTUCKET

The **Freedom Cruise Line** runs express ferries to Nantucket, allowing you to take a day trip from Harwich Port between May 15 and October 15 (☞ Nantucket A to Z *in Chapter 4). ⊠ Saquatucket Harbor, Harwich Port, ☎ 508/432–8999.*

### By Bus

**Bonanza Bus Lines** (☎ 508/548–7588 or 800/556–3815) offers direct service to Bourne, Falmouth, and the Woods Hole steamship terminal from Boston, Providence, Fall River, and New Bedford, as well as connecting service from New York and Connecticut. Another route travels between Boston, Wareham, and Buzzards Bay.

**Plymouth & Brockton Street Railway** (☎ 508/775–5524 or 508/746–0378) travels to Provincetown from Boston and Logan Airport, with stops en route.

### By Car

From Boston (60 mi), take Route I–93 South to Route 3 South, across the Sagamore Bridge which becomes Route 6, the Cape's main artery that leads to Hyannis and Provincetown. From Western Massachusetts, Northern Connecticut, and Northeastern New York State, take I–84 East to the Massachusetts Turnpike (I–90 East) and take I–495 to the Bourne Bridge. From Washington, D.C., Philadelphia, New Jersey,

New York City, and all other points south and west, take I–95 North toward Providence, where you'll pick up I–195 East (toward Fall River/New Bedford) to Route 25 East to the Bourne Bridge. From the Bourne Bridge, you can take Route 28 south to Falmouth and Woods Hole (about 15 mi), or go around the rotary, following the signs to Route 6; this will take you to the Lower Cape and central towns more quickly. On summer weekends, when more than 100,000 cars a day cross each bridge, make every effort to avoid arriving in the late afternoon, especially on holidays. Routes 6, 6A, and 28 are heavily congested eastbound Friday evenings and westbound Sunday afternoons.

## By Plane

**Barnstable Municipal Airport** (⊠ 480 Barnstable Rd., Rte. 28 rotary, Hyannis, ☎ 508/775–2020) is the region's main air gateway.

**Provincetown Municipal Airport** (⊠ Race Point Rd., ☎ 508/487–0241) has year-round Boston service through Cape Air. Both airports are just a few minutes from the town center.

CARRIERS

Airline service is extremely unpredictable because of the seasonal nature of Cape travel—carriers come and go, while others juggle their routes. The Barnstable Municipal Airport will always know which carriers fly in, should you encounter difficulty in making reservations.

**Business Express** (☎ 508/862-0556 or 800/345–3400), a subsidiary of Delta, flies into Hyannis nonstop from Boston from May through September, with connecting service on American, Northwest, and Delta to major cities.

**Cape Air/Nantucket Airlines** (☎ 508/771–6944 or 800/352–0714) flies direct from Boston year-round. Cape Air also flies from New Bedford to Martha's Vineyard and Nantucket. It also has joint fares with Continental, Delta, Midwest Express, and US Airways and ticketing-and-baggage agreements with eight major U.S. airlines and KLM.

**Colgan Air** (☎ 800/272–5488) flies from Newark or New York to Hyannis and Nantucket year-round.

**Continental Express** (☎ 800/525–0280) offers seasonal service from Newark to Hyannis.

**US Airways Express** (☎ 800/428–4322) flies nonstop from Boston and New York, including Westchester, year-round. Connect in Boston with the airline's other routes.

For charters, contact **Cape Air** (☞ *above*). **Westchester Air** (☎ 800/759–2929) offers a charter that flies from White Plains, New York to Hyannis.

## By Train

Citing dwindling ridership, **Amtrak** (☎ 800/872–7245) stopped offering train service to the Cape as of 1997. Amtrak's continuing fiscal woes make it improbable service will be restored, but it's possible it could start up again if there is enough interest; call for updates.

# Getting Around

## By Bicycle

The Cape will satisfy both the avid and the occasional cyclist. There are many flat back roads, as well as a number of well-developed and scenic bike trails. The 30-mi Cape Cod Rail Trail follows the paved right-of-way of the old Penn Central Railroad line between South Dennis and South Wellfleet, with many access points.

## By Bus

The **Cape Cod Regional Transit Authority** (☎ 508/385–8326 or 800/ 352–7155 in MA) operates several bus services that link Cape towns. The SeaLine operates along Route 28 Monday–Saturday between Hyannis and Woods Hole. Its many stops include Mashpee Commons, Falmouth, and the Woods Hole steamship docks. The SeaLine connects in Hyannis with the Plymouth & Brockton line, as well as the Villager, another bus line that runs along Route 132. (Average fare: $3.50, one-way, from Hyannis to Woods Hole.) The driver will stop when signaled along the route, and all buses have lifts for people with disabilities.

The **b-bus** is comprised of a fleet of minivans that will transport passengers door-to-door between any towns on the Cape. Service runs 7 days a week, year-round, though reservations must be made in advance. The cost is $2.00 per ride, plus 10¢ per mile.

The **H2O Line** offers regularly scheduled service, year-round and 7 days a week, between Hyannis and Orleans along Route 28. The Hyannis–Orleans fare is $3.50; shorter trips are less. All buses are wheelchair-accessible and equipped with bike racks. Buses connect in Hyannis with the SeaLine, the Villager, Bonanza, and Plymouth & Brockton lines.

**Plymouth & Brockton Street Railway** (☎ 508/775–5524 or 508/746–0378) has service between Boston and Provincetown, with stops at many towns in between.

**Bonanza** (☎ 508/548–7588 or 800/556–3815) runs between Bourne, Falmouth, Woods Hole, and Hyannis. All service is year-round.

## By Car

Traffic on Cape Cod in summer can be maddening, especially on Route 28, which traces the populous south shore. Route 6 is the main artery, a limited-access (mostly divided) highway running the entire length of the Cape. On the north shore, the Old King's Highway, Route 6A, parallels Route 6 and is a scenic country road passing through occasional towns. When you're in no hurry, use back roads—they're less frustrating and much more rewarding. Good to know: Massachusetts permits a **right turn on a red light** (after a stop) unless a sign says otherwise. Also, when approaching one of the Cape's numerous rotaries (traffic circles), note that the vehicles already in the rotary have the right of way, and that those vehicles entering the rotary *must* yield.

## By Limousine

The following companies provide 24-hour Cape-wide limo service: **Aristocrat Limousine** (☎ 508/420–5466 or 800/992–6163). **Black Tie Limousine** (☎ 508/862–6303 or 888/862–0405). **Cape Escape Tours & East Coast Limousine Co.** (☎ 508/430–0666 or 800/540–0808 in MA). **John's Taxi & Limousine** (☎ 508/394–3209).

## By Taxi

There are taxi stands at the Hyannis airport, the bus station, and at the Cape Cod Mall. In Hyannis, call **Hyannis Taxi** (☎ 508/775–0400 or 800/773–0600) and **Checker Taxi** (☎ 508/771–8294). Elsewhere on the Cape, call **All Village Taxi** (Falmouth, ☎ 508/540–7200), **Eldredge Taxi** (Chatham, ☎ 508/945–0068), or **Cape Cab** (Provincetown, ☎ 508/487–2222).

## By Trolley

The **Cape Cod Regional Transit Authority** (☎ 508/385–8326 or 800/ 352–7155 in MA) runs **seasonal trolleys** in Falmouth, Mashpee, Hyannis, Yarmouth, and Dennis. Fares and times vary; call for more information. In Sandwich, the **Glasstown Trolley** (☎ 508/428–9973) runs from the train station at Jarves Street to Heritage Plantation.

# Contacts and Resources

### B&B Reservation Agencies

In summer, lodgings should be booked as far in advance as possible—several months for the most popular cottages and B&Bs. Assistance with last-minute reservations is available at the Cape Cod Chamber of Commerce information booths. Off-season rates are much reduced, and service may be more personalized.

**Bed and Breakfast Cape Cod** (⊠ Box 341, W. Hyannis Port 02672–0341, ☎ 508/775–2772 or 800/686–5252, FAX 508/775–2884) lists about 115 B&Bs on the Cape and islands.

**DestINNations** (⊠ 572 Rte. 28, Suite 3, W. Yarmouth 02673, ☎ 508/790–0577 or 800/333–4667, FAX 508/790–0565) handles a limited number of hotels and B&Bs on the Cape and islands but will arrange any and all details of a visit.

**Orleans Bed & Breakfast Associates** (⊠ Box 1312, Orleans 02653, ☎ 508/255–3824 or 800/541–6226, FAX 508/255–2863) lists 70 no-smoking B&Bs on the Lower Cape from Harwich to Truro.

**Provincetown Reservations System** (⊠ 293 Commercial St., Province-town 02657, ☎ 508/487–2400 or 508/487–6515) makes reservations year-round for accommodations, shows, restaurants, transportation, and more.

### Camping

The Cape has many **private campgrounds,** as well as camping at state parks and forests. Call the **Cape Cod Chamber of Commerce** (☞ *below*) for its listing.

**Eastern Mountain Sports** (⊠ 1513 Rte. 132, Hyannis, ☎ 508/362–8690) rents tents and sleeping bags.

**Nickerson State Park** (⊠ Rte. 6A, Brewster, ☎ 508/896–4615) is popular with nature lovers.

**Sandy Terraces** (⊠ Box 98, Marstons Mills 02648, ☎ 508/428–9209), a surprise in this very traditional area, is a seasonal family nudist campground.

Although private campgrounds serve the area, the only camping permitted on the Cape Cod National Seashore itself is in nonrental, self-contained RVs at **Race Point Beach** (⊠ Provincetown, ☎ 508/487–1256).

### Car Rentals

Rental cars are available at the airport in **Hyannis** from: **Avis** (☎ 508/775–2888 or 800/331–1212), **Budget** (☎ 508/771–2744 or 800/527–0700), **Hertz** (☎ 508/775–5825 or 800/654–3131), and **National** (☎ 508/771–4353 or 800/227–7368). **Budget** also offers seasonal rentals out of **Provincetown Municipal Airport** (☎ 508/487–4405).

### Children's Activities

Each town has a recreation program open to visitors. The morning activities, including sports, trips, and crafts, provide a good opportunity for your kids to meet others.

The **Cape Cod YMCA** (⊠ Box 188, Rte. 132, W. Barnstable 02668, ☎ 508/362–6500) offers summer one-week "fun clubs" (sports, crafts, nature), kids' evenings, summer day camps, swimming classes, and more.

**Hyannis Public Library** (⊠ 401 Main St., Hyannis, ☎ 508/775–2280) has a new children's multicultural center with books, tapes, and videos (some in foreign languages), programs, a play area with puzzles and games, and a children's reading club in summer.

Libraries usually offer regular children's story hours or other programs—check them out on a rainy day. Hours are listed in the newspapers each week. The wonderful **Wellfleet Public Library** (⊠ W. Main St., ☎ 508/349–0310), in a town visited by numerous children's book authors, has a story hour.

The Cape has a number of other day and residential summer camps. For more information, write to the **Cape Cod Association of Children's Camps** (⊠ Box 38, Brewster 02631).

**Cape Cod Baseball Camp** (⊠ Box S, Buzzards Bay 02532, ☎ 508/432–6909) offers sessions of a week or more, either residential or day camp, for children 8–19.

The **Cape Cod Community College** (⊠ Rte. 132, W. Barnstable 02668, ☎ 508/362–2131, ext. 4365) also offers summer day camps.

**Cape Cod Museum of Natural History** (⊠ Rte. 6A, Brewster, ☎ 508/896–3867) has a full program of children's and family activities in summer, including overnights and one- and two-week day camps of art and nature classes for preschoolers through grade 6.

**Cape Cod Sea Camps** (⊠ Box 1880, Brewster 02631, ☎ 508/896–3451) teach sailing and water sports.

Children's concerts, plays, and programs are offered by the **Cape Cod Melody Tent** (☎ 508/775–9100), the **Cape Cod Symphony Orchestra** (☎ 508/362–1111), and the **Cape Playhouse** (☎ 508/385–3911).

**Massachusetts Audubon Wellfleet Bay Sanctuary** (⊠ Rte. 6, Box 236, S. Wellfleet 02663, ☎ 508/349–2615) has daily and weekly nature programs and camps for children.

## Emergencies
**Police, fire and ambulance** (☎ 911).

For rescues at sea, call the **Coast Guard** (Woods Hole, ☎ 508/548–5151; Sandwich, ☎ 508/888–0335; Chatham, ☎ 508/945–0164; Provincetown, ☎ 508/487–0070). Boaters should use channel 16 on their radios.

**Massachusetts Poison Control Center** (☎ 800/682–9211).

DENTISTS

**Dental Associates of Cape Cod** (⊠ 262 Barnstable Rd., Hyannis, ☎ 508/778–1200) accepts emergency walk-ins.

HOSPITALS

**Cape Cod Hospital** (⊠ 27 Park St., Hyannis, ☎ 508/771–1800).

**Falmouth Hospital** (⊠ 100 Ter Heun Dr., Falmouth, ☎ 508/548–5300).

LATE-NIGHT PHARMACIES

Most of the Cape's **CVS** stores are open seven days a week. Two are open 24 hours a day, with the pharmacies open until midnight during the summer: in Falmouth (⊠ 64 Davis Straits, ☎ 508/540–4307) and Dennis (⊠ Patriot Square Mall, Rte. 134, ☎ 508/398–0724). They usually accept out-of-town prescription refills with the prescribing doctor's phone verification. Most pharmacies post emergency numbers on their doors.

WALK-IN CLINICS

**Falmouth Walk-in Medical Center** (⊠ 309 Main St., Rte. 28, Teaticket, ☎ 508/540–6790).

**Mashpee Family Medicine** (⊠ 800 Falmouth Rd., Mashpee, ☎ 508/477–4282).

**Medicenter Five** (⊠ 525 Long Pond Dr., Harwich, ☎ 508/432–5936).

**Mid Cape Medical Center** (⊠ 489 Bearses Way, at Rte. 28, Hyannis, ☎ 508/771–4092).

**Outer Cape Health Services** (⊠ Rte. 6, Wellfleet, ☎ 508/349–3131; Harry Kemp Way, Provincetown, ☎ 508/487–9395).

## Guided Tours

CRUISES

**Art's Dune Tours** are hour-long narrated auto tours through the National Seashore and the dunes around Provincetown. ☎ 508/487–1950 or 508/487–1050; 800/894–1951 in MA only. ☞ $9 daytime, $10 sunset. Tours mid-Apr.–late Oct.

The gaff-rigged schooner **Bay Lady II** makes two-hour sails, including a sunset cruise, across Provincetown Harbor into Cape Cod Bay. ⊠ MacMillan Wharf, Provincetown, ☎ 508/487–9308. ☞ $10–$15.

**Cape Cod Canal Cruises** (two or three hours, narrated) leave from Onset, just northwest of the Bourne Bridge. A Sunday jazz cruise, sunset cocktail cruises, and Friday and Saturday dance cruises are available. Kids 12 and under cruise free on Family Discount Cruises, daily at 4 (except Sunday). ⊠ Onset Bay Town Pier, ☎ 508/295–3883. ☞ $8.

**Cape Cod Duck Mobile** offers the Cape's only amphibious land-and-sea vehicle tour, leaving from downtown Hyannis several times a day, weather-permitting, 7 days a week. You'll spend 30 minutes on land, 30 on sea. Refreshments are available. ⊠ 447 Main St., Hyannis, ☎ 508/362–1117 or 800/685–9111. ☞ $12.

**Cape Cod Tours** travels the main streets and back roads of the Mid Cape and gives a good overview of the history of Hyannis and the surrounding areas. The 29 passenger bus offers hotel pick-up and drop-off service, and coupons for resorts and activities. There is also an amphibious duck-vehicle tour leaving from downtown Hyannis. Refreshments are available. ⊠ Hyannis, ☎ 508/362–1117, or 800/685–9111. ☞ $12 tour; purchase ticket ½ hr before departure.

**Hy-Line** runs one-hour narrated tours of Hyannis Harbor, including a view of the Kennedy compound. Sunset and evening cocktail cruises are available. ⊠ Ocean St. dock, Pier 1, ☎ 508/778–2600. ☞ $12.

**Patriot Boats** offers two-hour day and sunset cruises between Falmouth and the Elizabeth Islands on the 68-ft schooner Liberté. Charters are available. ⊠ 227 Clinton Ave., Falmouth, ☎ 508/548–2626 or 800/734–0088 in MA. ☞ $20.

**Starfish River Cruise** offers 1½-hour water safari tours of the Bass River, past windmills, marshlands, and old captains' houses, on a 32-ft aluminum boat with an awning. ⊠ Rte. 28, W. Dennis, just east of the Bass River Bridge, ☎ 508/362–5555. ☞ $12.

FLIGHTSEEING

Sightseeing by air is offered by **Cape Cod Flying Service** (⊠ Cape Cod Airport, 1000 Race La., Marstons Mills, ☎ 508/428–8732), **Chatham Municipal Airport** (⊠ George Ryder Rd., W. Chatham, ☎ 508/945–9000), **Cape Flight** (⊠ Barnstable Municipal Airport, ☎ 508/775–8171), and **Cape Air** (⊠ Provincetown Municipal Airport, ☎ 800/352–0714). Among Cape Air's offerings are helicopter tours and tours in a 1930 Stinson. **Cape Cod Soaring Adventures** (☎ 508/420–4201 or 800/660–4563) offers glider flights and lessons out of Marstons Mills.

RAIL TOURS

**Cape Cod Scenic Railroad** runs 1¾-hour excursions (round-trip) between Hyannis and Sagamore with a stop in Sandwich. The train passes ponds, cranberry bogs, and marshes. A Dinner Train begins and ends in Hyannis (3 hours); the five-course gourmet meal is elegantly served as you watch the passing scenery, floodlighted after sunset. Men must wear jackets or ties. Ecology Discovery Tours (by reservation), conducted in conjunction with the Museum of Natural History, are narrated by a naturalist and include a nature walk. ⊠ *Main and Center Sts., Hyannis,* ☎ *508/771–3788 or 800/872–4508. Excursions: several departures per day (no service Mon.) in each direction mid-June–Oct. (weekends starting in May);* ⊟ *$11.50. Scenic Train: Feb.–Dec., departs various evenings (reservations required at least 24 hrs. in advance); ecology tour: call the Museum of Natural History (*☎ *508/896–3867) for times;* ⊟ *$43.95, Sat. $51.30. Dinner Train: call for times;* ⊟ *Same as excursions.*

TROLLEY TOURS

The **Provincetown Trolley** leaves from the town hall, with pickups at other locations, on the hour from 10 to 7 and on the half hour from 10:30 to 3:30. Points of interest on the 40-minute narrated tours include the downtown area and the Cape Cod National Seashore. Riders can get on and off at four locations. ☎ *508/487–9483.* ⊟ *$8. Tours May–Oct.*

## House Rentals

Many real-estate agencies can assist with house or apartment rentals; be sure to check local chambers' guidebooks and area telephone book yellow pages. Most agencies deal with only a segment of the Cape, and properties are often rented a year in advance, so plan ahead. Often, rental opportunities are greater in the smaller, quieter towns, such as Yarmouth, Chatham, Wellfleet, and Truro.

Mid Cape rentals are handled by **Century 21, Sam Ingram Real Estate** (⊠ 938 Rte. 6A, Yarmouth 02675, ☎ 508/362–1191 or 800/697–3340).

**Commonwealth Associates** (⊠ 551 Main St., Harwich Port 02646, ☎ 508/432–2618, ℻ 508/432–1771) can assist in finding vacation rentals in the Harwiches, Brewster, and Chatham, including waterfront properties.

**Compass Real Estate** (⊠ 2 Academy Pl., at Rte. 28, Orleans 02653, ☎ 508/240–0022 or 800/834–0061; branch office at 282 Main St., Wellfleet 02667, ☎ 508/349–1717 or 800/834–0703) is one of many realtors with a listing of available apartments and houses on the Lower Cape.

**Donahue Real Estate** (⊠ 850 Main St., Falmouth 02540, ☎ 508/548–5412) lists both apartments and houses on the Upper Cape.

**Great Vacations Inc.** (☎ 2660 Rte. 6A, Brewster 02631, ☎ 508/896–2090) specializes in locating vacation rentals in Brewster, Dennis, and Orleans.

**Peter McDowell Associates** (⊠ 585 Main St., Rte. 6A, Dennis 02638, ☎ 508/385–9114 or 888/385–9114) offers a wide selection of properties for rent by the week, month, or season.

**Real Estate Associates** (⊠ Rte. 151 and 28A, Box 738, N. Falmouth 02556, ☎ 508/563–7173) lists more expensive houses on the Upper Cape.

**Roslyn Garfield Associates** (⊠ 115 Bradford St., Provincetown 02657, ☎ 508/487–1308) lists rentals for Wellfleet, Truro, and Provincetown.

**Waterfront Rentals** (✉ 20 Pilgrim Rd., W. Yarmouth 02673, ☎ 508/778–1818) covers Bourne to Truro, listing everything from one-bedroom condos to a 7-bedroom, 7-bath beachfront estate.

## Libraries

Several area libraries have special collections. The **Centerville Library** (✉ 585 Main St., ☎ 508/775–1787) has a 42-volume noncirculating set of transcripts of the Nuremberg Trials. **Cotuit Library** (✉ 871 Main St., ☎ 508/428–8141) houses a noncirculating set of luxurious leather-bound classics. **Hyannis Public Library** (✉ 401 Main St., ☎ 508/775–2280) has a case full of books on JFK. Barnstable's **Sturgis Library** (✉ 3090 Rte. 6A, ☎ 508/362–6636) has extensive Cape genealogical and maritime materials. Other Cape libraries, though without special collections, are no less worthwhile. The **Wellfleet Public Library** (✉ W. Main St., ☎ 508/349–0310) is a particularly nice one.

## Outdoor Activities and Sports

### BICYCLING

**Local Guidebooks.** The Dennis Chamber of Commerce's guidebook includes bike tours and maps, and the Wellfleet chamber's pamphlet "Bicycling in Wellfleet" includes an annotated map. Check with local chambers for free maps, guidebooks and pamphlets.

For a brochure on the **Claire Saltonstall Bikeway** between Boston and Provincetown (135 mi) or Woods Hole (85 mi), using mostly bike paths and little-traveled roadways, send $2.10 and a business-size SASE to American Youth Hostels (✉ 1020 Commonwealth Ave., Boston 02215). It is also available in area bike shops and bookstores.

### FISHING

Charter boats and party boats (per-head fees, rather than the charters' group rates) fish in season for bluefish, tuna, marlin, and mako and blue sharks. Throughout the year there's bottom fishing for flounder, tautog, scup, fluke, cod, and pollack.

The Cape Cod Chamber's *Sportsman's Guide* gives fishing regulations, surf-fishing access locations, a map of boat-launching facilities, and more.

The state Division of Fisheries and Wildlife has a new book with dozens of maps of Cape ponds. Remember, you'll need a license for freshwater fishing, available for a nominal fee at bait and tackle shops.

Molly Benjamin's fishing column in the Friday *Cape Cod Times* tells the latest in fishing on the Cape—what's being caught and where.

### GOLF

The Cape Cod Chamber of Commerce has a "Golf Map of Cape Cod," locating dozens of courses on the Cape and islands. Summer greens fees range from $25 to $50.

### RUNNING

Many road races are held in season, including the world-class **Falmouth Road Race** in August and the **Cape Codder Triathlon** (✉ Box 307, W. Barnstable 02668) at Craigville Beach in early summer.

## Visitor Information

The **Cape Cod Chamber of Commerce** (Junction of Rtes. 6 and 132, Hyannis, ☎ 508/362–3225, 800/332–2732) is open year-round, weekdays 8:30–5; Memorial Day–Columbus Day, also weekends 10–4. There's also a year-round visitors information center, open daily 9–5, on Route 25 on the way to the Bourne Bridge.

**Local chambers of commerce,** many open only in season, put out literature on their area.

# Pick up the phone.
# Pick up the miles.

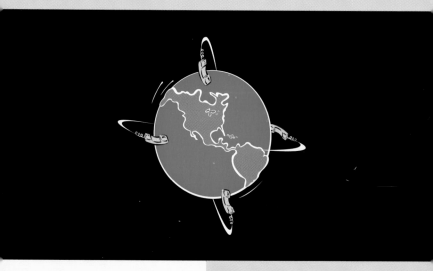

Is this a great time, or what? :-)

Now when you sign up with MCI you can receive up to 8,000 bonus frequent flyer miles on one of seven major airlines.

Then earn another 5 miles for every dollar you spend on a variety of MCI services, including MCI Card® calls from virtually anywhere in the world.*

You're going to use these services anyway. Why not rack up the miles while you're doing it?

# Speakeasy.

Fodor's and Living Language® bring you the most useful, up-to-date language course for travelers. Phrasebook/dictionary, two cassettes, booklet.

**Brewster Board of Trade** (✉ Box 1241, 02631, ☎ 508/896–8088; information center at the rear of the old town hall, Rte. 6A).

**Brewster** (✉ 70 Locust La., 02631, ☎ 508/255–7045) shares the information center with the Brewster Board of Trade.

**Cape Cod Canal Region,** for Sandwich and Bourne and Wareham (✉ 70 Main St., Buzzards Bay 02532, ☎ 508/759–6000; information centers at the Sagamore rotary, at the train depot in Buzzards Bay, and on Rte. 130 in Sandwich).

**Chatham** (✉ Box 793, 02633, ☎ 508/945–5199; information center at 553 Main St.).

**Dennis** (✉ Junction of Rtes. 28 and 134, W. Dennis; Box 275, S. Dennis 02660, ☎ 508/398–3568 or 800/243–9920).

**Eastham** (✉ Rte. 6 at Fort Hill Rd., Box 1329, 02642, ☎ 508/255–3444).

**Falmouth** (✉ Academy La., Box 582, 02541, ☎ 508/548–8500 or 800/526–8532).

**Harwich** (✉ Rte. 28, Box 34, Harwich Port 02646, ☎ 508/432–1600 or 800/441–3199).

**Hyannis** (✉ 1481 Rte. 132, 02601, ☎ 508/362–5230 or 800/449–6647).

**Mashpee** (✉ Rte. 151, Box 1245, 02649, ☎ 508/477–0792).

**Orleans** (✉ Box 153, 02653, ☎ 508/255–1386; information booth on Eldredge Pkwy., off Rte. 6A, ☎ 508/240–2484).

**Provincetown** (✉ MacMillan Wharf, Box 1017, 02657, ☎ 508/487–3424).

**Provincetown Business Guild** (✉ 115 Bradford St., Box 421–89, 02657, ☎ 508/487–2313; gay tourism).

**Truro** (✉ Rte. 6 at Head of the Meadow Rd., Box 26, N. Truro 02652, ☎ 508/487–1288).

**Wellfleet** (✉ Box 571, 02667, ☎ 508/349–2510; information center off Rte. 6 in S. Wellfleet).

**Yarmouth Area** (✉ 657 Rte. 28, Box 479, W. Yarmouth 02673, ☎ 508/778–1008 or 800/732–1008; information center on Rte. 6 heading east between Exits 6 and 7).

## Other Information

**Army Corps of Engineers 24-hour recreation hot line** (☎ 508/759–5991) for canal-area events, tide, and fishing information. **Arts and entertainment events** are listed in the *Cape Cod Times*'s "CapeWeek" section on Friday and in its daily editions. Also check out listings in *The Barnstable Patriot* and the Tuesday and Friday editions of *The Cape Codder*. In Provincetown, look for *The Advocate* and *The Banner,* both weeklies.

The **Cape Cod Jazz Society** operates a 24-hour hot line (☎ 508/394–5277) on jazz events throughout the Cape.

A recorded **tide, marine, and weather forecast hot line** is sponsored by local radio station WQRC (☎ 508/771–5522).

# 3 Martha's Vineyard

*On today's star-studded Vineyard, the summertime bustle and crush of Vineyard Haven, Oak Bluffs, and Edgartown continue to belie the quieter feeling that off-season visitors have come to love. In season, you can step back into rural time Up-Island at the wonderful West Tisbury Farmers' Market or on a conservation area's pine woods or rolling meadows. At all times some superb beach—that perennial favorite of island vacationing—beckons nearby.*

Updated by
Alan W.
Petrucelli

Dining
updated by
Seth Rolbein

**B**ARTHOLOMEW GOSNOLD charted Martha's Vineyard for the British Crown in 1602 and is credited with naming it, supposedly after his infant daughter or mother-in-law (or both) and the wild grapes he found growing in profusion. Later, a Massachusetts Bay Colony businessman, Thomas Mayhew, was given a grant to the island, along with Nantucket and the Elizabeth Islands, from King Charles of England. Mayhew's son, Thomas Jr., founded the first European settlement here in 1642 at Edgartown, finding the resident Wampanoags good neighbors. Among other survival skills, they taught the settlers to kill whales on shore. When moved out to sea, this practice would bring the island great prosperity, for a time. Historians estimate a Wampanoag population of 3,000 upon Mayhew's arrival. Today there are approximately 300. The tribe is now working hard to reclaim and perpetuate its cultural identity, and it has managed to take back ancestral lands in the town of Gay Head.

Europeans settled as a community of farmers and fishermen, and both occupations continue to flourish. In the early 1800s, the island made a decided shift to whaling as the basis of its economy. Never as influential as Nantucket or New Bedford, Martha's Vineyard nonetheless held its own, and many of its whaling masters returned home wealthy men. Especially during the industry's golden age, between 1830 and 1845, captains built impressive homes with their profits, and these, along with many graceful houses from earlier centuries, still line the streets of Vineyard Haven and Edgartown, both former whaling towns. The industry went into decline after the Civil War, but by then revenue from tourism picked up, and those dollars just keep flooding in.

The story of the Vineyard's development begins in 1835, when the first Methodist Camp Meeting—a two-week gathering of far-flung parishes for group worship and a healthy dose of fun—was held in the Oak Bluffs area, barely populated at the time. From the original meeting's nine tents, the number grew to 250 by 1857. Little by little, returning campers built permanent platforms, arranged around the central tent of the preachers. Then the odd cottage popped up where someone would have put a tent. By 1880, Wesleyan Grove, named for Methodism's founder, John Wesley, was a community of about 500 tiny cottages. Lacy filigree insets began to appear on the facades, and the ornamentation came to be known as Carpenter Gothic, a hybrid of Gothic Revival styles imported from Europe fitted with jigsaw-cut detail work.

Meanwhile, burgeoning numbers of cottagers coming to the island each summer helped convince speculators of its desirability as a resort destination, and in 1867 a separate secular community was laid out alongside the Camp Ground. Steamers from New Bedford, Boston, New York, and elsewhere began bringing in fashionable folk for the bathing and the sea air, for picking berries or playing croquet. Grand hotels sprung up around Oak Bluffs Harbor. A railroad followed, connecting the town with the beach at Katama. The Victorian seaside resort became known as Cottage City. Later, the name was changed to Oak Bluffs.

More than 300 of the Camp Ground cottages remain. And just as Edgartown and Vineyard Haven attest to their origins as whaling ports, so Oak Bluffs—with its porch-wrapped beach houses, and its village green and gazebo, where families still gather to hear the town band play—evokes the days of Victorian summer ease, of flowing white dresses and parasols held languidly against the sun.

Far less developed than Cape Cod—thanks to a few conservation organizations—yet more cosmopolitan than neighboring Nantucket,

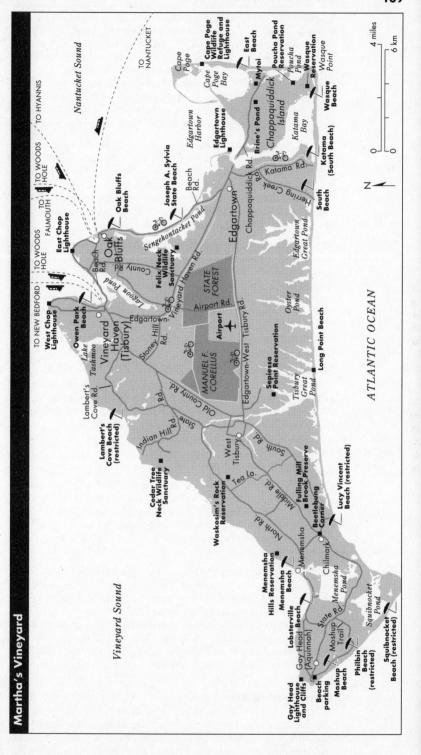

Martha's Vineyard is an island with a double life. From Memorial Day through Labor Day the quieter, some might say real, Vineyard is shaken into a vibrant, star-studded place. Edgartown is flooded with seekers of chic who've come to wander the tiny streets lined with boutiques, stately whaling captains' homes, and charming inns. The busy main port, Vineyard Haven, welcomes day-trippers fresh off a ferry or private yacht to browse in its own array of shops. Oak Bluffs, where pizza and ice-cream emporiums reign supreme, has a boardwalk-town air and several nightspots that cater to high-spirited, carefree youth.

Summer regulars include a host of celebrities, among them William Styron, Art Buchwald, Walter Cronkite, Beverly Sills, Patricia Neal, Spike Lee, and Sharon Stone. President Clinton and his wife Hillary have also returned here often since he took up office in Washington. Concerts, theater and dance performances, and lecture series draw top talent to the island, while a county agricultural fair, weekly farmers' markets, and miles of walking trails provide earthier pleasures.

Most people know this summer persona, but in many ways the Vineyard's other self is even more appealing, for the off-season island becomes a place of peace and simple beauty. On drives along country lanes through the agricultural center of the island, there's time to linger over pastoral and ocean vistas without deference to a throng of other cars, bicycles, and mopeds. In nature reserves, the voices of summer are gone, leaving only the sounds of birdsong and the crackle of leaves underfoot. Private beaches open to the public, and the water sparkles under crisp, blue skies.

Locals are at their convivial best off season. After the craziness of their short moneymaking months, they reestablish contact with friends and take up pastimes previously crowded out by work. The result for visitors—besides the extra dose of friendliness—is that cultural, educational, and recreational events continue year-round.

## Pleasures and Pastimes

### Beaches

On the Vineyard's south shore, the Atlantic Ocean side, surf crashes and the water is chillingly refreshing—a great place for bodysurfing. Beaches on the Nantucket or Vineyard sounds tend to be protected, a little warmer and calmer, perfect for swimming and for families. A few freshwater beaches offer a change of pace from the salty sea.

### Bicycling

Martha's Vineyard is a great place for cycling. Up-Island roads do get hilly, and during summer and fall the roads can be very crowded. Still, cycling is a pleasant and often more practical way to tour the island. There are well-maintained, flat paved paths along the coast road from Oak Bluffs to Edgartown—very scenic—and inland from Vineyard Haven to Edgartown to South Beach. These connect with paths—potholed in some places—that weave through the State Forest. Middle Road in Chilmark is a lovely, winding country road with less traffic than the main roads.

### Conservation Areas

There are a great many conservation areas on the Vineyard in which you can engage that sense of biophilia (love of nature) that we all have. Many are of the do-it-yourself variety, with beautiful walking trails meandering through a variety of habitats, while others offer bird walks or special kids' programs.

## Dining

Upscale, downscale, flamboyant and funky: The Vineyard offers an amazing variety of culinary choices, without the wash of pretense that increasingly characterizes neighboring Nantucket's dining scene. In general, the most interesting and varied cooking is going on in and around Oak Bluffs, although Edgartown continues to dominate the haute side of things, and Down-Island there are wonderful little outposts that feel like the food equivalent of a secluded, lovely beach.

It seems impossible in this generation, but the rumors are true: Much of the Vineyard is dry. That means no liquor stores, and no liquor service in restaurants, in the towns of Vineyard Haven, West Tisbury, Chilmark, Aquinnah (Gay Head), and Menemsha. Oak Bluffs and Edgartown are "wet," and their economies benefit from it. Virtually all restaurants in the "dry" towns allow you to bring your own beer or wine, and many of them now charge a "set up" or "corkage" fee of as little as one dollar per table, or as much as three dollars per bottle. Many travelers keep a cooler handy, and stop at a small package store like Our Market in Oak Bluffs as they begin their tour to Vineyard Haven and beyond.

If all this sounds quaint, rest assured that, foodwise, the Vineyard is no typical backwater. Celebrities and high-powered executives may come here to relax in their cutoffs and sneakers, but *no one* takes a vacation without packing their appetite. You don't have to leave this island to sample the cuisines of France, Italy, New Orleans, or New York. And if for some strange reason you want to eat New England seafood, you'll find plenty of it. Locals complain that much of the fish for sale these days is more likely to come from a New Bedford trawler than a local dayboat, but what reaches the plate still ranks among the best and freshest in the country.

**Lunches and Clambakes.** If you feel no need to make it yourself, box or picnic lunches are available at **Vineyard Gourmet** (⊠ Main St., Vineyard Haven, ☎ 508/693–5181). **Bill Smith** (☎ 508/627–8809 or 800/828–6936) prepares clambakes to go, and **New England Clambake Co.** (☎ 508/627–7462) will cater them for parties of 10 or more.

For price ranges, *see* dining Chart A in On the Road with Fodor's.

## Fishing

Huge trawlers unload their daily catches at the docks in Vineyard Haven and Menemsha, attesting to the richness of the waters surrounding the island. But some of the most zealous fishing is done by amateurs— the striped bass and bluefish derby in the fall is very serious business. One of the most popular spots for sport fishermen is Wasque Point on Chappaquiddick. Another is South Beach and the jetty at the mouth of the Menemsha Basin. Striped bass and bluefish are island stars. There are several outfits that have deep-sea fishing trips if surf fishing is not your thing.

## Hiking and Walking

The nature preserves and conservation areas are laced with well-marked, scenic trails through varied terrains and ecological habitats— and the island's miles of uninterrupted beaches are perfect for stretching your legs.

## Lodging

Martha's Vineyard offers a variety of lodging places, from historical whaling captains' mansions filled with antiques to sprawling ocean-front hotels to cozy cottages in the woods. When choosing your accommodations, keep in mind that each town has a different "personality": Oak

Bluffs tends to cater to a younger, active, nightlife-oriented crowd while Edgartown is more proper and reserved. Chilmark has beautiful beaches and miles of conservation lands, but not much of a downtown shopping area.

For price ranges, *see* the lodging chart in On the Road with Fodor's.

## Shopping

A specialty of the island is wampum—black, white, or purple beads made from shells that are fashioned into jewelry and sold at the cliffs and elsewhere. Antique and new scrimshaw jewelry, and jewelry incorporating Vineyard and island-specific designs such as a bunch of grapes or lighthouses, are also popular. The ultra-expensive Nantucket lightship baskets can be found at island shops. Many Vineyard shops close for the winter, though quite a few in Vineyard Haven and a few elsewhere remain open—call ahead before making a special trip.

The three main towns have the largest concentrations of shops. In Vineyard Haven, shops line Main Street. Edgartown's are clustered together within a few blocks of the dock, on Main, Summer, and Water streets. Primarily fun, casual clothing and gift shops crowd along Circuit Avenue in Oak Bluffs. At Gay Head Cliffs, you'll find touristy Native American crafts and souvenirs in season.

The West Tisbury Farmers' Market, the largest farmers' market in Massachusetts, offers a different, more earthy sort of shopping experience, and some of the antique stores hidden along the back roads are full of the interesting and the unusual.

## Water Sports

Martha's Vineyard is an ideal place for windsurfing. With the many bays and inlets there is always a patch of protected water for neophytes. The ocean-side surf provides plenty of action for experts. Swimming, sailing, sea-kayaking and canoeing are also favorite pastimes, with both ocean and freshwater locales to choose from.

# Exploring Martha's Vineyard

The island is roughly triangular, with maximum distances of about 20 mi east to west and 10 mi north to south. The west end of the Vineyard, known as Up-Island—from the nautical expression of going "up" in degrees of longitude as you sail west—is more rural and wild than the eastern Down-Island end, comprising Vineyard Haven, Oak Bluffs, and Edgartown. Almost a quarter of the island is conservation land, and more is being acquired all the time by organizations. The Land Bank, funded by a 2% tax on real estate transactions, is a leading group, set up in order to preserve as much of the island in its natural state as is possible and practical.

*Numbers in the text correspond to numbers in the margin and on the Vineyard Haven, Oak Bluffs, Edgartown, and Up-Island maps.*

## Great Itineraries

The itineraries below cover all of the towns on the island, but you don't have to. You might want to spend a short time in Vineyard Haven before getting rural Up-Island or heading for a beach. Or you might prefer to go straight to Edgartown to stroll around looking at the abundance of pristine white houses, popping into a museum or two, and shopping. Or you may just want to have fun, in which case Oak Bluffs, its harbor scene, and nearby beaches will be the place to go. In essence, pick and choose what you'll like best from what follows.

IF YOU HAVE 2 DAYS

Start your trip in **Vineyard Haven.** Historic houses line **William Street,** and you'll find shops and eateries along Main Street. Take a quick jaunt out to **West Chop** ⑥ for a great view over Vineyard Sound from the lighthouse, then head back through town toward **Oak Bluffs** via Beach Road. Spend some time wandering the streets of the **Oak Bluffs Camp Ground** ⑩, where tightly packed pastel-painted Victorian cottages vie with one another for the fanciest gingerbread trim. Then head into the center of Oak Bluffs for a ride on the Flying Horses, the oldest continuously operating carousel in the country. Play miniature golf. Have an ice cream at Mad Martha's after lunch.

Instead of going to Oak Bluffs from Vineyard Haven, you could head straight to **Edgartown.** Take the On-Time ferry to **Chappaquiddick Island** ㉕. Visit the **Mytoi** preserve or have a picnic at Brine's Pond. If conservation areas are your thing, the **Cape Poge Wildlife Refuge** on the island is a must. You'll need a few hours on Chappaquiddick to make the visit worthwhile. If you spend the afternoon in Edgartown, take your pick of shops and interesting museums. After dinner, attend a performance at the Old Whaling Church. Spend the night in Edgartown or back in Oak Bluffs.

On the second day head all the way out to **Gay Head** ㉟, one of the most spectacular spots on the Island. Go to the lookout at the cliffs for the view, or take the boardwalk to the beach and walk back to see the cliffs and the Gay Head Light from below. In the afternoon, **West Tisbury** ㉖ has great nature preserves and a public beach on the south shore. The Field Gallery in the town center has whimsical statues set about the lawn.

You could also spend the afternoon in the fishing village of **Menemsha** ㉞. As the day comes to an end, pick up some seafood and bring it to Menemsha Beach for a sunset picnic. Spend the night Up-Island.

IF YOU HAVE 4 DAYS

Follow the two-day tour at a slower pace. If conservation areas appeal, **Felix Neck Wildlife Sanctuary** northwest of Edgartown is a great spot. Chappaquiddick conservation areas are also wonderful. Start your tour in **Vineyard Haven,** but spend two nights in **Edgartown,** with jaunts into Oak Bluffs to sample the nightlife. If schedules permit, catch a show at the Vineyard Playhouse or at the amphitheater at Tashmoo Overlook, or a performance at the Wintertide Coffeehouse.

Spend two days Up-Island as well, where the beaches are some of the best on the Vineyard. Visit one of the **West Tisbury** ㉖ farms, some of which have pony rides for kids or fruit picking, or the **Winery at Chicama Vineyards** ㉘. The **Mayhew Chapel and Indian Burial Ground** are interesting and offer an almost eerie look into the past. Because you'll have more time, don't miss the **Gay Head Cliffs** ㊳ via **Moshup Beach,** and spend some time sunning and swimming at this breathtaking spot. (A note to the modest: The beach attracts nude sunbathers; though it's illegal, the officials usually look the other way.) Spend two nights Up-Island. If you choose West Tisbury, be sure to take time on its gorgeous beach.

IF YOU HAVE 6 DAYS

With nearly a week on your hands, follow the suggestions mentioned in the four-day tour, allowing plenty of time in each place. Give yourself whole days on the beach. Take a fishing trip from one of the harbors. Bike the trails from town to town or through the State Forest. You'll have plenty of time for shopping.

Up-Island, bird-watchers can enjoy leisurely walks around **Sepiessa Point Reservation** or **Wompesket Preserve**'s wet meadow—the conservation areas here have much to offer. The **Menemsha Hills Reservation, Cedar Tree Neck Wildlife Sanctuary,** and **Waskosim's Rock Reservation** are also quite good. Get out of the car and walk around the rural towns of West Tisbury and Chilmark. Since you'll be on the island for Wednesday and/or Saturday, you won't miss the **West Tisbury Farmers' Market.** If you are a John Belushi fan, stop by the **Chilmark Cemetery** to see the booze bottles and cigarette collections left beside the rock bearing his name. (It is a memorial stone only; Belushi's real grave is unmarked to deter overzealous fans.) Travel the back roads to find some interesting, out-of-the-way antiques shops. And be sure to catch sunset from this side of the island—from the cliffs or from a Menemsha beach or its harbor.

### When to Tour Martha's Vineyard

Summer is the most popular season on the Vineyard, the time when everyone is here and everything is open and happening. The weather is perfect for all kinds of activities, and there are special events, including the Martha's Vineyard Agricultural Fair and the Edgartown Regatta. Fall is also busy, with harvest celebrations and fishing derbies. Tivoli Day, an end of summer–start of fall celebration, includes a street fair. The island does tend to curl up in winter, and many shops and restaurants close. However, for the weeks surrounding the Hanukkah–Christmas–New Year's holidays, the Vineyard puts bells on and there are all kinds of special events and celebrations, most notably in Edgartown and Vineyard Haven. Spring sees the island awakening from its slumber with a burst of flowers and garden and house tours as islanders prepare for the busy season.

# DOWN-ISLAND

The three towns that comprise Down-Island Martha's Vineyard—**Vineyard Haven, Oak Bluffs,** and **Edgartown**—are the most popular and the most populated. Here you'll find the ferry docks, the shops, and a concentration of things to see and do, and here is where the history of Martha's Vineyard is documented in the centuries-old houses and churches that line the main streets. A stroll through any one of these towns allows a look into the past while enjoying the pleasures of the present.

## Vineyard Haven (Tisbury)

*3.3 mi west of Oak Bluffs, 8 mi northwest of Edgartown by the inland route.*

Most people call this town Vineyard Haven for the name of the port where the ferry pulls in, but its official name is in fact Tisbury. Not as high-toned as Edgartown, nor as honky-tonk as Oak Bluffs, Vineyard Haven blends the past and the present with a graceful touch of the bohemian. Settled in the mid-1600s when the Island's first governor-to-be purchased rights to the land from local Wampanoags, it is the Island's busiest year-round community. Many visitors arriving by ferry dock at this port to be greeted by the bustle of the harbor and the shops that line Main Street.

➊ If you need to stock up on maps or information on the island, **Martha's Vineyard Chamber of Commerce** is a good place to get your bearings. It is around the corner from the steamship terminal (which also has an information booth) on Beach Road. ✉ *Beach Rd.,* ☎ *508/693-0085.* ⊙ *Weekdays 9–5; also Sat. 10–2 Memorial Day–Labor Day.*

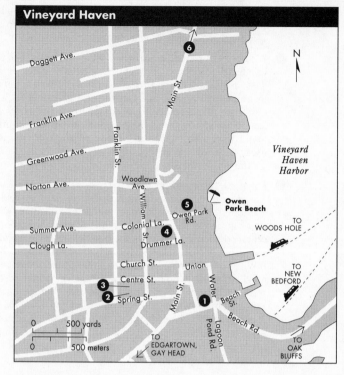

**Vineyard Haven**

Association Hall, **2**

Centre Street Cemetery, **3**

Martha's Vineyard Chamber of Commerce, **1**

Old Schoolhouse Museum, **4**

Owen Park, **5**

West Chop, **6**

NEED A BREAK?

The **Black Dog Bakery**'s delicious breads, baked goods, and quick-lunch items are simply not to be missed—it's a popular stop for good reason (☞ Dining, below). ⌧ Water St., ☎ 508/693-4786.

The stately, neoclassic 1844 **Association Hall** houses the town hall and the **Katharine Cornell Memorial Theatre**, created in part with funds that Cornell—one of America's foremost stage actresses in the 1920s, '30s, and '40s, and a longtime summer resident—donated in her will. The walls of the theater on the second floor are painted with murals depicting such island scenes as whaling and a Native American gathering, and there is a blue sky with seagulls overhead. Island artist Stan Murphy painted the murals on the occasion of the town's tercentenary in 1971. In addition to theatrical performances, concerts and dances are held here. ⌧ 51 Spring St., ☎ 508/696-4200.

The **Centre Street Cemetery** stands as a reminder of the town's past. Tall pine trees shade grave markers dating from as far back as 1817. Some stones are simple gray slate slabs, others are carved with such motifs as the death's-head—a skull, common on early tombstones. A more recent grave is that of the actress Katharine Cornell, who died in 1974 and whose largesse helped to build the theater (housed in the Association Hall, ☞ above) named for her. ⌧ Centre St.

A stroll down **William Street,** a quiet stretch of white picket fences and Greek Revival houses, many of them built for prosperous sea captains, lets you imagine the town as it was in the 19th century. Now a part of a National Historic District, the street was spared when the Great Fire of 1883 claimed much of the old whaling and fishing town.

**108 William Street,** set back on a wide lawn behind a wrought-iron fence, is an imposing monument to a later source of the town's prosperity:

tourism. The elegantly detailed three-story house was built in 1873 by Benjamin C. Cromwell—captain not of a whaling ship but rather a steamer that brought New Bedford folk to the island.

❹ Built in 1829, the **Old Schoolhouse Museum** was the first town school. Exhibits include items brought back from voyages during whaling days, including Inuit and Polynesian tools, as well as antique musical instruments, clothing, and records of 19th-century schoolchildren. Out front, the **Liberty Pole** was erected by the Daughters of the American Revolution in commemoration of three patriotic girls who blew up the town's liberty pole in 1776 to prevent it from being taken for use on a British warship. ⊠ *110 Main St.,* ☎ *508/696–7644.* 🎟 *$2.* ☉ *Mid-June–mid-Sept., Tues.–Fri. and Sun. noon–4.*

❺ For a little relaxation, head for the tree-shaded benches in **Owen Park,** a lovely spot for a picnic. In summer, band concerts are held at the bandstand. At the end of the lawn is a public beach with a swing set and a close-up view of the boats sailing in and out of the harbor. In the 19th century this harbor was one of the busiest ports in the world, welcoming thousands of coastwise vessels each year. The headlands—West Chop in Vineyard Haven and East Chop in Oak Bluffs—each came to have a lighthouse at its tip to help bring ships safely into port. Both areas were largely settled in the late-19th to early 20th centuries, when the very rich from Boston and Newport built expansive bluff-top "summer cottages." These shingle style houses, characterized by broad gable ends, dormers, and, of course, natural shingle siding that weathers to gray, were meant to eschew pretense, though they were sometimes gussied up with a turret or two.

❻ Beautiful and green, **West Chop** retains its exclusive air and can claim some of the island's most distinguished residents. An approximate 2-mi walk, drive, or bike ride along Vineyard Haven's Main Street, which becomes increasingly residential on the way, will take you there.

One of two lighthouses that mark the opening to the harbor, the 52-ft white-and-black **West Chop Lighthouse** was built in 1838 of brick to replace an 1817 wood building. It has been moved back twice from the edge of the eroding bluff. Just beyond the lighthouse, on the point, is a scenic overlook with a landscaped area and benches.

**West Chop Woods** is an 85-acre conservation area with marked walking trails through pitch pine and oak.

**Martha's Vineyard Shellfish Group** grows seed clams, scallops, and oysters to stock lagoons and beds throughout the county. From spring through fall, tours of the solar shellfish hatchery on Lagoon Pond in Vineyard Haven can be arranged with advance notice. ⊠ *Lagoon Pond,* ☎ *508/693–0391.* 🎟 *Free.* ☉ *Call for appointment.*

## Dining and Lodging

**$$$**  ✕ **Black Dog Tavern.** Time was, people went to the Black Dog to eat and to enjoy the handsome, simple dining room overlooking the harbor. Now, it seems people go there to buy T-shirts, hats, bandannas, and coffee cups, order from the Black Dog catalog, and wait in line. The logo is ubiquitous, and has bred a generation of funny knockoffs like the Bad Dog and the Dead Dog. The room hasn't changed, but locals have generally adopted the old Yogi Berra line: It's so crowded, no one goes there anymore. Regardless, prices seem to go up every year. The menu has all the usual suspects, and the truth is that the food (if that still matters) is just fine breakfast, lunch, or dinner. Even the bakery out front has gotten so popular that a bakery-café has opened down State Road (it's about a fifteen-minute walk). ⊠ *Beach St. Ext.,* ☎ *508/693–9223. Reservations not accepted. AE, D, MC, V. BYOB. No smoking.*

$$$  ✕ **Dry Town Café.** Despite its café name and the large plate-glass windows that offer a front-row seat on Main Street traffic, this little restaurant has a formal feel, imparted in large part by its chic, white interior. Thanks to its recent success, Dry Town has a more upscale menu than it had in the past. Appetizers now include such refined fare as salad with grilled marinated quail on endive and radicchio. Main courses such as the paella and the grilled lobster and saffron polenta are equally ambitious. Dry Town definitely wins the award for best restaurant logo: an inverted codfish holding a bottle with its tail, allowing the last, lonely drop of red wine in this dry town to drip into its mouth. ✉ *70 Main St.,* ☎ *508/693–1484. MC, V. BYOB. No lunch.*

$$$  ✕ **Le Grenier.** Up narrow stairs, away from the street bustle, a life-sized rendering of a stereotypical, goateed French chef greets you at the door. A truly French restaurant in a dry town is certainly a stretch, but what is served here is authentic, expert, and always aims for the Continental and classic. For frogs' legs, sweetbreads, tournedos, and calves' brains, Le Grenier is the clear choice. For veal, scallops, and lobster, the preparations are not as inventive, but certainly time-honored and sophisticated. ✉ *Upper Main St.,* ☎ *508/693–4906. AE, MC, V. BYOB. No lunch.*

$  ✕ **Diodati's Restaurant & Clam Bar.** A red-and-white-striped awning is a nice invitation to grab a table on Main Street's sidewalk and have a quick bite. "Dio's" has just about everything that a food joint should, from good breakfasts to 10-inch subs to fried clams. Next door is the Get a Life coffee shop, where, if you already have a life, you can at least wake up and smell the coffee. ✉ *55 Main St.,* ☎ *508/696–7448. Reservations not accepted. No credit cards. BYOB.*

$$$$  🏨 **Thorncroft Inn.** Set on 3½ acres of woods about 1 mi from the ferry,
★  the main inn, a 1918 Craftsman bungalow, combines fine Colonial and richly carved Renaissance Revival antiques with tasteful reproductions to create an environment that is somewhat formal but not fussy. Ten of the rooms have working fireplaces. Deluxe rooms have mini-refrigerators, and some have whirlpools or canopy beds; three rooms have two-person whirlpool baths, two have private hot-tub spas. Set apart from the main house and reached via a breezeway, the private ultra-deluxe room has a king-size bed and a whirlpool tub. Owners Karl and Lynn Buder serve gourmet breakfasts in two seatings as well as afternoon tea. ✉ *460 Main St., Box 1022, 02568,* ☎ *508/693–3333 or 800/332–1236,* 🖷 *508/693–5419. 13 rooms. No smoking. AE, D, DC, MC, V. BP.*

$$$  🏨 **Tisbury Inn.** At the center of the shopping district, this hotel—dating back to 1794—offers tiled bathrooms with tub showers, firm beds, and amenities that include a well-equipped health club, cable TV/HBO, room phones, and ceiling fans. Pastel colors, floral fabrics, and simple decor lend the rooms a pleasant, islandy feel; on the down side, they are small and those facing the main street tend to be noisy. In summer, stay three nights early in the week and the fourth is free. ✉ *9 Main St., Box 428, 02568,* ☎ *508/693–2200 or 800/332–4112,* 🖷 *508/ 693–4095. 31 rooms, 4 suites. Restaurant, indoor pool, health club. AE, D, DC, MC, V. CP.*

$$  🏨 **The Hanover House.** Set on a half-acre of beautifully landscaped lawn within walking distance of the ferry, this charming inn offers spotlessly clean rooms decorated in casual country style with a combination of antiques and reproduction furniture. Each room has an individual decorative flair, some with floral wallpaper and quilts, others tastefully sponge-painted and filled with whimsical, creative furnishings, such as an antique sewing machine that serves as a TV stand. All have a queen or two double beds, and some have private entrances that open

onto one of two spacious sundecks. Three suites in the separate carriage house are roomy, with private decks or patios; two have kitchenettes. The owners attend to every detail, including touches like Hemingway paperbacks on night tables. Homemade breads and muffins and a special house cereal are served each morning on a lovely sunporch with fresh flowers from the gardens. ⊠ *28 Edgartown Rd., Box 2107, 02568,* ☎ *508/693–1066 or 800/339–1066,* FAX *508/696–6099. 12 rooms, 3 suites. No pets. AE, D, MC, V. Closed Dec.–Mar. CP.*

$ ▥ **Captain Dexter House.** An 1843 sea captain's house at the edge of the shopping district is the setting for this intimate B&B. Small guest rooms are beautifully appointed with period-style wallpapers, velvet wingback chairs, and 18th-century antiques and reproductions, including several four-poster canopy beds with lace or fishnet canopies and hand-sewn quilts. The Captain Harding Room is larger, with the original wood floor, fireplace, bay windows, canopy bed, desk, and bright bath with claw-foot tub. The charming, helpful innkeepers tell wonderful anecdotes about the history of the house. There is a common refrigerator and TV. Afternoon tea and evening sherry are available. ⊠ *92 Main St., Box 2457, 02568,* ☎ *508/693–6564.*FAX *508/693–8448. 7 rooms, 1 suite. No pets. No smoking. AE, MC, V. Closed Dec.–Apr. CP.*

$ ⛺ **Martha's Vineyard Family Campground.** Wooded sites, tent-trailer
★ rentals, a recreation hall, ball field, playground, camp store, bicycle rentals, and electrical and water hookups are among the facilities here. No dogs or motorcycles are allowed. There are also three cabins with electricity, refrigerators, and gas grills. ⊠ *569 Edgartown–Vineyard Haven Rd., Box 1557, 02568,* ☎ *508/693–3772,* FAX *508/693–5767. Picnic area. Closed mid-Oct.–mid-May.*

## Nightlife and the Arts

The **Vineyard Playhouse** (⊠ 24 Church St., ☎ 508/693–6450 or 508/696–6300) has a year-round schedule of community theater and Equity productions. Mid-June through early September, a mostly Equity troupe performs drama, classics, and comedies on the air-conditioned main stage, and summer Shakespeare and other productions at the natural amphitheater at Tashmoo Overlook on State Road in Vineyard Haven (bring insect repellent and a pillow). Children's programs and summer camp are also featured. Local art exhibitions are held throughout the year. One performance of each summer main-stage show is interpreted in American sign language.

**Island Theater Workshop** (☎ 508/693–5290), the island's oldest year-round company, performs at various venues.

**Wintertide Coffeehouse** (⊠ Five Corners, Vineyard Haven, ☎ 508/693–8830) has live folk, blues, jazz, and other music featuring local and national talent, including open-mike nights, in a homey alcohol- and smoke-free environment. Light meals, desserts, and freshly ground coffees and cappuccino are served at candlelit tables year-round.

Sunday night **Vineyard Haven Town Band concerts** take place on alternate weeks in summer at 8 PM at Owen Park in Vineyard Haven and at the gazebo in Ocean Park on Beach Road in Oak Bluffs.

## Outdoor Activities and Sports

### BEACHES

**Lake Tashmoo Town Beach,** at the end of Herring Creek Road in Vineyard Haven, has swimming in the warm, relatively shallow brackish lake or in the cooler, gentle Vineyard Sound. There is a lifeguarded area and some parking.

**Owen Park Beach,** a small, sandy harbor beach off Main Street in Vineyard Haven, is a convenient spot with a children's play area, lifeguards, and a harbor view.

### BIKING

**Martha's Bike Rental** (✉ Five Corners, ☎ 508/693–6593) rents bicycles for $10 per day. They also do repairs.

**Martha's Vineyard Scooter and Bike Rental** (✉ 24 Union St., ☎ 508/693–0782) rents scooters and a variety of bicycles. They also do repairs.

### BOATING

**Wind's Up!** (✉ 199 Beach Rd., ☎ 508/693–4252) rents day sailers, catamarans, surfboards, sea kayaks, canoes, and Sunfish, as well as Windsurfers and boogie boards, and offers lessons.

### BOWLING

**Spinnaker Lanes** (✉ State Rd., ☎ 508/693–9691) offers 12 lanes of candlepin bowling and six tournament pool tables. Lanes are wheelchair-accessible.

### GOLF

The semi-private **Mink Meadows Golf Course** (✉ Golf Club Rd. at Franklin St., ☎ 508/693–0600), on West Chop, has nine holes and ocean views. Reservations must be made 48 hours in advance.

**Island Cove Mini Golf** (✉ State Rd., ☎ 508/693–2611) offers an 18-hole course featuring bridges, a cave, rocklike obstacles, sand traps, and a stream that powers a water mill.

### HEALTH AND FITNESS CLUBS

The **Health Club at the Tisbury Inn** (✉ 9 Main St., Vineyard Haven, ☎ 508/693–7400) has Lifecycle, Nautilus, Universal, and StairMaster machines, bikes, free-weight rooms, tanning facilities, aerobics classes, and personal trainers. There are also a large heated pool, a hot tub, and a sauna. Day memberships are available.

### TENNIS

Tennis is very popular on the island, and at all times reservations are strongly recommended. Public clay courts are on Church Street in Vineyard Haven; they're open in season only, and a small fee is charged (reserve a court with the attendant the previous day).

## Shopping

**All Things Oriental** (✉ 123 Beach Rd., ☎ 508/693–8375) has jewelry, porcelains, paintings, furniture, and more.

**Black Dog General Store** (✉ Between the Black Dog Tavern and the Bakery on Water St., ☎ 508/696–8182) sells T-shirts, sweatshirts, beach towels, and many other gift items, all emblazoned with the telltale Black Dog.

**Bramhall & Dunn** (✉ 19 Main St., ☎ 508/693–6437) carries 19th-century English country pine furniture, as well as fine crafts, linens, and housewares, and hand-knit sweaters and original women's accessories and shoes.

**Brickman's** (✉ 8 Main St., ☎ 508/693–0047) sells beach and sports gear, such as camping, fishing, and snorkeling equipment and boogie boards. It also carries sportswear, surfer-type clothing, and major-label footwear for the family.

**Bunch of Grapes Bookstore** (✉ 68 Main St., ☎ 508/693–2291) carries a wide selection of new books, including many island-related titles, and sponsor book signings—watch the papers for announcements.

**C. B. Stark Jewelers** (⊠ 126 Main St., Vineyard Haven, ☎ 508/693–2284) creates one-of-a-kind pieces for both women and men, including island charms. It also carries other fine jewelry and watches.

**Crispin's Landing** (⊠ 80 Main St., ☎ 508/693–6758) has shops selling leather, pottery, jewelry, and crafts.

**Cronig's Market** (⊠ 109 State Rd., ☎ 508/693–4457) supports local farmers and specialty item producers—and carries whatever food items you might need.

**C. W. Morgan Marine Antiques** (⊠ Beach Rd., next to MV Shipyard, ☎ 508/693–3622) offers a wide range of museum-quality nautical items—instruments, sea chests, ship models, paintings, prints, and so forth.

**Island Children—Althea and Emily Designs** (⊠ 94 Main St., ☎ 508/693–6130) has children's and women's clothing in 100% cotton, hand-block-printed with unique African- and Caribbean-inspired and classic designs.

**Lorraine Parish** (⊠ 18 S. Main St., ☎ 508/693–9044) sells sophisticated, upscale women's dresses (some with a summery island feel), suits, and blouses in natural fibers, all designed by Lorraine Parish, the designer of Carly Simon's wedding dress.

**Murray's of the Vineyard** (⊠ 72 Main St., ☎ 508/693–2640), sister shop of Nantucket's Murray's Toggery, has classic men's and women's fashions, shoes, and accessories, from such names as Ralph Lauren and Liz Claiborne.

**Paper Tiger** (⊠ 29 Main St., ☎ 508/693–8970) carries a wide array of handmade paper, cards, and gift items, plus wind chimes, pottery, writing utensils, and works by local artists.

**Pyewacket's** (⊠ 135 Beach Rd., ☎ 508/696–7766) carries an interesting conglomeration of antiques, island crafts, handmade soaps and candles, and jewelry.

**Rainy Day** (⊠ 86 Main St., ☎ 508/693–1830) is the ultimate home store, with high-quality cookware and home furnishings, including reasonably priced kilim rugs.

**Sioux Eagle Designs** (⊠ 29 Main St., ☎ 508/693–6537) sells unusual handmade pieces of jewelry from around the world.

**Union St. Clothing** (⊠ 180 Union St., ☎ 508/693–4534) is a boutique filled with fun and funky clothes by Kiko and Esprit as well as hats and jewelry.

**Wind's Up!** (⊠ 199 Beach Rd., ☎ 508/693–4340) sells swimwear, windsurfing and sailing equipment, boogie boards, and other outdoor gear.

## Oak Bluffs

*6 mi northwest of Edgartown, 22 mi northeast of Gay Head.*

Purchased from the Indians in the 1660s, Oak Bluffs was a farming community that did not come into its own until the Methodists began holding summer revivalist camp meetings in a grove of oaks known as Wesleyan Grove. As the camp meetings gained attendees, small cottages were built in place of tents, then the general population took notice and the area became a popular summer vacation spot. Hotels, a dance hall, a roller-skating rink and other shops and amusements were built to accommodate the flocks of summer visitors.

Circuit Avenue is the bustling center of today's Oak Bluffs action, with most of the town's shops, bars, and restaurants. Oak Bluffs Harbor, once the setting for a number of grand hotels—the 1879 Wesley Hotel on Lake Avenue is the last of them—is still colorful with its gingerbread-trimmed guest houses and food and souvenir joints.

**❼ East Chop** is one of two points of land that jut out into the Nantucket–Vineyard sound, creating the sheltered harbor at Vineyard Haven.

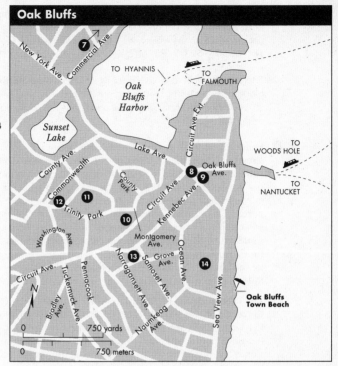

You can loop out to the point on your way from Vineyard Haven by taking Highland Drive off Beach Road after crossing the drawbridge.

The **East Chop Lighthouse** was built of cast iron in 1876 to replace an 1828 tower—used as part of a semaphore system of visual signaling between the island and Boston—that burned down. The 40-ft tower stands high atop a bluff with spectacular views of Nantucket Sound.

**⑧** The **information booth** (⌧ At Lake and Oak Bluffs Aves.) will help you get your bearings and point the way to the not-to-be-missed spots in Oak Bluffs, with some good tips for gingerbread-trim lovers.

**✋ ⑨** The **Flying Horses Carousel** is the nation's oldest continuously operating carousel, and a National Historic Landmark. The carousel was handcrafted in 1876 (the horses have real horse hair and glass eyes) and extensively renovated in 1992, and offers children entertainment from a Nintendo-free time. While waiting in line, you can munch on popcorn or cotton candy or slurp a slush. In the waiting area there are a number of arcade games. ⌧ *Oak Bluffs Ave.*, ☎ *508/693–9481.* 🖂 *Rides: $1; $8 for a book of 10.* ☺ *mid-June–Labor Day, daily 9:30 AM–10 PM; spring and fall, weekends only; closed mid-Oct.–Easter.*

In summer Oak Bluffs has all kinds of attractions to keep kids entertained. The **Game Room,** in a large white building across the street from the Flying Horses Carousel, has 75 arcade games, as well as electric bumper cars, air hockey, and Skee-ball. ⌧ *Oak Bluffs Ave.*, ☎ *508/693–5163.*

✋ At **Dockside Minigolf,** each of the 18 holes—half indoors, half in the open—has an island motif, such as a ferryboat, a lighthouse, or a gingerbread house. ⌧ *Upstairs at Dockside Marketplace, Oak Bluffs Harbor,* ☎ *508/693–3392.*

NEED A
BREAK?

**The Coop de Ville** is often teeming with people eager to sample the delectables from the raw bar and the simple fried seafood. Eat out on the patio deck overlooking the water—the oysters are fantastic. ⊠ *Oak Bluffs Harbor,* ☎ *508/693-3420. MC, V.* ☺ *May–Columbus Day, daily 11–10.*

Some like it sweet, and with all of its ice cream **Mad Martha's** is just the place for sweet teeth. There's a great jukebox, too. ⊠ *117 Circuit Ave.,* ☎ *508/693-9151.* ☺ *May–June and Sept.–Oct., daily noon–9; July–Aug., daily 11 AM–midnight.*

★ ⑩ Don't miss a look at **Oak Bluffs Camp Ground,** a 34-acre warren of streets tightly packed with more than 300 Carpenter Gothic Victorian cottages gaily painted in pastels with wedding-cake trim. As you wander through this fairy-tale setting, imagine it at night, lit by the warm glow of hundreds of Japanese paper lanterns hung from every cottage porch. This is what happens each summer on Illumination Night, when the end of the Camp Meeting season—attended these days by some fourth- and fifth-generation cottagers—is marked as it has been for more than a century, with lights, singing, and open houses for families and friends. Note that because of overwhelming crowds of onlookers in seasons past, the date is not announced until the week before.

⑪ The **Tabernacle,** an impressive open-air structure of iron and wood at the center of Trinity Park, is the original site of the Methodist services. On Wednesdays at 8 PM in season, visitors are invited to join in on an old-time community sing-along. If you know tunes like "The Erie Canal" or just want to listen in, drop by the Tabernacle and take a seat. Also, music books are available for a donation. Sunday services are held in summer at 9:30 AM. The 1878 **Trinity Methodist Church** is also in the park.

⑫ For a glimpse at life in Cottage City during its heyday, visit the **Cottage Museum,** in an 1867 Creamsicle cottage near the Tabernacle. The museum exhibits cottage furnishings from the early days, including photographs, hooked rugs, quilts, and old bibles. The gift shop offers Victorian and nautical items. ⊠ *1 Trinity Park,* ☎ *508/693-0525.* ⊠ *$1 donation requested.* ☺ *Mid-June–Sept., Mon.–Sat. 10–4.*

Worth at least a glance, the **Wooden Valentine** is quite a sight—just think pink. ⊠ *25 Washington Ave.*

⑬ An octagonal, nonsectarian house of worship, **Union Chapel** was constructed in 1870 for the Cottage City resort folk who lived outside the Camp Ground's 7-ft-high fence. In summer, concerts are held here, as are 10 AM Sunday services. ⊠ *Corner of Kennebec and Samoset Aves.,* ☎ *508/693-1093.*

⑭ A long stretch of green facing the sea, **Ocean Park** is half circled by a crescent of large shingle-style cottages with numerous turrets, breezy porches, and pastel facades. Band concerts take place at the gazebo here on summer nights, and in August the park hosts hordes of island families and visitors for a grand fireworks display over the ocean. ⊠ *Sea View Avenue.*

## Dining and Lodging

$$$–$$$$  ✕ **Sweet Life Café.** Aptly named, this lilting, lovely new arrival has
★ quickly become an island favorite. Its subdued, intimate interior, with its warm tones and low lighting, will make you feel like you've entered someone's home. An appetizer of Maine crab over young green beans with a lemon butter sauce is very appealing, if a bit skimpy. But the main courses more than make up for it, particularly a roasted red pepper cod that is served with a fresh corn polenta and sweet peas. The

desserts are all homemade, from gingerbread to lemon ice cream. La dolce vita—and a meal here will confirm it. ⊠ *Upper Circuit Ave., at the far end of town,* ☎ *508/696–0200. Reservations essential. AE, D, MC, V. Closed Jan.–Mar.*

**$$$**    ✕ **Lola's.** "Where Oak Bluffs meets Edgartown," jokes Lola Domitrovich, and in a way that's true. On the beach road between the towns, Lola's actually is closer to Oak Bluffs in every respect. Boisterous, open 364 days a year (closed only on Christmas), with a huge kitchen cranking out Louisiana-style fare, this is the spot for the Mardi Gras crowd. Neither of the two signature dishes, étouffée and jambalaya, is particularly inspired, but both are generous and boast huge crayfish flown up, says Lola, straight from Bayou Country. The wine list is strong on America, the bar is big and welcoming, the wall mural full of familiar local faces. Year-round residents very much appreciate the live music into the dead of winter. ⊠ *Beach Rd., just over 1 mi from Oak Bluffs,* ☎ *508/693-5007. D, MC, V.*

**$$–$$$**    ✕ **Zapotec.** Southwest gingerbread describes the decor, with plenty of color, flowerboxes and Santa Fe funk. The food follows suit: corn tortillas, grilled chicken breast with chili sauce, and of course nachos and quesadillas. The steamed mussels with *chipotle* chili (a dried, smoked jalapeño), cilantro, lime, and cream breaks the mold a bit and is a favorite appetizer. There is a nice, simple, cheaper children's menu, which is a welcome sight about a week into the vacation. ⊠ *10 Kennebec Ave.,* ☎ *508/693–6800. Reservations not accepted. AE, MC, V. Closed mid-Oct.–late Apr.*

**$$**    ✕ **City, Ale and Oyster.** Architect Robert Skydell opened the Vineyard's first brew pub in the summer of 1997 (he also owns the Dry Town Café; ☞ Dining and Lodging in Vineyard Haven, *above*). The beer is made according to traditional German style; brews include the Oak Bluffberry, made with pure clover honey, and a Hazelnut Porter. As the name suggests, oysters get pride of place, but plenty of other fresh, local seafood is also on the menu, as are pizzas, burgers, fried chicken, and steak. ⊠ *30 Kennebec Ave.,* ☎ *508/693–2626. Reservations not accepted. AE, MC, V.*

**$$**    ✕ **Smoke 'n Bones.** Opened in 1997, this is the island's only rib joint, with a real-life smoker out back and a cord of hickory, apple, oak, and mesquite wood stacked up around the lot. Co-owner Ed Jigarjian makes no bones about his hope to franchise, and the place has a cookie-cutter, pre-fab feeling, with all the appropriate touches like neon flames around the kitchen and marble bones for doorknobs. But it's fun, with details kids can really enjoy, like a hole in each tabletop for a bucket to hold discarded ribs. The allusions to pot-smoking don't seem bother anybody, the beer is cold and the mugs are big (although there is no bar), and, like in any good rib spot, there is a thriving takeout business. Like the menu says, Bone appetit. ⊠ *Siloam Rd., about 7 blocks from Oak Bluffs,* ☎ *508/696–7427 (RIBS). Reservations not accepted. No credit cards.*

**$**    ✕ **Café Luna.** Right on the marina in Oak Bluffs, Café Luna is a great
★    late-night spot, one of the few places open on the island till midnight. It has scaled up its menu with an exciting Italian antipasto (small and large sizes available) to go along with their excellent pizzas and focaccia sandwiches. Upstairs is the place for tapas, dessert, and coffee; downstairs has more seating and is more formal, but you won't go wrong on either floor. ⊠ *Oak Bluffs Harbor,* ☎ *508/693–8078. Reservations not accepted. D, MC, V.*

**$**    ✕ **Giordano's.** Consistently good, affordable food is what this spot's all about. Portions are notoriously huge and traditionally Italian, with fried fish and other seafood thrown in for good measure. A great place to dine with kids, Giordano's has been family-run for nearly 70 years and has a sizable children's menu. The summertime wait is long but at least entertaining, thanks to its downtown corner location and the brightness and noisiness of it all. ☒ *107 Circuit Ave.,* ☎ *508/693–0184. Reservations not accepted. No credit cards. Closed late-Sept.– early June.*

$    ✕ **Linda Jean's.** A classic local hangout. Want to eat breakfast at 6 AM?
★    No problem. Want to eat breakfast at 11:30 AM? No problem. Tired of the gourmet world and want something close to good diner food, with comfortable booths, friendly waitresses, and few frills? No problem. The only problem? You may have to wait. ☒ *34 Circuit Ave.,* ☎ *508/693–4093. Reservations not accepted. No credit cards.*

$$$    ▦ **Oak House.** The wraparound veranda of this pastel-painted, 1872
★    Victorian looks across a busy street to the beach. Several rooms have private terraces; if you're bothered by noise, ask for a room at the back. Inside, the reason for the inn's name becomes clear: Everywhere you look, you see richly patinated oak, in ceilings, wall paneling, wainscoting, and furnishings. All this well-preserved wood makes a solid backdrop for the choice antique furniture and nautical-theme accessories. An elegant afternoon tea with cakes and cookies, baked by the innkeeper, Betsi Luce, a Cordon Bleu–trained pastry chef, is served in a glassed-in sunporch—with its white wicker, plants, floral print pillows, and original stained-glass window accents, a lovely place to while away the end of the day. ☒ *Sea View Ave., Box 299, 02557,* ☎ *508/693–4187,* FAX *508/696–7385. 8 rooms, 2 suites. AE, D, MC, V. Closed mid-Oct.–mid-May. CP.*

$$    ▦ **Admiral Benbow Inn.** On a busy road between Vineyard Haven and Oak Bluffs harbor, the Benbow is in need of a little sprucing up, but is nonetheless endearing in its funkiness. The small, homey B&B was built for a minister at the turn of the century, and it is decked out with elaborate woodwork, a comfortable hodgepodge of antique furnishings, and a Victorian parlor with a stunning tile-and-carved-wood fireplace. Joyce Dodge, who took over the inn in 1996, serves a Continental-plus breakfast that includes seasonal fruit, yogurt, and homemade cereal. ☒ *520 New York Ave., Box 2488, 02557,* ☎ *508/693–6825. 6 rooms. No smoking. AE, D, MC, V. CP.*

$$    ▦ **Martha's Vineyard Surfside Motel.** These two buildings, the newest built in 1989, are right in the thick of things, where it tends to get noisy in summer. Rooms are spacious, bright (corner rooms more so), and well maintained, each with typical motel furnishings, carpeting, and table and chairs. Deluxe rooms have water views. Four new suites are nicely decorated and have Jacuzzi baths. Two rooms are wheelchair-accessible. ☒ *Oak Bluffs Ave., Box 2507, 02557,* ☎ *508/693–2500 or 800/537–3007,* FAX *508/693–7343. 34 rooms, 6 suites. Hot tub. AE, D, MC, V.*

$$    ▦ **Sea Spray Inn.** In 1989, artist and art restorer Rayeanne King con-
★    verted her Victorian, porch-wrapped summer house into a B&B, and it still *feels* like a summer house. It's set on a quiet drive circling an open park that borders an ocean beach; public tennis and golf are within walking distance. The decor is simple and restful. In the Honeymoon Suite (one room), an iron-and-brass bed is positioned for viewing the sunrise through bay windows draped in lacy curtains, and the cedar-lined bath includes an extra-large shower. The Garden Room has a king-size bed with gauze canopy and a private enclosed porch. The common living room is large and airy. ☒ *2 Nashawena Park, Box 2125, 02557,* ☎ *508/693–9388. 7 rooms. No smoking. MC, V. Closed mid-Nov.–mid-Apr. CP.*

$    ▦ **Attleboro House.** This guest house across from bustling Oak Bluffs harbor is a big 1874 gingerbread Victorian with wraparound verandas on two floors. It has small, simple rooms, some with sinks, powder-blue walls, lacy white curtains, and a few antiques. Singles have three-quarter beds. Linen exchange but no chambermaid service is provided during a stay. The shared baths are rustic and old but clean. ⊠ *42 Lake Ave., Box 1564, 02557, ☎ 508/693–4346. 9 rooms share 5 baths. MC, V. Closed Oct.–mid-May. CP.*

**$** △ **Webb's Camping Area.** Set on 84 acres, this campground is woodsy and private, with some water-view sites and a store. Swimming is permitted in Lagoon Pond, and the campground offers bathrooms, showers, laundry facilities, playgrounds, and RV hookups. ⊠ *Barnes Rd., R.F.D. 3, Box 100, 02568, ☎ 508/693–0233. MC, V. Closed day after Labor Day–mid-May.*

## Nightlife and the Arts

**Atlantic Connection** (⊠ 124 Circuit Ave., ☎ 508/693–7129) offers fancy light and sound systems (including a strobe-lit dance floor topped by a glitter ball) and live reggae, R&B, funk, and blues.

At **Island House** (⊠ 118–120 Circuit Ave., ☎ 508/693–4516), renowned pianist David Crohan entertains with popular and classical music throughout dinner in season. Other musical guests perform nightly in the lounge, where a light menu is served.

The **Ritz Café** (⊠ 109 Circuit Ave., ☎ 508/693–9851), a popular bar with a pool table (off-season) and a jukebox, has live blues and jazz every weekend, more often in season. The **Lamppost** and the **Rare Duck** (⊠ 111 Circuit Ave., ☎ 508/696–9352) are popular (and rowdy) spots with younger crowds.

**The Tabernacle** is the scene of a popular Wednesday evening community sing-along at 8, as well as other family-oriented entertainment. For a schedule, contact the **Camp Meeting Association** (⊠ Box 1176, Oak Bluffs 02557, ☎ 508/693–0525).

Sunday night summer **town band concerts** alternate weeks between the gazebo in Ocean Park on Beach Road in Oak Bluffs and Owen Park in Vineyard Haven.

## Outdoor Activities and Sports

BEACHES

**Joseph A. Sylvia State Beach,** between Oak Bluffs and Edgartown, is a 6-mi-long sandy beach with a view of Cape Cod across Nantucket Sound. The calm, warm water and food vendors make it popular with families. There's parking along the roadside, and the beach is accessible by bike path or shuttle bus.

**Oak Bluffs Town Beach,** between the steamship dock and the state beach, is a crowded, narrow stretch of calm water on Nantucket Sound, with snack joints, lifeguards, parking, and rest rooms at the steamship office.

BIKING

**Anderson's** (⊠ Circuit Ave. Ext., ☎ 508/693–9346), on the harbor, rents several different styles of bicycle.

**DeBettencourt's** (⊠ Circuit Ave. Ext., ☎ 508/693–0011) rents bikes, mopeds, scooters and Jeeps.

**King's Rentals** (⊠ Circuit Ave. Ext., ☎ 508/693–1887) rents several different types of bicycle, as well as scooters and mopeds.

**Ride-On Mopeds** (⊠ Circuit Ave. Ext., ☎ 508/693–2076) rents bicycles and mopeds.

**Sun 'n' Fun** (⊠ Lake Ave., ☎ 508/693–5457) rents bikes, cars, and Jeeps.

FISHING

**Dick's Bait and Tackle** (⊠ New York Ave., ☎ 508/693–7669) rents gear, sells accessories and bait, and has a copy of the current listing of fishing regulations.

The party boat **Skipper** (☎ 508/693–1238) leaves for deep-sea fishing trips out of Oak Bluffs Harbor in July and August. Reservations are mandatory.

GOLF

**Farm Neck Golf Club** (⊠ Farm Neck Way, ☎ 508/693–3057), a semi-private club on marsh-rimmed Sengekontacket Pond, has 18 holes in a championship layout and a driving range. Reservations are required 48 hours in advance.

ICE-SKATING

**Martha's Vineyard Ice Arena** (⊠ Edgartown–Vineyard Haven Rd., Oak Bluffs, ☎ 508/693–5329) is open mid-July through March.

SAILING AND BOATING

**Vineyard Boat Rentals** (⊠ Dockside Marketplace, Oak Bluffs Harbor, ☎ 508/693–8476) rents Boston Whalers, Bayliners, and Jet Skis.

SCUBA DIVING

Vineyard waters hold a number of sunken ships, among them several schooners and freighters off East Chop and Gay Head.

**Vineyard Scuba** (⊠ S. Circuit Ave., Oak Bluffs, ☎ 508/693–0288) has diving information and equipment rentals. Certification classes are also given.

TENNIS

Tennis is very popular on the island, and at all times reservations are strongly recommended. Hard-surface courts in Niantic Park in Oak Bluffs cost a small fee and are open year-round.

**Farm Neck Tennis Club** (⊠ County Rd., Oak Bluffs, ☎ 508/693–9728) is a semiprivate club with four clay courts, lessons, and a pro shop. It's open mid-April through mid-November; reservations are required.

**Island Inn Tennis Club** (⊠ Beach Rd., Oak Bluffs, ☎ 508/693–6574) has three Har-Tru courts and a pro shop; it's open May–Columbus Day.

## Shopping

**Book Den East** (⊠ New York Ave., ☎ 508/693–3946) is an amazing place to browse and buy, with 20,000 out-of-print, antiquarian, and paperback books housed in an old barn.

**Laughing Bear** (⊠ 138 Circuit Ave., ☎ 508/693–9342) has fun children's and women's wear made of Balinese or Indian batiks and other unusual materials, plus jewelry and accessories from around the world.

**Michaela Ltd. Gallery of American Crafts** (⊠ 124 Circuit Ave., ☎ 508/ 693–8408) has pottery, blown glass, jewelry, soaps, candles, baskets, woven goods, and more.

**The Secret Garden** (⊠ 148 Circuit Ave., ☎ 508/693–4759), set in a yellow gingerbread cottage, sells lace, baby gifts, wicker furniture, and prints.

**Tisbury Marketplace** (⊠ Beach Road between Oak Bluffs and Vineyard Haven) has crafts and gift shops, a toy store, a music store, a sporting-goods shop, and a pizza parlor.

# Edgartown and Chappaquiddick Island

*9.3 mi southeast of Vineyard Haven via Beach Rd., 8.6 mi east of West Tisbury.*

Edgartown has long been the Vineyard's most prestigious town. Thomas Mayhew Jr. landed here in 1642 and became the Vineyard's first governor, and the town has remained the county seat ever since. Though plenty of settlers inhabited the area, making it the Island's first colonial settlement, the town was not officially named until 1652, when it was called Great Harbour. For political reasons, the town was renamed some 30 years later, after the three-year-old son of the Duke of York.

Once a well-to-do whaling town, Edgartown has managed to preserve the elegance of that wealthy era. Lining the streets are 17th- and 18th-century sea captains' houses, ensconced in well-manicured gardens and lawns. The uniformity—of both the quality of houses originally built and their current condition—sets Edgartown apart from the two other major Vineyard towns, making it a sort of dignified museum piece. There are plenty of shops here, as well as other interests to occupy the crowds who walk the streets to see and be seen.

★ ⑮ To orient yourself historically before making your way around town, you might want to stop off at a complex of buildings and lawn exhibits that constitute the **Vineyard Museum.** The following opening hours apply to all buildings and exhibits, which are detailed below. The museum is part of the Dukes County Historical Society, and it sells an excellent Edgartown walking tour booklet ($4.95) that is full of anecdotes and history of the people who have lived in the old houses over the past three centuries. ⊠ *Cooke St., corner of School St.,* ☎ *508/627–4441.*  *Mid-June–mid-Sept.: $5; mid-Sept.–mid-June: $3.* ☉ *July–Labor Day, daily 10–4:30; Labor Day–June, Wed.–Fri. 1–4, Sat. 10–4.*

The one Vineyard Museum property open in summer only is the **Thomas Cooke House,** set in the 1765 home of a customs collector. The house itself is part of the display, including its low doorways, wide-board floors, original raised-panel woodwork with fluted pilasters, and hearths in the summer and winter kitchens. Docents conduct tours of the 12 rooms, whose exhibits explore the island's history through furniture, tools, costumes, portraits, toys, crafts, and various household objects. One room is set up as a 19th-century parlor, illustrating the opulence of the golden age of whaling with such period pieces as a pianoforte. Upstairs are ship models, whaling paraphernalia, old customs documents, and a room tracing the evolution of the Camp Meeting through photographs and objects. ⊠ *School St.,* ☎ *508/627–4441.*

The **Francis Foster Museum** houses a small collection of whaling implements, scrimshaw, navigational instruments, and lots of old photographs. One interesting exhibit is a collection of 19th-century miniature photographs of 110 Edgartown whaling masters, grouped by family. The **Gale Huntington Reference Library** is also in the building, with genealogical records, rare island books, and ships' logs from the whaling days, as well as some publications for sale. ⊠ *School St.,* ☎ *508/ 627–4441.*

The **Capt. Francis Pease House,** an 1850s Greek Revival, houses a permanent exhibit of Native American, prehistoric, pre-Columbian, and later artifacts, including arrowheads and pottery, and changing exhibits from the collection. The Children's Gallery displays changing exhibits created by children, and the museum shop sells books, maps, jewelry, and island crafts. ⊠ *School St.,* ☎ *508/627–4441.*

The **Carriage Shed** displays a number of vessels and vehicles, including a whaleboat, a snazzy 1855 fire engine with stars inlaid in wood,

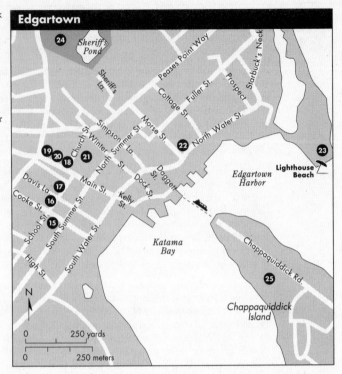

**Edgartown**

and an 1830 hearse, considerably less ornate than the fire engine, and rightly so. The shed also houses some peculiar gravestones that mark the eternal resting places of an eccentric poet's strangely beloved chickens. In the yard outside are a replica of a 19th-century **brick tryworks,** used to process whale oil from blubber aboard ship, and the 1,008-prism **Fresnel lens** installed in the Gay Head Lighthouse in 1854 and removed when the light was automated in 1952. Each evening the lens lamp is lighted briefly after sundown. A new addition is the **Tool Shed,** which contains harvesting tools used both on land and at sea in the early 19th century. ⊠ *School St.,* ☎ *508/627–4441.*

A **stroll** along the streets of Edgartown is the best way to get a good look at all of the beautiful old houses, reminders of the prosperity the town once enjoyed.

⑯ **60 Davis Lane** is a particularly handsome white clapboard Greek Revival with black shutters and fan ornament, surrounded by gardens. It was built in 1825 as a private school.

⑰ A monumental, white Greek Revival fronted with four solid Doric columns, **20 School Street** was built as a Baptist church in 1839. It is now a private residence.

⑱ The **Old Whaling Church** began in 1843 as a Methodist church and is now a performing-arts center. The massive building has a six-column portico, unusual triple-sash windows, and a 92-ft clock tower that can be seen for miles. The simple, graceful interior is brightened by light from 27-ft-tall windows and still contains the original box pews and lectern. Aside from attending performances, you can get inside only as part of the historical walking tours offered by Liz Villard (☞ *Martha's Vineyard A to Z, below*). ⊠ *Main St.,* ☎ *For tour: 508/627–8619.*

⑲ A truly elegant sight is the graceful **Dr. Daniel Fisher House,** with a wraparound roof walk, a small front portico with fluted Corinthian columns, and an architecturally economical, elegant side portico with thin fluted columns. It was built in 1840 for one of the island's richest men, who was a doctor, the first president of the Martha's Vineyard National Bank, and the owner of a whale-oil refinery, a spermaceti (whale-oil) candle factory, and a gristmill, among other things. It is interesting to note that the good doctor came to a portion of his fortune through marriage—as a wedding gift, his father-in-law presented him with the bride's weight in silver. The house is now used for functions and office space, and you can get inside only as part of Liz Villard's historical walking tours. ⊠ *Main St.,* ☎ *For tour: 508/627–8619.*

**20** The island's oldest dwelling is the 1672 **Vincent House.** It was moved to its present location behind the Fisher House in 1977, restored, and furnished with pieces that date from the 17th to the 19th century. A tour of this weathered, shingle farmhouse works like a time line, beginning with the sparse furnishings of the 1600s and traveling to a Federal-style parlor of the 1800s. This is another Liz-Villard-tour-only building. ⊠ *Main St.,* ☎ *For tour: 508/627–8619.*

**21** A good place to stop for directions or suggestions is the **Edgartown Visitors Center,** with information, rest rooms, and snacks. ⊠ *Church St., no phone.*

NEED A BREAK?

If you need a pick-me-up, pop into **Espresso Love** (⊠ 3 S. Water St., ☎ 508/627-9211) for a cappuccino and a homemade raspberry scone or blueberry muffin. If you prefer something cold, Fruit Smoothies are made with fruit and apple juice. Light lunch fare is also served: bagel sandwiches, soups, and delicious pastries and cookies—all homemade, of course. (The First Family has been known to make breakfast stops here.)

The architecturally pristine, much-photographed upper part of North Water Street is lined with many fine captains' houses. There's always an interesting detail on this stretch that you never noticed before—like a widow's walk with a mannequin poised, spyglass in hand, watching for her seafaring husband to return. The 1832 house where this piece **22** of whimsy can be seen is at **86 North Water Street,** which the Society for the Preservation of New England Antiquities maintains as a rental property.

**23** The **Edgartown Lighthouse,** surrounded by a public beach, offers a great view (but seaweedy bathing). The original light guarding the harbor was built in 1828 and set on a little island made of granite blocks. The island was later connected to the mainland by a bridge. By the time the 1938 hurricane made a new light necessary, sand had filled in the gap between the island and the mainland. The current white, cast-iron tower was floated by barge from Ipswich, Massachusetts in 1938. This area, called Starbuck's Neck, is a good place to wander about, with views of ocean, harbor, a little bay, and moorland.

A pleasant walking trail circles an old ice pond that is at the center of **24** **Sheriff's Meadow Sanctuary,** 17 acres of marsh, woodland, and meadow. One of the area's many wildlife preserves, Sheriff's Meadow offers a variety of habitats to observe.

★ ☙ The Vineyard's conservation areas are a good way to get acquainted with local flora and fauna. The 350-acre **Felix Neck Wildlife Sanctuary,** a Massachusetts Audubon Society preserve 3 mi out of Edgartown toward Oak Bluffs and Vineyard Haven, has 6 mi of hiking trails traversing marshland, fields, oak woods, seashore, and waterfowl and reptile ponds. There are also nesting osprey and barn owls in the sanctuary. A full schedule of events is offered throughout the year, including sunset hikes along the beach, exploration of the salt marsh, stargazing, snake or bird walks, snorkeling, canoeing, and more, all led by trained naturalists. An exhibit center has trail maps, aquariums, snake cages, and a gift shop. A bit of summer fun and learning experience combined, the sanctuary's **Fern & Feather Day Camp** is a great way for children to learn about wildlife, plants, and the stars. There are one- or two-week summer sessions that include overnight camping expeditions. Early registration is advised and begins in February. ⊠ *Off Edgartown–Vineyard Haven Rd.,* ☎ *508/627–4850.* ☞ *$3.* ☼ *Center: Mid-June–mid-Sept., daily 8–4; mid-Sept.–mid-June, Tues.–Sun. 9–4. Trails: sunrise–sunset.*

**㉕** **Chappaquiddick Island,** a sparsely populated area with a great number of nature preserves, makes for a pleasant day trip or bike ride on a sunny day. If you are interested in covering a lot of it, cycling is an excellent way to save time getting from point to point. Chappaquiddick Island is actually connected to the Vineyard by a long sand spit from South Beach in Katama—a spectacular 2¾-mi walk if you have the energy. If not, the On Time ferry makes the trip across from 7 AM to midnight in season. The ferry makes frequent trips but posts no schedule, and therefore earns its name because, technically, it cannot be late.

The Land Bank's 41-acre **Brine's Pond** is a popular, scenic picnicking spot. Mown grasses surround a serpentine pond that has an island in its center and a woodland backdrop behind—a truly lovely setting.

★ The Trustees of Reservations' 14-acre **Mytoi** preserve comprises a tranquil Japanese park with a creek-fed pool, spanned by a bridge and rimmed with Japanese maples, azaleas, and irises. The garden was created in 1958 by a private citizen. Rest room facilities are available. ⊠ *Dyke Rd., .2 mi from its intersection with Chappaquiddick Rd.,* ☎ *508/693–7662.* ☞ *Free.* ☼ *Daily sunrise–sunset.*

At the end of Dyke Road (¼ mi farther) is the **Dyke's Bridge,** infamous as the scene of the 1969 accident in which a young woman died in a car driven by Ted Kennedy. The rickety bridge has been replaced, after having been dismantled in 1991, but for ecological reasons, vehicle access over it is limited. The **Cape Poge Wildlife Refuge** (☞ *below*), which includes the spectacular **East Beach** and the **Cape Poge Light** are across the bridge.

The **Poucha Pond Reservation,** near the southeast corner of the Vineyard, totals 99 acres, through which trails wander among shady pitch pine and oak forests around a marshy pond on one-time farmland. One trail-end has a great view of the pond, the Dyke Bridge, and the East Beach dunes in the distance. Bring binoculars for the birds—terns, various herons, gulls, plovers—and repellent for the mosquitoes. ⊠ *3.8 mi from Chappaquiddick ferry landing,* ☎ *508/627–7141.* ☞ *Free.* ☼ *Daily sunrise–sunset.*

A conglomeration of habitats where you can swim, walk, fish, or just sit and enjoy the surroundings, the **Cape Poge Wildlife Refuge,** on the easternmost shore of Chappaquiddick Island, is more than 6 mi of wilderness—dunes, woods, cedar thickets, moors, salt marshes, ponds, tidal flats, and barrier beach—as well as an important migration stopover and nesting area for numerous sea and shore birds. The best way to get to the refuge is as part of a naturalist-led Jeep drive (☎ 508/627–3599.) You can also get there from **Wasque Reservation** (☞ *below*) on the south shore of the island. You'll need a four-wheel-drive vehicle to do that and to get to much of the acreage. The Trustees of Reservations require an annual permit ($60–$110) for a four-wheel drive, available on-site or through Coop's Bait and Tackle. ⊠ *East end of Dyke Rd., 3 mi from the Chappaquiddick ferry landing.*

★ The 200-acre **Wasque Reservation** (pronounced *wayce*-kwee) connects Chappaquiddick "Island" with the mainland of the Vineyard in Katama and forms Katama Bay. It is mostly a vast beach. Fish, sunbathe, take the trail by Swan Pond, walk to the island's southeasternmost tip at Wasque Point, or dip into the surf—with caution, of course. **Wasque Beach** is accessed by a flat boardwalk with benches overlooking the west end of Swan Pond. It's a pretty walk skirting the pond, with ocean views on one side and osprey poles on the other. Atop a bluff there is a pine-shaded picnic grove with a spectacular, practically 180-degree panorama. Swan Pond lies below the bluff, rich in bird life, including the requisite swans, and surrounded by marsh and beach grasses. Beyond that: beach, sky, and boat-dotted sea. From the grove, a long boardwalk leads down amid the grasses to **Wasque Point,** a prime surf-casting spot for bluefish and stripers. There's plenty of wide sandy beach here to sun on, but swimming is dangerous because of strong currents. ⊠ *Located at east end of Wasque Rd., 5 mi from Chappaquiddick ferry landing,* ☎ *508/627–7260.* 🖭 *$3 cars, plus $3 per adult, Memorial Day–mid-Sept.; free rest of year.* ☉ *Property: 24 hrs; gatehouse: Memorial Day–Columbus Day, daily 9–5. Rest rooms, drinking water.*

## Dining and Lodging

$$$$ ✕ **L'étoile.** Perhaps the Vineyard's finest traditional restaurant, L'étoile
★ carries on a long history of quality. Both the food and the setting in the stunning Charlotte Inn are unforgettable. The glass-enclosed dining room reminds you why hunter green, dark wood, and glass became so popular—and imitated. Quiet Edwardian charm and beautiful decorative touches perfectly anticipate the food. Preparations are at once classic and creative. Not to be missed are a terrine of grilled vegetable appetizer, roasted ivory king salmon with a wonderful horseradish and scallion crust, and black angus sirloin with zinfandel and oyster sauce. L'étoile has an outstanding wine list with selections from California and Europe that are solid, if a little pricey. An outdoor patio is ideal for brunch ($24 prix fixe). ⊠ *27 S. Summer St.,* ☎ *508/627–5187. Reservations essential. AE, MC, V. Closed Jan.–mid-Feb. and weekdays spring and fall. No lunch.*

$$$–$$$$ ✕ **Savoir Fare.** From William Styron to Bill Clinton, Savoir Faire has
★ carved out a reputation as the Vineyard's celebrity favorite. At first glance there's no obvious reason: The big black-and-white diamonds on the floor are nice but not so special, the buff walls and white linen can be found everywhere, even the view of the small garden and patio is island typical. What is it, then? The food! The chefs bring a Mediterranean sensibility to terrific ingredients, fusing combinations of spices and staples with astonishing complexity and mastery. The experience of a dish like lobster *en brodo* (in broth), made with corn, roasted peppers, and pancetta, served with spicy Sicilian mashed potatoes and lobster vinaigrette with chive oil, can make you begin to reconsider your culinary beliefs. ⊠ *14 Church St., in courtyard opposite Main St.'s town hall,* ☎ *508/627–9864. AE, MC, V. Closed Nov.–Apr. No lunch.*

$$$ ✕ **Daggett House.** With its low ceilings, ancient beams, wide-plank floors, and open fireplace, Daggett House is one of a few Vineyard establishments that uses its decor and atmosphere to transport diners back to the Colonial era. Speaking of transport, both the dining room, down a flight of stairs, and the handsome backyard have a lovely view of the harbor and the always-on-time Chappaquiddick ferry. Though the Puritans might have been at home in the surroundings, they would not recognize the menu, with fare such as shiitake mushrooms, roasted orange-pepper aioli, and seared sirloin in a Bass ale sauce. ⊠ *59 N. Water St.,* ☎ *508/627–4600. Reservations essential. AE, MC, V. No smoking. No lunch.*

$$$  ✕ **The Navigator.** This is Edgartown's version of big: A big room with a big view of big boats, with a big menu, a big patio, and a big bar. None of this comes, of course, without your paying big prices. Probably the best time to visit is for a big Bloody Mary late in the afternoon. ⊠ *2 Main St.,* ☎ *508/627–4320. AE, D, DC, MC, V. Closed mid-Oct.–Apr.*

$$  ✕ **The Beeftender.** After a week of lobster and clams, many diners end up screaming, "Enough already! I want red meat and I want it now!" The Beeftender is ready and at your service, with traditional steaks and chops. A number of the most popular items are available in children's portions. Steaks are offered in three sizes, which allows for the possibility of sampling an appetizer. ⊠ *Upper Main St.,* ☎ *508/627–8344. Reservations not accepted. AE, D, DC, MC, V.*

$–$$  ✕ **The Newes from America.** Sometimes a nearly subterranean, darkened scene feels right on a hot summer afternoon, in which case the Newes is the best place in town for an informal lunch or dinner. The food is Americana all the way (burgers and fries), and the beer selection is microbrewed but massively inclusive. Inside, there's plenty of wood and greenery, and many things "olde." You're likely to have to wait, but it'll give you an opportunity to consider the multiple choice questions on the menu, such as, Who is the clothing-optional Lucy Vincent Beach in Chilmark named after? A pioneering nudist? A 19th century stripper? A prudish Chilmark librarian? All answers lie within. ⊠ *23 Kelly St.,* ☎ *508/627–4397. Reservations not accepted. AE, D, MC, V.*

$–$$  ✕ **The Sand Bar.** Much more bar than sand, the Sand Bar is an official fan-club watering hole for the New England Patriots football team. The front of the restaurant has one of the nicest little second-floor porches in town. One table deep, shaded, it looks over busy Main Street. The food is secondary to the sports on the tube out back, or the breeze on the porch in front. Downstairs, the same owners run CJ's on Main Street, and send up their award-winning clam chowder for lunch. ⊠ *Main St.,* ☎ *508/627-9027. Reservations not accepted. AE, D, MC, V.*

$$$$  ✕🖬 **Charlotte Inn.** Tastefully and intelligently decorated inside and out—
★ in parlors, guest rooms, suites, and a maze of garden spaces—the Charlotte consummately realizes the values of a bygone era. Its exquisite and fascinating antique furnishings, objects, paintings, and books, come together with energy and life—that which the caring and attentive staff and the inn's owners, Gery and Paula Conover, pour into it every day. Come to the inn for an Edwardian fantasy, for an escape in one of the luxurious suites in the Carriage or Coach houses, for an utterly tranquil winter holiday with the island nearly to yourself, or for a sumptuous meal at L'étoile (☞ *above*). Every detail is considered. From the Scottish barrister's desk at check-in to a lovely plant-filled window niche upstairs to the decanter of sherry waiting in each uniquely decorated room, elegant touches bespeak the Charlotte Inn's everyday timelessness. ⊠ *27 S. Summer St., 02539,* ☎ *508/627–4751,* 𝔽𝔸𝕏 *508/627–4652. 22 rooms and 3 suites in 5 buildings. Restaurant. AE, MC, V. CP.*

$$  ✕🖬 **Daggett House.** The flower-bordered lawn that separates the main
★ house from the harbor makes a great retreat after a day of exploring town, a minute away. All three inn buildings—the main 1660 Colonial house, the Captain Warren house across the street, and a three-room cottage between the main house and the water—are decorated with fine wallpapers, antiques, and reproductions (some canopy beds). Much of the buildings's historical ambiance has been preserved, including a secret stairway that's now a private entrance to an upstairs guest room. The Widow's Walk Suite has a full kitchen and a private roof walk with a superb water view and a hot tub. Two other rooms have kitchenettes, and another has a hot tub. Breakfast and dinner are served in the 1750

tavern (☞ *above*). ✉ *59 N. Water St., Box 1333, 02539,* ☎ *508/627–4600 or 800/946–3400,* fax *508/627–4611. 21 rooms, 4 suites. No smoking. AE, MC, V.*

**$$$$** 🏨 **Harbor View Hotel.** This historic hotel, centered in the 1891 gray-shingled main building with wraparound veranda and a gazebo, is now a complex of buildings in a residential neighborhood a few minutes from town. Town houses have cathedral ceilings, decks, kitchens, and large living areas with sofa beds. Rooms in other buildings, however, very much resemble upscale motel-style rooms, so ask for main building or town house rooms for more unique lodgings. A good beach for walking stretches ¾ mi from the hotel's dock, from which there's good fishing for blues. Children enjoy swimming in the sheltered bay. Packages and theme weekends, as well as rooms with kitchenettes, are also available. ✉ *131 N. Water St., 02539,* ☎ *508/627–7000 or 800/225–6005,* fax *508/627–7845. 124 units. Restaurant, piano bar, room service, pool, golf privileges, 2 tennis courts, volleyball, baby-sitting, laundry service, concierge. AE, DC, MC, V.*

**$$$$** 🏨 **Kelley House.** At the center of town, this sister property of the Harbor View combines services and amenities with a country-inn feel, through complimentary Continental breakfasts, afternoon tea, and evening cookies and milk. The 1742 white clapboard main house and the adjacent Garden House are surrounded by pink roses; inside, the decor is an odd mix of country French and Shaker. Large suites in the Chappaquiddick House and the two spacious town houses with full kitchens in the Wheel House have porches (most with harbor views) and living rooms. All guest rooms have cable TVs, phones, and air-conditioning. The 1742 pub, with original hand-hewn timbers and ballast-brick walls, serves light fare and microbrewed beers on tap until 11 PM. ✉ *23 Kelly St., 02539,* ☎ *508/627–7900 or 800/225–6005,* fax *508/627–8142. 42 rooms, 9 suites, 2 town house units. Pool, 2 tennis courts, baby-sitting, laundry service. AE, DC, MC, V. Closed Nov.–Apr. CP.*

**$$$$** 🏨 **Mattakesett.** This community of individually owned three- and four-bedroom homes and condominiums is within walking distance of South Beach. All units are spacious, sleep eight, and have phones, full kitchens with dishwashers, washer/dryers, freestanding fireplaces, and decks with or without bay and ocean views. The staff provides plenty of service, and the children's program, pool, and barbecue grills add to the family atmosphere. Usually there's a one-week minimum stay, and it's best to book for summer by January 15. ✉ *Katama Rd., 02539,* ☎ *508/627–8920; reservations c/o Stanmar Corp., 130 Boston Post Rd., Sudbury, MA 01776,* ☎ *508/443–1733,* fax *508/627–7015. 92 units. Pool, 8 tennis courts, aerobics, bicycles, children's programs. No credit cards. Closed Columbus Day–Memorial Day.*

**$$$$** 🏨 **Tuscany Inn.** Inspired by the region for which it was named, the Tuscany Inn manages to combine European sophistication and elegance with casual comfort. Owners Rusty Scheuer and Laura Sbrana-Scheuer transformed the property into an Italian villa with a Victorian flair. Rooms are tastefully decorated with quality furnishings, fine fabrics, and carefully selected treasures that the owners have collected in their travels. Choose from king, queen, or twin accommodations; some baths have whirlpool tubs and harbor views. Mingle with other guests in front of the fire in the parlor, read a book out on the wicker-furnished veranda, or sip cappuccino and munch on homemade biscotti on the patio in the side yard, surrounded by lovely gardens and white lights. Breakfast is a highlight, as the owner is also an accomplished chef who teaches cooking classes off-season. Dining in the intimate restaurant will make you feel like you're in Italy! ✉ *22 N. Water St., Box*

2428, 02539, ☎ 508/627–5999 or 508/627–8999, ℻ 508/627–
6605. 8 rooms. AE, MC, V. CP.

**$$$** 🖬 **Edgartown Commons.** This condominium complex of seven build-
ings, which include an old house and motel units around a busy pool,
is just a couple of blocks from town. Studios and one- or two-bedroom
units all have full kitchens, and some are very spacious. Each has been
decorated by its owner, so the decor varies—some have an older look,
while some are new and bright. Definitely family-oriented, the place
has lots of kids to keep other kids company. Units away from the pool
are quieter. ⊠ 20 Peases Point Way, 02539, ☎ 508/627–4671, ℻ 508/
627–4271. 35 units. Picnic area, pool, shuffleboard, playground, coin
laundry. AE, D, MC, V. Closed Nov.–Apr.

**$$$** 🖬 **Shiverick Inn.** Innkeepers Denny and Marty Turmelle add warmth
★ to the elegance of this inn, set in a striking 1840 house with mansard
roof and cupola. Rooms are airy and bright, with high ceilings, lots of
windows, American and English antiques, rich fabrics and wallpapers,
and antique art. Beds are mostly queen-size with canopies or carved
four-posters. Several rooms have fireplaces or woodstoves. Breakfast
is served in a lovely summerhouse-style room with a wood-burning fireplace. There's a library with cable TV and a stereo, and a flagstone
garden patio. ⊠ Corner of Peases Point Way and Pent La., Box 640,
02539, ☎ 508/627–3797 or 800/723–4292, ℻ 508/627–8441. 10
rooms. Bicycles. No smoking. AE, D, MC, V. CP.

**$$** 🖬 **Colonial Inn.** Part of a busy downtown complex of shops, the inn
is hardly a tranquil escape, but if you like modern conveniences and
being at the center of the action, it may be for you. Rooms are decorated
in soft floral peaches, with white pine furniture, wall-to-wall carpeting,
and brass beds and lamps. Each has a good-size bath, and suites come
with a sofa bed. One common fourth-floor deck has a superb view of
the harbor, as do some rooms. Holiday theme weekends, celebrating
Halloween and Thanksgiving, are good for the whole family. ⊠ N. Water
St., Box 68, 02539, ☎ 508/627–4711 or 800/627–4701, ℻ 508/627–
5904. 43 rooms, 2 suites, 2 efficiencies. Restaurant. AE, MC, V. Closed
Dec.–Apr. CP.

**$$** 🖬 **Harborside Inn.** Right on the harbor, with boat docks at the end of
a nicely landscaped lawn, the inn offers a central town location, harbor-
view decks, and plenty of amenities. Seven two- and three-story buildings are set around a wide lawn with formal rose beds, brick walkways,
a brick patio, and a pool. Rooms have Colonial-style furnishings
(which could use some updating), brass beds and lamps, and textured
wallpapers. ⊠ 3 S. Water St., Box 67, 02539, ☎ 508/627–4321 or
800/627–4009, ℻ 508/627–7566. 89 rooms, 3 suites. Grills, in-room
VCR rental, pool, hot tub, sauna. AE, MC, V.

## Nightlife and the Arts

Throughout the year, lectures, classic films, concerts, plays, and other
events are held at the **Old Whaling Church** in Edgartown; watch the
papers, or check the kiosk out front. ⊠ 89 Main St., ☎ 508/627–4442.

## Outdoor Activities and Sports

BEACHES

**Bend-in-the-Road Beach,** Edgartown's town beach, is a protected area
marked by floats adjacent to the state beach. Backed by low, grassy dunes
and wild roses, Bend-in-the-Road has calm, shallow waters, some parking, and lifeguards. It is on the shuttle bus and bike routes. ⊠ Beach Rd.
**East Beach,** on Chappaquiddick Island, one of the area's best beaches, is
accessible only by boat or Jeep from the Wasque Reservation. It has heavy
surf, good bird-watching, and relative isolation in a lovely setting.
**South Beach,** also called Katama Beach, is the island's largest and most
popular. It is a 3-mi ribbon of sand on the Atlantic with strong surf

and occasional riptides, so check with the lifeguards before swimming. There is limited parking. ⊠ *Katama Rd.*

**Wasque Beach,** at the Wasque Reservation on Chappaquiddick, is an uncrowded ½-mi sandy beach with sometimes strong surf and currents, a parking lot, and rest rooms. ⊠ *Chappaquiddick Island.*

### BIKING

Several bike paths lace through the Edgartown area, including a path to Oak Bluffs that has a spectacular view of Sengekontacket Pond on one side and Nantucket Sound on the other.

**R. W. Cutler Bike** (⊠ 1 Main St., ☎ 508/627–4052) rents and repairs all types of bicycle.

**Triangle** (⊠ Upper Main St., ☎ 508/627–7099) has bicycles available for rent.

**Wheelhappy** (⊠ 8 S. Water St., ☎ 508/627–5928) rents bicycles and will deliver them to you.

### FISHING

The annual **Martha's Vineyard Striped Bass & Bluefish Derby** (⊠ Box 2101, Edgartown 02539, ☎ 508/693–0728), from mid-September to mid-October, offers daily, weekly, and derby prizes for striped bass (which were reintroduced in the '93 derby), bluefish, bonito, and false albacore catches, from boat or shore. The derby is a real Vineyard tradition, cause for loyal devotion among locals who drop everything to cast their lines at all hours of day and night in search of that prizewinning whopper.

One of the most popular spots for surf casting is **Wasque Point** on Chappaquiddick.

**Big Eye Charters** (☎ 508/627–3649) offers fishing charters that leave from Edgartown Harbor.

**Coop's Bait and Tackle** (⊠ 147 W. Tisbury Rd., ☎ 508/627–3909) sells accessories and bait, rents fishing gear, and has a copy of the current listing of fishing regulations.

**Larry's Tackle Shop** (⊠ 141 Main St., ☎ 508/627–5088) rents gear and sells accessories and bait, and has a copy of the current listing of fishing regulations.

*Slapshot II* (☎ 508/627–8087) is available for fishing charters from Edgartown Harbor.

### RECREATION AREAS

**Edgartown Recreation Area** (⊠ Robinson Rd., ☎ 508/627–6110) has three tennis courts, a basketball court, a softball field, a roller hockey court, a picnic area, and playground equipment. All areas are lighted some nights in summer. Activities (published in local papers) include tennis round-robins, softball and basketball games, arts and crafts, and rainy-day events. Also see the "Island Recreation" section of the *Vineyard Gazette*'s calendar for open Frisbee, rugby, and other games.

### TENNIS

Because tennis is so popular on the island, reservations are usually necessary for the hard-surface courts on Robinson Road in Edgartown, which cost a small fee and are open year-round.

## Shopping

**Bickerton & Ripley Books** (⊠ Main St., ☎ 508/627–8463) carries a wide selection of current and island-related titles.

**Brickman's** (⊠ 33 Main St., ☎ 508/627–4700) sells beach and sports gear, such as camping, fishing, and snorkeling equipment, and boogie boards. It also carries sportswear, surfer-type clothing, and major-label footwear for the family.

**The Colonial Inn Shops** (✉ 38 N. Water St., ☎ 508/627–4711) sell art, crafts, pottery, and sports gear.

**Edgartown Scrimshaw** (✉ 17 N. Water St., ☎ 508/627–9439) carries a large collection of scrimshaw, including some antique pieces, as well as Nantucket lightship baskets and 14-karat lightship-basket jewelry.

**The Fligors** (✉ 27 N. Water St., ☎ 508/627–8811) is the closest thing to a department store on the island, with varied offerings, including preppy clothing and a Christmas shop.

**The Gallery Shop** (✉ 20 S. Summer St., ☎ 508/627–8508) has 19th- and 20th-century oils and watercolors, including English sporting prints, marine art, and works by major local artists, plus small English antiques.

**Nevin Square** (✉ Winter St., no phone) sells leather, clothing, art, antiques, and crafts.

**Optional Art** (✉ 35 Winter St., ☎ 508/627–5373) carries fine jewelry mostly in 18-karat gold, handcrafted by 30 award-winning American artisans.

**Vivian Wolfe Antiques** (✉ 42 Main St., ☎ 508/627–5822) has antique and estate jewelry, as well as antique silver tea services and so forth.

**Willoughby's** (✉ 12 N. Water St., Edgartown, ☎ 508/627–3369) displays mostly island landscapes and scenes by Vineyard and Cape artists, including limited-edition prints and drawings.

# UP-ISLAND

Much of what makes the Vineyard special is found in its rural reaches, in the agricultural heart of the island and the largely undeveloped lands along the West Chop to the Edgartown line. Country roads meander through woods and tranquil farmland, and dirt side roads lead past crystalline ponds, abandoned cranberry bogs, and conservation lands. In **Chilmark**, **West Tisbury**, and **Gay Head**, nature lovers, writers, artists, and others have established close ongoing summer communities. In winter, the isolation and bitter winds send even many year-round Vineyarders from their Up-Island homes to places in the cozier Down-Island towns.

## West Tisbury

**26** *6.6 mi southwest of Vineyard Haven, 11.9 mi northeast of Gay Head.*

Founded in the 1670s by settlers from Edgartown, among them the son of Myles Standish and the son-in-law of the Aldens from the *Mayflower,* West Tisbury, known for its first 200 Westernized years as simply Tisbury, offered advantages not found Down-Island, most important a strong-flowing stream running into a pond that was perfect for a mill site—a rare thing on the Vineyard. A gristmill was built and a community developed. Farming, especially sheep farming, became Tisbury's mainstay.

West Tisbury has retained its rural appeal, and continues its agricultural tradition with several active horse and produce farms. The town center looks very much the small New England village, complete with a white, steepled church. And half of the 5,146-acre State Forest lies within the town limits.

Just outside the center of Vineyard Haven, on the way to West Tisbury, **27** the **Tashmoo Overlook** is a scenic viewpoint overlooking a meadow leading down to Lake Tashmoo and Vineyard Sound beyond. The meadow is public space. Across the lane from it is the amphitheater where summer Vineyard Playhouse productions are held. ✉ *State Rd. and Spring St.*

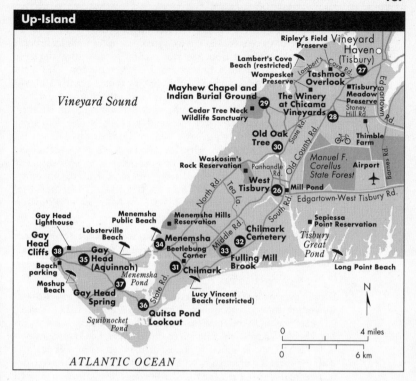

**NEED A BREAK?**

In a humble dale on the way out of Vineyard Haven, the **Scottish Bakehouse** has long been an island staple for breads, desserts, and pastries of all kinds. You can also pick up Scotch meat pies, jams and jellies, pickles, and chutneys, and the shortbread is to die for. It's a favorite spot for a light breakfast if you're bicycling Up-Island for the day—or looking for a treat to take on your way. ⊠ *State Rd.,* ☎ *508/693–1873. No credit cards. No seating.*

West Tisbury has a plethora of conservation areas, all of them unique. One of the rare meadows open for walking on the island, **Tisbury Meadow Preserve** isn't being farmed, but it is mowed to keep it from returning to woodland. An old farmstead sits on the property, and the back acres are wooded and cross an 18-century cart path. You can walk Tisbury Meadow in less than an hour or combine it with two others, the **Wompesket** and **Ripley's Fields** preserves across State Road, for a longer hike. ⊠ *Trailhead on east side of State Rd., .4 mi south of Tashmoo Overlook. 83 acres.*

**Ripley's Field Preserve** gives an idea of what the island must have looked like 200 years ago. The preserve spreads over glacier-formed undulating land, both meadows and woodland. A windmill and wildflowers are pleasant attractions. It is connected by old cart paths to **Tisbury Meadow** and **Wompesket** preserves in North Tisbury. ⊠ *John Hoft Rd., off north end of Lambert's Cove Rd., .7 mi from State Rd. Parking area and bike rack on left. 56 acres.*

Bordering part of Merry Farm, **Wompesket Preserve** includes an interesting wet meadow and ponds that are good for birding. The walk to the area overlooks the farm and the Atlantic in the distance. Follow marked dirt roads from Ripley's Field or Tisbury Meadow preserves to get to Wompesket. ⊠ *Red Coat Hill Rd. 18 acres.*

★ ㉘ A rather unique island undertaking, certainly appropriate considering the island's name, the **Winery at Chicama Vineyards** is the result of the labors of the Mathiesen family. From 3 acres of trees and rocks, George, a broadcaster from San Francisco, his wife Cathy, and their six children have built a fine vineyard. They started in 1971 with 18,000 vinifera vines, and today the winery produces nearly 100,000 bottles a year from chardonnay, cabernet, and other European grapes. Chenin Blanc, merlot, and a cranberry dessert wine are among their 10 or more tasty varieties. Free tours and tastings are given daily from Memorial Day to Columbus Day. A shop selling their wine, along with herbed vinegars, mustards, jellies, and other foods prepared on the premises, is open year-round. A Christmas shop with glassware, gift baskets, wreaths, and wine-related items is open mid-November through New Year's Eve. ⊠ *Stoney Hill Rd.,* ☎ *508/693–0309.* ⊙ *Memorial Day–Columbus Day, Mon.–Sat. 11–5, Sun. 1–5; call for off-season hrs. and tastings.*

Pick your own strawberries and raspberries at **Thimble Farm,** where you can also buy pre-boxed fruit if you're not feeling quite so rural. The farm also sells cut flowers, melons, pumpkins, hydroponic tomatoes, and other produce. ⊠ *Stoney Hill Rd.,* ☎ *508/693–6396.*

㉙ Deep in the woods off a dirt road, the **Mayhew Chapel and Indian Burial Ground** are suffused with history. The tiny chapel, built in 1829 to replace an earlier one, and a memorial plaque are dedicated to the pastor Thomas Mayhew Jr., leader of the original colonists who landed at Edgartown in 1642. Mayhew was an enlightened man, noted for his fair dealings with the local Wampanoags. Within a few years, he had converted a number of them to Christianity. Called Praying Indians, they established a community here called Christiantown.

An overgrown wildflower garden grows near the chapel. Beyond the boulder with the plaque are rough-hewn stones marking Indian grave mounds—the dead are not named, for fear of calling down evil spirits. Behind the chapel is the beginning of a loop trail through the woods that leads to a lookout tower. There's a map at the first trail fork.

**Cedar Tree Neck Wildlife Sanctuary,** 300 hilly acres of unspoiled West Tisbury woods managed by the Sheriff's Meadow Foundation, consists of varied environments, including a sphagnum bog and a pond. The sanctuary has interesting flora, among them bayberry and swamp azalea bushes, tupelo, sassafras, and pygmy beech trees. Wooded trails lead to a stony but secluded North Shore beach (swimming, picnicking, and fishing prohibited), and from the summit of a headland there are views of Gay Head and the Elizabeth Islands. ⊠ *Follow Indian Hill Rd. off State Rd. for 2 mi, then turn right 1 mi on an occasionally steep, rocky dirt road to the parking lot.* ☎ *508/693–5207.* ⊡ *Free.* ⊙ *Daily 8:30–5:30.*

☾ The **Takemmy Farm** invites children to visit with its llamas, raised as pets and breeding stock, and miniature donkeys. The farm sells eggs and yarn year-round; in season, there are vegetables, flowers, and honey for sale as well. ⊠ *State Rd., N. Tisbury,* ☎ *508/693–2486.* ⊡ *$3 per car.* ⊙ *Mon.–Sat., 1–5.*

The **Martha's Vineyard Glassworks** offers a chance to watch glass being blown—a fascinating process to observe—by glassmakers who have pieces displayed in Boston's Museum of Fine Arts. Their work is also for sale. ⊠ *State Rd., N. Tisbury,* ☎ *508/693–6026.*

㉚ Its limbs twisting into the sky and along the ground, the **Old Oak Tree** stands near the intersection of State and North roads. This massive, much-loved, member of the *quercus* family is thought to be about 150 years old, and it's a perennial subject for nature photographers.

NEED A BREAK? Step back in time with a visit to **Alley's General Store** (⊠ State Rd., ☎ 508/693–0088), the heart of town since 1858. Alley's sells a truly general variety of goods: everything from hammers and housewares and dill pickles to all those great things you find only in a country store. The Martha's Vineyard Preservation Trust purchased the building in 1993 and beautified it, preserving its rural character. Behind the parking lot, **Back Alley's** (☎ 508/693–7367) serves tasty sandwiches and pastries to go year-round.

An unusual sight awaits at the **Field Gallery,** where Tom Maley's ponderous white sculptures, such as a Colonial horse and rider, or a whimsical piper, are displayed on a wide lawn. Inside there are changing summer exhibitions of island artists' work. ⊠ *State Rd.,* ☎ *508/693–5595.*

The weekly **West Tisbury Farmers' Market**—Massachusetts' largest—is held Wednesdays and Saturdays in summer at the 1859 **Agricultural Hall,** near the town hall. The colorful stands overflowing with fresh produce—most of it organic—offer a refreshing return to life before fluorescent-lit supermarkets (☞ Shopping, *below*). ⊠ *South Rd.*

The Agricultural Society purchased a nearby parcel of land in 1992, and construction on a new, larger **Ag Hall** was completed early in 1996. The new hall, about a mile from the original, is the setting for various shows, speakers, dances, and potluck dinners. A yearly county fair—including a woodsman contest, dog show, games, baked goods and jams for sale, and, of course, livestock- and produce-judging—is held at the new hall in late August (☞ Festivals and Seasonal Events *in* Chapter 1). ⊠ *35 Panhandle Rd.*

**Music Street,** named for a preponderance of pianos bought with whaling profits and played perpetually by the prodigy of sea captains, boasts the oldest houses in West Tisbury.

The **Mill Pond** is a lovely spot to stop to watch the swans—at the right time of year you might see a cygnet with the swan couple. The small building nearest the pond has been a grammar school, an ice house, and a police station until a larger building opened for the cops to occupy. The mill itself is across the road. Originally a gristmill, it was opened in 1847 to manufacture island wool for peacoats, which whalers wore. The Martha's Vineyard Garden Club uses the building now. The pond is just around the corner from the town center on the way toward Edgartown.

A paradise for bird-watchers, **Sepiessa Point Reservation,** a Land Bank–Nature Conservancy property, consists of 164 acres on splendid Tisbury Great Pond, with expansive pond and ocean views, walking trails around coves and saltwater marshes, bird-watching, horse trails, swimming, and a boat launch. On the pond beach, watch out for razor-sharp oyster shells. Beaches across the pond along the ocean are privately owned. ⊠ *1.2 mi on right down New La./Tiah's Cove Rd., off W. Tisbury Rd., W. Tisbury,* ☎ *508/627–7141.* ☜ *Free.* ☉ *Daily sunrise–sunset.*

For a lovely walk with the promise of a refreshing swim at its end, **Long Point,** a 633-acre Trustees of Reservations preserve, is an open area of grassland and heath bounded on the east by the freshwater Homer's Pond, on the west by the saltwater West Tisbury Great Pond, and on the south by a mile of fantastic South Beach on the Atlantic Ocean. Tisbury Great Pond and Long Cove Pond, a sandy freshwater swimming pond, are ideal spots for bird-watchers. Drinking water and a rest room facility are available. Arrive early on summer days if you're coming by car, since the lot fills quickly. ⊠ *Mid-June–mid-Sept., turn left onto the unmarked dirt road (Waldron's Bottom Rd., look for mailboxes) ³⁄₁₀ mi west of airport on Edgartown–W. Tisbury Rd., at end, follow signs to Long Point parking lot. Mid-Sept.-mid-June, follow unpaved Deep Bottom Rd. (1 mi west of airport) 2 mi to lot.* ☎ *508/ 693–3678.* 🎟 *Mid-June–mid-Sept., $7 per vehicle, $3 per adult; free rest of year.* ☉ *Daily 10–6.*

★ At the center of the island, the **Manuel F. Correllus State Forest** is a 2,000-acre pine and scrub oak forest crisscrossed with hiking trails and circled with a paved but rough bike trail (mopeds are prohibited). There's a 2-mi nature trail, a 2-mi par course, and horse trails. The West Tisbury side of the state forest joins with an equally large Edgartown parcel to virtually surround the airport. ⊠ *Headquarters on Barnes Rd. by the airport,* ☎ *508/693–2540.* 🎟 *Free.* ☉ *Daily sunrise–sunset.*

A memorial to Thomas Mayhew Jr., called **Place on the Wayside**, stands along the Edgartown–West Tisbury Road, just east of the airport entrance on the south side of the road. A plaque identifies the spot where Mayhew had his "last worship and interview with them before embarking for England" in 1657, never to return: The ship was lost at sea. Wampanoags passing this spot would leave a stone in Mayhew's memory, and the stones were later cemented together to form the memorial.

## Dining and Lodging

$$ ★ ✕ **The Red Cat.** When compared to the town offerings, this restaurant feels like the culinary equivalent of a lovely, less-touristed beach. The atmosphere is casual and comfortable, and the cooking has a decidedly island lilt to it, with "Bleu on Blue," a signature dish of fresh bluefish baked with creamy bleu cheese. To avoid disappointment, it's always wise to reserve a table a few days ahead in season. ⊠ *688 State Rd., near North Rd.,* ☎ *508/693–9599. MC, V. BYOB. No smoking. Closed Mon. No lunch. Call for off-season hrs.*

$$–$$$ ✕🖃 **Lambert's Cove Country Inn.** A narrow road winds through pine woods to this secluded inn, set on a lawn surrounded by gardens and old stone walls. In 1996, after a brief period of neglect, Louis and Katherine Costabel took over the inn, sprucing up the lodging facilities and the property overall. Rooms in the 1790 farmhouse have light floral wallpapers and a sweet country feel. Rooms in outbuildings have screened porches or decks. Among the common areas is a library with fireplace. At the restaurant, the soft candlelight and excellent Continental cooking make it another destination for a special dinner, whether you stay the night or not. The chef's delicate, refined creations rely on local produce and seafood. Especially good are the crisp-baked soft-shell crab appetizer, red and yellow Belgian endive salad, and grilled Muscovy duck breast on caramelized onions. Reservations are essential at the restaurant, and it's BYOB. ⊠ *Off Lambert's Cove Rd., W. Tisbury; R.R. 1, Box 422, Vineyard Haven 02568,* ☎ *508/693–2298,* 🖷 *508/ 693–7890. 16 rooms. Restaurant, tennis court. AE, MC, V. BP.*

$ 🖃 **Hostelling International–Martha's Vineyard.** The only budget, roof-over-your-head alternative in season, this hostel is one of the country's best. You'll catch wind of local events from the bulletin board, and there is a large common kitchen and a fireplace in the common room. The hostel offers summer programs on island history and nature tours. Morning chores are required in summer, after AYH custom. It is near a bike path and about 3 mi from the nearest beach. The Island Shuttle makes a stop out front. ✉ *Edgartown–W. Tisbury Rd., Box 158, 02575,* ☎ *508/693–2665. 78 dorm-style beds. MC, V. 11 PM curfew June–Aug. Closed daytime 10–5 and completely Nov.–Apr.*

## Nightlife and the Arts

**Granary Gallery** (✉ Red Barn Emporium, Old County Rd., W. Tisbury, ☎ 508/693–0455 or 800/472–6279) showcases sculptures and mostly representational paintings by island and international artists, including the photographs of the late Alfred Eisenstaedt. Biweekly shows spotlighting a local artist are preceded by Sunday-evening receptions.

**Hermine Merel Smith Fine Art** (✉ 548 Edgartown Rd., W. Tisbury, ☎ 508/693–7719) specializes in paintings and drawings by contemporary American realists and impressionists.

**Hot Tin Roof** (✉ Martha's Vineyard Airport, ☎ 508/693–1137), opened by Carly Simon in 1976, then closed, was re-opened by the singer in the summer of 1996, and it is now the island's hottest club. Simon sometimes shows up to perform (unannounced), but it's usually other big names—Jimmy Cliff and Jerry Lee Lewis, for instance—crowding the stage.

## Outdoor Activities and Sports

### BEACHES

★ **Lambert's Cove Beach** (✉ Lambert's Cove Rd., W. Tisbury), one of the island's prettiest, has fine sand and very clear water. On the Vineyard Sound side, it has calm waters good for children and views of the Elizabeth Islands. In season it is restricted to residents and those staying in West Tisbury.

**Long Point** has a beautiful beach on the Atlantic, as well as freshwater and saltwater ponds for swimming, including the brackish Tisbury Great Pond. There are rest rooms.

**Uncle Seth's Pond** is a warm freshwater pond on Lambert's Cove Road in West Tisbury, with a small lifeguarded beach right off the road. Car parking is very limited.

### HORSEBACK RIDING

**Manuel F. Correllus State Forest** has horse trails running through it that are open to the public. ✉ *Access off Barnes Rd., Old County Rd., and Edgartown–W. Tisbury Rd.*

**Arrowhead Farm** (✉ Indian Hill Rd., ☎ 508/693–8831) has riding lessons for adults and children year-round, as well as children's summer horsemanship programs. The farm has an indoor ring and leases horses, but does not offer trail rides.

**Misty Meadows Horse Farm** (✉ Old County Rd., ☎ 508/693–1870) offers trail rides. Be sure to call ahead to reserve.

### TENNIS

Stop by to reserve hard-surface courts at the grammar school on Old County Road in West Tisbury. There is a small fee for using the courts, which are open year-round.

## Shopping

**Alley's General Store** (✉ State Rd., ☎ 508/693–0088), in business since 1858, deals in everything from fresh fruit and preserves to shoelaces, suntan lotions, and vintage tablecloths and cookie jars (though horrendously pricey). There's even a post office. The scale out front will give you your weight for a penny.

**Chilmark Pottery** (✉ Fieldview La., off State Rd., W. Tisbury, ☎ 508/693–6476) is a workshop and gallery selling hand-formed stoneware, porcelain, and Raku by island potters.

**Cronig's** (✉ State Rd., ☎ 508/693–2234) supports local farmers and specialty food producers and carries most food items that you'll need.

★ **West Tisbury Farmers' Market**—with booths selling fresh flowers, plants, fruits and vegetables, homemade baked goods and jams, and honey—is held at the Agricultural Hall (☎ 508/693–9549) on South Road in West Tisbury mid-June to mid-October, Wednesdays and Saturdays from 9 to noon.

# Chilmark

**㉛** *12 mi southwest of Vineyard Haven, 5.4 mi southwest of West Tisbury.*

**Chilmark** is a rural, unspoiled village whose scenic ocean-view roads, rustic woodlands, and lack of crowds have drawn chic summer visitors, and, hard on their heels, stratospheric real estate prices. Laced with ribbons of rough roads and winding stone fences that once separated fields and pastures, Chilmark is a reminder of what the Vineyard was like in an earlier time, before developers took over.

After years of fighting for it, the Land Bank bought **Waskosim's Rock Reservation** in 1990 from a developer who'd planned to build 40 houses on it. Today the reservation comprises diverse habitats—rolling green hills, wetlands, oak and beetlebung (black gum) woods, and 1,500 ft of frontage on Mill Brook—as well as the ruins of an 18th-century homestead. Waskosim's rock itself was deposited by the retreating glacier 10,000 years ago and is said to resemble the head of a breaching whale. It is one of the highest points on the Vineyard, on a ridge above the valley, from which there is a panoramic view of more than 1,000 acres of protected land. At the trailhead off North Road there's a map outlining a 3-mi hike throughout the 166 acres. ⊠ *Parking areas in Chilmark on North Rd.,* ☎ *508/627–7141.* ☜ *Free.* ☉ *Daily sunrise–sunset.*

**㉜** **Chilmark Cemetery** offers an interesting look into the Island's past. Longtime summer resident and writer Lillian Hellman, one of many who continued the Vineyard's tradition of liberal politics, is buried here. John Belushi is also buried here, in an unmarked grave. The boulder, deeply engraved with the comedian's name and sitting near the entrance, is a decoy placed to deter overzealous fans from finding the actual burial site. Visitors often leave odd tokens of remembrance. ⊠ *South Rd., Chilmark.*

The view of Chilmark's **Allen Farm**—an impressive vista from the road across pastureland protected by the Land Bank—is breathtaking. A shop here sells handwoven blankets and knitted items made from the farm's wool. ⊠ *South Rd.,* ☎ *508/645–9064.*

★ A dirt road leads off South Road to **Lucy Vincent Beach,** perhaps the island's most beautiful. In summer it is open only to Chilmark residents and those staying in town. Off-season a stroll here is the perfect getaway.

**Beetlebung Corner,** a crossroads named for the stand of beetlebung trees that grow here, marks Chilmark's town center. Here are the town's public buildings, including the firehouse and the post office, as well as the **Chilmark Community Center,** where events including everything from town meetings to auctions to chamber music concerts are held. In summer, a general store, a clothing boutique, a restaurant and breakfast café, a gallery, and a bank turn the little crossroads into a minimetropolis. ⊠ *Middle, State, South, and Menemsha Cross Rds.*

NEED A BREAK?     The **Chilmark Store** has the basics that you'd expect as well as Primo's Pizza, a solid take-out lunch spot. If you've bicycled into town, the wooden rockers on the porch may be just the place to take a break—or find a picnic spot of your own nearby to enjoy a fishburger or a slice of the pizza of the day. ⊠ *7 State Rd.,* ☎ *508/645-3655.* ☉ *May–*

*mid-Oct. Grocery open till 8 PM; Primo's Pizza open till 7 PM. No credit cards.*

.....................................................................................................

The **Chilmark Community Center** has activities for children of all ages in summer, including dances, concerts, and movies. It's a great place for kids to meet other kids. ⊠ *Beetlebung Corner,* ☎ *508/645–9484.*

Jointly owned by the Land Bank and the town, **Fulling Mill Brook** is a 50-acre conservation area. Its easy walking trail that slopes gently down toward the lowlands along the brook, where there are boulders to sit and sun on, to the property's edge at South Road, where there's a bike rack. ⊠ *Middle or South Rds.*

## Dining and Lodging

**$$$** ✕ **Feast of Chilmark.** After almost a decade in service, this restaurant ★ seems to keep getting stronger. Civilized, calm, and calming, the Feast is a welcome break from the busyness of the Vineyard's bigger-town joints. Photographer Peter Simon's work adds much to the ambiance in the bright and airy dining room, where the clientele is generally laid-back. Lots of seafood interpretations, light appetizers, and fresh salads play up summer tastes and flavors. For starters, panfried calamari is a good choice, as are the littleneck clams casino. The whole menu takes advantage of local produce, and the mixed green salad is simple and perfect. One unfortunate annoyance is the $3 corkage fee—per bottle—when you bring your own wine. ⊠ *Beetlebung Corner,* ☎ *508/ 645–3553. AE, MC, V. BYOB. No lunch.*

**$$$** ✕🖫 **Inn at Blueberry Hill.** Exclusive and secluded, this unique property comprising 56 acres of former farmland puts you in the heart of the rural Vineyard. The restaurant is relaxed and elegant, and the fresh, innovative, and health-conscious food is superb. Three meals are available on the property if you don't feel like leaving during the day (the restaurant is open to the public for dinner by reservation only). Guest rooms are simply and sparsely decorated with Shaker-inspired, island-made furniture, handmade mattresses with all-cotton sheets and duvets, and fresh flowers. All have large baths. Most rooms have glass doors that open onto terraces or private decks set with Adirondack chairs. Common areas are inviting—a large parlor-library with a fireplace, a casual den with fireplace in the health club building, and lawns, gardens, and nature trails in the preserve outside. Be aware that less expensive rooms may be small, and consider whether you'd prefer staying in one of the cottages or the main building. ⊠ *74 North Rd., R.R. 1, Box 309, Chilmark 02535,* ☎ *508/645–3322 or 800/356– 3322.* 🖷 *508/645–3799. 25 rooms, 1- to 3-room suites available. Pool, beauty salon, massage, tennis court, exercise room, meeting room, airport shuttle. Closed Dec.–Apr. BP.*

## Nightlife and the Arts

**Chilmark Community Center** (⊠ Beetlebung Corner, Chilmark, ☎ 508/ 645–9484) holds square dances and other events for different age groups throughout the summer; watch for announcements in the papers.

**Martha's Vineyard Chamber Music Society** (☎ 508/696–8055), formerly called the Chilmark Chamber Players, performs 10 summer concerts and three in winter at various venues.

**Nancy Slonim Aronie** (☎ 508/645–9085), a former visiting writer at Trinity College and NPR commentator, leads weeklong summer writing workshops in Chilmark.

**The Yard** (⊠ Off Middle Rd., Chilmark, ☎ 508/645–9662)—a colony of dancers and choreographers, formed in 1973—gives several performances throughout the summer at its 100-seat Barn Theater in a wooded setting. Artists are selected each year from auditions that are held in New York. Dance classes are available to visitors.

## Outdoor Activities and Sports
BEACHES

**Lucy Vincent Beach,** on the south shore, is a beautiful, wide strand of fine sand backed by high clay bluffs and fronted by the Atlantic surf. Keep walking to the left for the unofficial nude beach. In season, Lucy Vincent is restricted to town residents and visitors with passes.

**Squibnocket Beach,** on the south shore, offers an appealing boulder-strewn coastline; a narrow beach that is part smooth rocks and pebbles, part fine sand; and gentle waves. During season, this beach is restricted to residents and visitors with passes.

## Shopping
**Chilmark Chocolates** (⊠ State Rd., ☎ 508/645–3013) sells superior chocolates and the world's finest butter crunch that you can sometimes watch being made in the back room. Don't forget to pick up a catalog—mail orders are available except during warm months.

# Menemsha

★ ③④  *1.6 mi northwest of Chilmark, 3.5 mi east of Gay Head.*

A fishing village unspoiled by the "progress" of the 20th century, Menemsha is a jumble of weathered fishing shacks, fishing and pleasure boats, drying nets and lobster pots, and dads and kids pole-fishing from the jetty. Though the picturesque scene is not lost on myriad photographers and artists, this is very much a working village. The catch of the day taken off boats returning to port is sold at markets along Dutcher's Dock. Romantics bring picnic suppers to the public **Menemsha Beach** for catching perfect sunsets over the water. If you recognize this charming town and feel you've seen it before, you probably have: It was used for location shots in the film *Jaws.*

Very different from most other island conservation areas, **Menemsha Hills Reservation,** a 210-acre Trustees of Reservations property, includes a mile of rocky shoreline and high sand bluffs along Vineyard Sound, hilly walking trails through scrub oak and heathland with interpretive signs at viewpoints, and the 309-ft Prospect Hill, the island's highest, from which there are excellent views of the Elizabeth Islands and beyond. ⊠ *Off North Rd., Chilmark, 1 mi east of Menemsha Cross Rd.,* ☎ *508/693–7662. Call ahead about naturalist-led tours.* ☒ *Free.* ☉ *Daily sunrise–sunset.*

## Dining and Lodging

$$$–$$$$     ✕ **Homeport.** What began as a homestyle restaurant has become an institution, and institutionalized. Unless you get a broiled chicken (and why come to Menemsha for chicken?), the prices are as high as at the fanciest French joint. Of course, you pay for the location—away from the bustle of the towns, with outdoor seating overlooking the harbor. The Homeport has expanded over the years, pretty much obliterating the handsome lines of the original building, and symbolically that says much. What should be a great island treat, from raw bar to lobster, take out or sit down, winds up feeling vaguely like someone's taking advantage of you. ⊠ *At the end of North Rd.,* ☎ *508/645–2679. Reservations essential. MC, V. BYOB. Closed mid-Oct.–mid-Apr. No lunch.*

$     ✕ **The Bite.** Fried everything—clams, fish-and-chips, you name it—are on the menu at this simple, roadside shack, where two outdoor picnic tables are the only seating options. Small, medium, and large are the three options: All of them are perfect if you're craving that classic seaside fried lunch. Closes at 3 PM weekdays, 7 PM weekends. ⊠ *Basin Rd., no phone. No credit cards.*

$ ✕ ★ **Larson's.** Dutcher's Dock would not be the same without this authentic, funky port-side stopover. Basically a retail fish store, with reasonable prices for the island and superb quality anywhere, Larson's also will open some oysters for you, rustle up fish cakes, stuff quahogs, and cook a lobster to order. Then again, you can also grab some fish stock for a stew at home, or a handful of herring roe, and even bait squid for the bluefish you want to catch. The best deal is a plate of littlenecks or cherrystones for $6.50 a dozen. They taste especially good with a bottle of your own wine, sitting at one of the two outdoor picnic tables or on a blue bench, back against the wall, with upturned fish crates for tables. Maybe the nicest view of all is late in the afternoon, not just for sunset, but to contemplate the hardworking rust-bucket fishing trawlers moored at the back door. If it's lobster without fanfare that you want, it's here too, but be sure to call ahead (it sometimes runs out by dinnertime). Closes at 6 PM weekdays, 7 PM weekends. ⊠ *Dutcher's Dock,* ☎ *508/645–2680. MC, V. No seating. BYOB. Closed mid-Oct.–mid-May.*

$ ✕ **Menemsha Deli.** What fills the bill in Menemsha would not qualify for the title "deli" in New York City, but the sandwiches are thick, fresh, and ready to go. Don't mind the celebrity-naming of various combinations; the celebrities don't, and it doesn't affect the taste. ⊠ *Basin Rd.,* ☎ *508/645–9902. No credit cards. No dinner.*

$$$$ ✕🛏 **Beach Plum Inn.** A woodland setting, a panoramic view of the ocean and Menemsha Harbor, and a wonderful romantic restaurant with spectacular sunset views: The main draws of this 10-acre retreat are superb. What a pleasure to have the feeling that you're in good hands—whether those of an attentive waiter or of an owner who will tell you that the clams that have just come off the boat may be a little sandy. And they turn out to be the best clams you've ever tasted. Smoked bluefish and caper pâté served with crusty bread is also excellent. You can't go wrong if you choose roasted loin of pork with chili, coconut, and basil, or bouillabaisse in white wine saffron broth. And crème brûlée doesn't come too much closer to perfection than it does here. The romantic dining room overlooks Menemsha Harbor. Reservations are essential; it's BYOB, and lunch is not served. The inn's cottages are decorated in casual beach style. Inn rooms—some with private decks offering great views—have modern furnishings and small new baths; all are air-conditioned and have TVs and mini-refrigerators. Optional MAP rates include breakfast and dinner as well as afternoon cocktails and hors d'oeuvres served on the terrace. Owners Paul and Janie Darrow create a casual feel here, making it a great place to bring children. ⊠ *Beach Plum La., 02552,* ☎ *508/645–9454,* ℻ *508/645–2801. 5 inn rooms, 4 cottages. Restaurant, tennis court, croquet. No pets. No smoking. AE, D, MC, V. Closed mid-Oct.–mid-May. BP.*

$$ 🛏 ★ **Menemsha Inn and Cottages.** For 40 years the late *Life* photographer Alfred Eisenstaedt returned to his cottage on the hill for the panoramic view of Vineyard Sound and Cuttyhunk beyond the trees below. All with screened porches, fireplaces, and full kitchens, the cottages are nicely spaced on 10 acres and vary in privacy and water views. You can also stay in the 1989 inn building, or in the pleasant Carriage House, both of which have white walls, plush blue or sea green carpeting, and Appalachian light pine reproduction furniture. All rooms and suites have private decks, most with fine sunset views. Suites have sitting areas, desks, mini-refrigerators, and big tiled baths. The Continental breakfast is available at the inn and Carriage House only.

✉ *North Rd., Box 38, Menemsha 02552,* ☎ *508/645–2521. 9 rooms, 6 suites, 12 cottages. No pets. No credit cards. Closed Nov.–Apr. CP.*

### Outdoor Activities and Sports

BEACHES

**Menemsha Public Beach,** adjacent to Dutcher's Dock, is a pebbly beach with gentle surf on Vineyard Sound. Located on the western side of the island, it is a great place to catch the sunset. Fishing boats and anglers on the jetty add atmosphere. There are rest rooms and lifeguards. Snack stands and restaurants are a short walk from the parking lot.

FISHING

**North Shore Charters** (✉ Menemsha Harbor, ☎ 508/645–2993) has boats for charter.

### Shopping

**Pandora's Box** (✉ Basin Rd., ☎ 508/645–9696) has unique women's clothing and accessories.

## Gay Head (Aquinnah)

③⑤ *11.9 mi southwest of West Tisbury, 20 mi west of Edgartown.*

Gay Head is an official Native American township. In 1987, after more than a decade of struggle in the courts, the Wampanoag tribe won guardianship of 420 acres of land, which are being held in trust by the federal government in perpetuity and constitute the Gay Head Native American Reservation. Also in Gay Head is the 380-acre estate of the late Jacqueline Onassis. The "center" of Gay Head consists of a combination fire and police station, the town hall, a Tribal Council office, and a public library, formerly the little red schoolhouse. Because the town's year-round population hovers around 650, Gay Head children attend schools in other towns. In 1997, the town voted to change Gay Head back to its original Native American name, **Aquinnah** (pronounced a-*kwih*-nah), which is Wampanoag for "land under the hill"; the official change will take several years to fully implement.

③⑥ **Quitsa Pond Lookout** has a good view of the adjoining Menemsha and Nashaquitsa ponds, the woods, and the ocean beyond. ✉ *State Rd.*

③⑦ **Gay Head spring,** from a roadside iron pipe, gushes water cold enough to slake a cyclist's thirst on the hottest day. Feel free to fill a canteen. Locals come from all over the island to fill jugs. The spring is just over the town line. ✉ *State Rd.*

★ ③⑧ Don't miss the spectacular **Gay Head Cliffs,** a National Historic Landmark and part of the Wampanoag reservation land. These dramatically striated walls of red clay are the island's major tourist attraction, as evidenced by the tour bus–filled parking lot. Native American crafts and food shops line the short approach to the overlook, from which you can see the Elizabeth Islands to the northeast across Vineyard Sound and Noman's Land Island, part wildlife preserve, part military bombing-practice site, 3 mi off the Vineyard's southern coast.

If you've reached a state of quiet, vacationland bliss, keep in mind that this *is* a heavily touristed spot, and it might turn out to be a shock to your senses. When you come, consider going to the lighthouse first, then down and around to the beach and cliffs. If you're famished when you arrive, you can eat at L'Osteria at the Aquinnah (☞ *below*), at the top of the loop by the lighthouse.

There is no immediate access to the beach from the light—nothing like an easy staircase down the cliffs to the sand below. To reach the cliffs, park in the **Moshup Beach** (☞ Beaches *in* Outdoor Activities and

Sports, *below*) lot by the lighthouse loop, walk five-plus minutes south on the boardwalk, then continue another 20 or more minutes on the sand to get back around to the lighthouse. The cliffs themselves are pretty marvelous, and you should plan ahead if you want to see them. It takes a while to get to Gay Head from elsewhere on the island, and in summer the parking lot and beach fill up, so start early to get a jump on the throngs.

Adjacent to the cliffs overlook is the **Gay Head Lighthouse,** the largest of the Vineyard's five, precariously stationed atop the rapidly eroding cliffs. In 1799 a wooden lighthouse was built here—the island's first—to warn ships of Devil's Bridge, an area of shoals ¼ mi offshore. The current incarnation, built in 1856 of redbrick, carries on with its alternating pattern of red and white flashes. Despite the light, the Vineyard's worst wreck occurred here in January 1884, when the *City of Columbus* sank, taking with it into the icy waters more than 100 passengers and crew. The original Fresnel lens was removed when the lighthouse was automated in 1952, and it is preserved at the Vineyard Museum in Edgartown (☞ *above*). The lighthouse is open to the public summer weekends for sunsets, weather permitting; private tours can also be arranged. ☎ *508/645–2211.* 🎫 *$2.*

A scenic route worth the trip is West Basin Road, which takes you along the Vineyard Sound shore of Menemsha Bight and through the **Cranberry Lands,** an area of cranberry bogs gone wild that is a popular nesting site for birds. No humans can nest here, but you can drive by and look. At the end of the road, with marshland on the right and low dunes and grasses and the long blue arc of the bight on the left, you get a terrific view of the quiet fishing village of Menemsha, across the water. **Lobsterville Road Beach** here is public, but public parking is limited to—literally—three spots, so get here early if you want one of them.

### Dining and Lodging

**$$** ✕ **L'Osteria at the Aquinnah.** At the far end of a little row of shops owned and operated by members of the Gay Head Wampanoag tribe, L'Osteria at the Aquinnah is currently under lease from an old Gay Head family to the owners of Diodatti's (in Vineyard Haven). The result is straight-ahead, generic seaside servings, better for lunch than for dinner. Most imaginative is a bluefish burger, served just like a hamburger, which suddenly makes perfect sense. If the feel is generic, the view is not. The deck is world class, high on the Gay Head bluffs. The ocean rolls to the horizon, while clay and sand along the shore make for Caribbean colors far below. The owners always add one serious request: Please don't feed the seagulls. ✉ *Gay Head bluffs,* ☎ *508/645-9654. AE, MC, V.*

**$$$$** ✕🏠 **Outermost Inn.** In 1990 Hugh and Jeanne Taylor converted the
★ home they built years ago by the Gay Head Cliffs into a B&B. Their design takes full advantage of the superb location: Standing alone on acres of moorland, the house is wrapped with windows revealing breathtaking views of sea and sky in three directions. A wide porch that extends along another side of the house affords a great view of the Gay Head Lighthouse. The restaurant presents a lovely, romantic dinner. Crab cakes are very good for an appetizer (the menu calls them "The Best"), and the baked stuffed seafood platter is easy to recommend, especially if you feel like giving yourself a treat. The restaurant seats twice for dinner, at 6 and 8 PM four to six nights a week from spring through fall, and it is open to the public. Dinners are all prix fixe, from $38 to $48. (Reservations are essential, it's BYOB, and lunch is not served.) The overall decor of the inn is clean and contemporary, with white walls, local art, and polished light-wood floors. Each room has a phone, and one has a whirlpool tub. The beach is a 10-minute walk

away. Hugh has sailed area waters since childhood and charters out his 50-ft catamaran for excursions to Cuttyhunk Island. ⊠ *Lighthouse Rd., R.R. 1, Box 171, 02535,* ☎ *508/645–3511,* F̅A̅X̅ *508/645–3514. 7 rooms. Restaurant. No pets. AE, D, MC, V. Closed Nov.–May. BP.*

## Outdoor Activities and Sports

BEACHES

**Moshup Beach** is, according to the Land Bank, "probably the most glamorous" of its holdings. It is also the most expensive, with a seasonal parking fee of $15 per day (Memorial Day–Labor Day). The reason for that is that Moshup provides access to the spectacular Gay Head Cliffs. The best of the cliffs and a view up to the lighthouse are a 25-plus-minute walk on boardwalk and beach. There is a drop-off close to the beach, but you still must park in the lot and walk to the sand. If you're not driving, bike racks are on the beach itself, and the island shuttle bus stops here. Come early in the day to ensure both a quieter experience and finding a parking spot. Keep in mind that climbing the cliffs is against the law—they're eroding much too quickly on their own. It's also illegal to take any of the clay with you. ⊠ *Parking lot: intersection of State Rd. and Moshup Dr.*

**Lobsterville Road Beach** comprises 2 mi of beautiful sand and dune beach on the Vineyard Sound. It is a seagull nesting area and a favorite fishing spot. Though the water tends to be cold, the beach is protected and suitable for children. There is no public parking.

**Lobsterville Beach,** off Lobsterville Road in Gay Head, is a shallow, calm, sandy beach on Menemsha Pond with resident-only parking and a lovely setting.

# MARTHA'S VINEYARD A TO Z

## Arriving and Departing

### By Bus
**Bonanza Bus Lines** (☎ 508/548–7588 or 800/556–3815) travels to the Woods Hole ferry port from Rhode Island, Connecticut, and New York year-round.

### By Ferry
Car-and-passenger ferries travel to Vineyard Haven from Woods Hole on Cape Cod year-round. In season, passenger ferries from Falmouth and Hyannis on Cape Cod, and from New Bedford, serve Vineyard Haven and Oak Bluffs. All provide parking lots for leaving cars overnight (🖃 $6–$10 per night).

FROM WOODS HOLE

The **Steamship Authority** runs the only car ferries, which make the 45-minute trip to Vineyard Haven year-round and to Oak Bluffs from late May through September.

If you plan to take a car to the island (or to Nantucket) in summer or weekends in the fall, you *must* have a reservation. (Passenger reservations are never necessary.) You should book your car reservation as far ahead as possible; in season, call weekdays from 5 AM to 9:55 PM for faster service. Stand-by car reservations to the island are only available on Tuesday, Wednesday and Thursday; there are no stand-by car reservations to Nantucket. Those with confirmed car reservations must be at the terminal 30 minutes (45 minutes in season) before sailing time. ☎ *508/477–8600 for information and car reservations; on the Vineyard, 508/693–9130 for information and car reservations; TTY 508/540–1394.* 🖃 *One-way in season (mid-May–mid-Oct.): $5 per* *adult, $44 cars; one-way off-season: $5 per adult, call for car rates. Bicycles: $3 one-way year-round.*

A number of **parking lots** in Falmouth hold the overflow of cars when the Woods Hole lot is filled, and free shuttle buses take passengers to the ferry, about 15 minutes away. Signs along Route 28 heading south from the Bourne Bridge direct you to open parking lots, as does AM radio station 1610, which can be picked up within 5 mi of Falmouth.

A **free Martha's Vineyard Chamber of Commerce reservations phone** at the ticket office in Woods Hole connects you with many lodgings and car- and moped-rental firms on the island.

### FROM HYANNIS
**Hy-Line** makes the 1¾-hour run to Oak Bluffs May–October. From June to mid-September, the "Around the Sound" cruise makes a one-day round-trip from Hyannis with stops at Nantucket and Martha's Vineyard ($31). The parking lot fills up in summer, so call to reserve a space in high season. ⊠ *Ocean St. dock,* ☏ *508/778–2600 or 508/778–2602 for reservations; in Oak Bluffs,* ☏ *508/693–0112.* ⊒ *One-way: $11, $4.50 bicycles.*

### FROM FALMOUTH
The **Island Queen** makes the 35-minute trip to Oak Bluffs from late May through Columbus Day. ⊠ *Falmouth Harbor,* ☏ *508/548–4800.* ⊒ *Round-trip: $10, $6 bicycles; one-way: $6, $3 bicycles.*

**Patriot Boats** (☏ 508/548–2626) allows passengers on its daily Falmouth Harbor–Oak Bluffs mail run in the off-season ($4 one-way) and operates a year-round 24-hour water taxi. You can also charter a boat—for about $300.

### FROM NEW BEDFORD
The **Schamonchi** travels between Billy Woods Wharf and Vineyard Haven from mid-May to mid-October. The 600-passenger ferry makes the 1½-hour trip at least once a day, several times in high season, allowing you to avoid Cape traffic. ☏ *508/997–1688 (New Bedford); Martha's Vineyard ticket office:* ⊠ *Beach Rd., Vineyard Haven,* ☏ *508/693–2088.* ⊒ *Round-trip: $16, $5 bicycles; one-way: $9, $2.50 bicycles.*

### FROM NANTUCKET
**Hy-Line** makes 2¼-hour runs to and from Oak Bluffs from early June to mid-September—the only inter-island passenger service. (To get a car from Nantucket to the Vineyard, you must return to the mainland and drive from Hyannis to Woods Hole.) ☏ *508/778–2600 in Hyannis,* ☏ *508/693–0112 in Oak Bluffs,* ☏ *508/228–3949 on Nantucket.* ⊒ *One-way: $11, $4.50 bicycles.*

## By Plane
**Martha's Vineyard Airport** (☏ 508/693–7022) is in West Tisbury, about 5 mi west of Edgartown.

**Air New England** (☏ 508/693–8899) offers an island-based charter service year-round.

**Cape Air/Nantucket Airlines** (☏ 508/771–6944 or 800/352–0714) connects the Vineyard year-round with Boston (including an hourly summer shuttle), Hyannis, Nantucket, and New Bedford. It offers joint fares and ticketing and baggage agreements with several major carriers.

**Continental Express** (☏ 800/525–0280) has seasonal nonstop flights to the Vineyard from Newark.

**Direct Flight** (☏ 508/693–6688) is a year-round charter service based on the Vineyard.

**US Airways Express** (☏ 800/428–4322) has service to the Vineyard out of Boston and New York City's LaGuardia Airport.

**Westchester Air** (☎ 800/759–2929) flies charters out of White Plains, New York.

## By Private Boat

Town harbor facilities are available at **Vineyard Haven** (☎ 508/696–4249), **Oak Bluffs** (☎ 508/693–4355), **Edgartown** (☎ 508/627–4746), and **Menemsha** (☎ 508/645–2846). Private companies include **Vineyard Haven Marina** (☎ 508/693–0720), **Dockside Marina** (☎ 508/693–3392) in Oak Bluffs, and **Edgartown Marine** (☎ 508/627–4388).

# Getting Around

## By Car

In season, the Vineyard is overrun with cars, and many innkeepers will advise you to leave your car at home, saying you won't need it. This is true if you are coming over for just a few days and plan to spend most of your time in the three main towns, Oak Bluffs, Vineyard Haven, and Edgartown, which are connected in summer by a shuttle bus. Otherwise, you'll probably want a car. Driving on the island is fairly simple. There are few main roads, and they are all well marked. You can book rentals through the Woods Hole ferry terminal free phone.

## By Ferry

The three-car **On Time** ferry—so named because it has no printed schedules and therefore can never be late—makes the five-minute run to Chappaquiddick Island June–mid-October, daily 7 AM–midnight, less frequently off-season. ⊠ *Dock St., Edgartown,* ☎ *508/627–9427.* ⊡ *Round-trip: $1 individual, $4.50 car and driver, $2.50 bicycle and rider, $3.50 moped or motorcycle and rider.*

## By Four-Wheel-Drive

Four-wheel-drive vehicles are allowed from Katama Beach to Wasque Reservation with $50 annual permits ($75 for vehicles not registered on the island) sold on the beach in summer, or anytime at the Dukes County Courthouse (⊠ Treasurer's Office, Main St., Edgartown 02539, ☎ 508/627–4250). Wasque Reservation has a separate mandatory permit and requires that vehicles carry certain equipment, such as a shovel, tow chains, and rope; call the rangers before setting out for the dunes.

Jeeps are a good idea for exploring areas approachable only by dirt roads, but going can be difficult in over-sand travel, and even Jeeps can get stuck—stay on existing tracks whenever possible, as wet sand by the water line can suck you in. Actually, most rental companies don't allow their Jeeps to be driven over sand for insurance reasons. Renting a four-wheel-drive vehicle costs $40–$140 per day (seasonal prices fluctuate widely).

## By Limousine

**Muzik's Limousine Service** (☎ 508/693–2212) provides limousine service on- and off-island.

**Holmes Hole Car Rental & Limo Service** (⊠ 5 Corners, Vineyard Haven, ☎ 508/693–8838) provides limo service on-island and occasionally travels off-island.

## By Shuttle Bus

From late June to early September, shuttles operate daily from 7 AM to 12:45 AM between Vineyard Haven (pickup on Union Street in front of the steamship wharf), Oak Bluffs (by the Civil War statue), and Edgartown (on Church Street). For the current bus schedule, call the shuttle hot line (☎ 508/693–1589) or flip through a Steamship Authority schedule. Shuttle tickets cost $1.50–$3 each way. An all-day, unlimited pass is $5, and weekly and monthly passes are also available.

Buses from the Down-Island towns to Gay Head—which stop at the airport, West Tisbury, Chilmark, and on demand wherever it's safe to do so—run every couple of hours between 9 AM and 5 PM in July and August. At other times, call to confirm. Cost is $1–$5 one-way.

## By Taxi

Taxis meet all scheduled ferries and flights, and there are taxi stands by the Flying Horses Carousel in Oak Bluffs, at the foot of Main Street in Edgartown, and by the Steamship office in Vineyard Haven. Fares range from $4 within a town to $35–$40 one-way from Vineyard Haven to Gay Head. Rates double after midnight.

Companies serving the island include **All Island Taxi** (☎ 508/693–3705), **Marlene's** (☎ 508/693–0037) and **Martha's Vineyard Taxi** (☎ 508/693–8660).

## By Minibus in Edgartown

The **Martha's Vineyard Transit Authority** has three shuttle bus routes, two in Edgartown and one in Tisbury. For information on weekly, monthly, or seasonal passes, call 508/627–9663 or 508/627–7448.

DOWNTOWN

From mid-May to mid-September, white-and-purple shuttle buses make a continuous circuit downtown, beginning at free parking lots on the outskirts—at the Triangle off Upper Main Street, at the Edgartown School on Robinson Road (a right off the West Tisbury Road before Upper Main), and on Mayhew Lane. It's well worth taking it to avoid parking headaches in town, and it's cheap (50¢ each way) and convenient. The shuttle buses run every 15 minutes from 7:30 AM to 11:30 PM daily (mid-May–June, 7:30–7) and can be flagged along the route. Service is also available between Vineyard Haven (free parking lot near Cronig's Market) and the Steamship Terminal. Minibuses run every 15 minutes from 5:30 AM–10:30 PM Monday–Thursday and Saturday, 5:30 AM–midnight Friday, Sunday, and holidays. Cost is 50¢ each way and seasonal passes are available; seniors and kids under 12 ride free.

SOUTH BEACH

Shuttle bus service to South Beach via Herring Creek and Katama roads is available mid-June–mid-September for $1.50 one-way (10-trip pass, $10). Pickup is at the corner of Main and Church streets. In good weather there are frequent pickups daily between 10 and 6:30, every half-hour in inclement weather, or you can flag a shuttle bus whenever you see one. In July and August, evening service to and from South Beach runs to 9:45 PM.

# Contacts and Resources

## B&B Reservation Agencies

You can reserve a room at many island establishments via the toll-free phone inside the waiting room at the Woods Hole ferry terminal. The Chamber of Commerce maintains a listing of availability in the peak tourist season, from mid-June to mid-September. During these months, rates are, alas, astronomical, and reservations are essential. In winter, rates go down by as much as 50%. **Martha's Vineyard and Nantucket Reservations** (✉ Box 1322, Lagoon Pond Rd., Vineyard Haven 02568, ☎ 508/693–7200 or 800/649–5671 in MA) book cottages, apartments, inns, hotels, and B&Bs. **DestINNations** (☎ 508/428–5600 or 800/333–4667) handles a limited number of Vineyard hotels and B&Bs, but the staff will arrange any and all details of a visit.

## Vineyard Activities

RESTRICTED BEACHES

Inns occasionally lend guests parking stickers for town beaches that are otherwise limited to residents. These restricted beaches are often much less crowded than public beaches, and some, such as Lucy Vincent and Lambert's Cove, are the most beautiful. These beaches are restricted only seasonally, roughly mid-June to Labor Day. At other times they allow public access.

BIKING

For information on unofficial group rides, call **Cycleworks** (☎ 508/ 693–6966).

CHILDREN'S ACTIVITIES

The following towns offer recreational programs for children: **Edgartown** (☎ 508/627–6145), **Oak Bluffs** (☎ 508/693–2303), **Vineyard Haven** (☎ 508/696–4200), and **West Tisbury** (☎ 508/696–0147).

**Storytelling hours** for children are offered by all the island's libraries. Call towns about days and times: **Chilmark** (✉ State Rd., ☎ 508/645– 3360), **Edgartown** (✉ N. Water St., ☎ 508/627–4221), **Oak Bluffs** (✉ Circuit Ave., ☎ 508/693–9433), **Vineyard Haven** (✉ Upper Main St., ☎ 508/696–4210), and **West Tisbury** (✉ State Rd., ☎ 508/693–3366).

CONSERVATION AREAS

A free map to the islands' conservation lands is available from the Martha's Vineyard Land Bank (✉ 167 Main St., Edgartown 02539, ☎ 508/627–7141) or from some town halls and libraries. It is one of the best general maps covering the Vineyard, and it includes detailed directions, parking information, and usages permitted in conservation areas.

Conservation area bird walks, kids' programs and a schedule of events is listed in newspapers year-round and in the *Best Read Guide,* available free in shops and hotels.

NIGHTLIFE AND THE ARTS

Both **island newspapers,** the *Martha's Vineyard Times* and the *Vineyard Gazette,* as well as the *Cape Cod Times,* publish weekly calendars of events. Also scan the *Best Read Guide,* free at many shops and hotels.

Dial 508/627–6689 to reach a **24-hour movie hot line** with schedules for the **Capawock** (✉ Main St., Vineyard Haven), **Island Theater** (✉ Circuit Ave., Oak Bluffs) and the **Strand** (✉ Oak Bluffs Ave. Ext., Oak Bluffs). **Entertainment Cinemas** (✉ Main St., Edgartown, ☎ 508/627– 8008) has two screens showing first-release films. Films are also shown at other locations from time to time, so check local papers.

SHELLFISHING

Each town issues **shellfish licenses** for the waters under its jurisdiction. Contact the town hall of the town in which you wish to fish for a permit, as well as information on good spots and a listing of areas closed because of seeding projects or contamination: **Chilmark** (☎ 508/645– 2100), **Edgartown** (☎ 508/627–6180), **Gay Head** (☎ 508/645–2300), **Oak Bluffs** (☎ 508/693–5511), **Vineyard Haven** (☎ 508/696–4200), and **West Tisbury** (☎ 508/696–0102).

## Car Rentals

Budget (☎ 508/693–1911), Hertz (☎ 508/693–2402), and All Island (☎ 508/693–6868) arrange rentals from their airport desks.

Atlantic Rent-A-Car (✉ 15 Beach Rd., Vineyard Haven, ☎ 508/693– 0480) rents cars, Jeeps, vans, and convertibles. Adventure Rentals (✉

Beach Rd., Vineyard Haven, ☎ 508/693–1959) rents cars (through Thrifty), as well as mopeds, Jeeps, and buggies, and offers half-day rates. Cost is $25–$65 per day for a basic model car. Vineyard Classic and Specialty Cars (☎ 508/693–5551) rents classic Corvettes, a '57 Chevy, and other cars, as well as mopeds and mountain bikes.

## Cash Machines

There are ATMs at the Edgartown National Bank (⊠ 2 S. Water St. and 251 Upper Main St.), at Park Ave. Mall (⊠ Oak Bluffs, ☎ 508/627–3343), and by the Compass Bank (⊠ Opposite steamship offices in Vineyard Haven and Oak Bluffs; ⊠ 19 Lower Main St., Edgartown; ⊠ Up-Island Cronig's Market, State Rd., W. Tisbury; ☎ 508/693–9400).

## Emergencies

Dial 911 for the hospital, physicians, ambulance services, police, fire departments, or Coast Guard.

Martha's Vineyard Hospital (⊠ Linton La., Oak Bluffs, ☎ 508/693–0410).

Vineyard Medical Services (⊠ State Rd., Vineyard Haven, ☎ 508/693–6399) provides walk-in care; call for days and hours.

## Guided Tours

### CRUISES

The 50-ft sailing catamaran **Arabella** (☎ 508/645–3511) makes day and sunset sails out of Menemsha to Cuttyhunk and the Elizabeth Islands with Captain Hugh Taylor, co-owner of the Outermost Inn.

The teakwood sailing yacht **Ayuthia** (☎ 508/693–7245) offers half-day, full-day, and overnight sails to Nantucket or the Elizabeth Islands out of Coastwise Wharf in Vineyard Haven.

Half- and full-day cruises and sunset cruises to Nantucket or Cuttyhunk on the 54-ft Alden ketch **Laissez Faire** (☎ 508/693–1646) leave from Vineyard Haven.

**Mad Max** (☎ 508/627–7500), a 60-ft high-tech catamaran, offers day sails and charters out of Edgartown.

The **Shenandoah** (☎ 508/693–1699), a square topsail schooner, offers six-day cruises including meals; passengers are ferried to ports, which may include Nantucket, Cuttyhunk, New Bedford, Newport, Block Island, or others. One or two weeks each summer, day sails with lunch are offered (call for schedule). Cruises depart from Coastwise Wharf in Vineyard Haven. "Kids Cruises" (throughout the summer) are for 10- to 18-year-olds only.

### FLIGHTSEEING

See the Vineyard by silent sailplane with Soaring Adventures of America (⊠ Katama Airfield, Herring Creek Rd., Edgartown, ☎ 508/627–3833 or 800/762–7464). Tours are given daily in summer. It's just you and the pilot, and you might even take a turn at the controls.

### ORIENTATION

Buses meet all ferries for two-hour narrated tours of the island from spring through fall, with a stop at the Gay Head Cliffs. They are crowded and claustrophobic, so see the island another way if at all possible.

### WALKING

Liz Villard (☎ 508/627–8619) leads walking tours of Edgartown's "history, architecture, ghosts, and gossip" that include visits to the historic Dr. Daniel Fisher House, the Vincent House, and the Old Whaling Church. Tours are given April–December; call for times. Liz also leads similar tours of Oak Bluffs.

## House Rentals

Sandcastle Realty (✉ Box 2488, 256 Edgartown Rd., Edgartown 02539, ☎ 508/627–5665) and Martha's Vineyard Vacation Rentals (✉ 51 Beach Rd., Box 1207, Vineyard Haven 02568, ☎ 508/693–7711) can help you find rentals for long-term visits.

## Late-Night Pharmacies

Leslie's Drug Store (✉ Main St., Vineyard Haven, ☎ 508/693–1010) is open daily year-round and has a pharmacist on 24-hour call for emergencies.

## Visitor Information

**Martha's Vineyard Chamber of Commerce** is two blocks from the Vineyard Haven ferry. There are town information booths by the Vineyard Haven Steamship terminal, on Circuit Avenue in Oak Bluffs, and on Church Street in Edgartown. ✉ *Beach Rd.,* ☎ *508/693–0085.* ☉ *Weekdays 9–5; also Sat. 10–2 Memorial Day–Labor Day.*

# 4  Nantucket

*Herman Melville may never have set foot on Nantucket, but he was right about its exuberance—in summer, the place brims over with activity. To the eye, Nantucket Town's museum-quality houses and the outlying beaches and rolling moors make the island an aesthetic world unto itself. So hop a ferry to the Gray Lady of the Sea to measure your gait on its historic streets, stand knee-deep in surf casting for stripers, pick up some old whaling lore, or set yourself up on the sand as the sun arcs its way across the sky.*

Updated by
Dorothy
Antczak

Dining
updated by
Seth Rolbein

**A**T THE HEIGHT OF ITS PROSPERITY in the early to mid-19th century, the little island of Nantucket was the foremost whaling port in the world. Its harbor bustled with the coming and going of whaling ships and coastal merchant vessels putting in for trade or outfitting. Along the wharves, a profusion of sail lofts, ropewalks, ship's chandleries, cooperages, and other shops stood cheek by jowl. Barrels of whale oil were off-loaded from ships onto wagons, then wheeled along cobblestone streets to refineries and candle factories. On strong sea breezes the smoke and smells of booming industry were carried through the town as inhabitants eagerly took care of business. It's no wonder that Melville's Ishmael felt the way he did about the place:

*My mind was made up to sail in no other than a Nantucket craft, because there was a fine, boisterous something about everything connected with that famous old island, which amazingly pleased me.*

But the island's boom years didn't last long. Kerosene came to replace whale oil, sought-after sperm whales were overhunted and became scarce, and a sandbar at the mouth of the harbor silted up. Before the prosperity ended, however, enough hard-won profits went into building the grand houses that remind us of the glory days. And wharves once again teem with shops of merchants who tend to the needs of incoming ships—ferry boats, mostly, bringing tourists who crave T-shirts and chic handbags rather than ropes and barrels.

Thanks in no small part to the island's isolation, 30 mi out in the open Atlantic—its original Indian name, *Nanticut,* means faraway land—and to an economy that experienced frequent depressions, Nantucket has managed to retain much of its 17th- to 19th-century character. The town itself hardly seems to have changed since whaling days—streets are still lined with hundreds of beautifully preserved houses and lit by old-fashioned lamps. This remarkable preservation also owes much to the foresight and diligence of people working to ensure that Nantucket's uniqueness can be enjoyed by generations to come. In 1955 legislation was initiated to designate the island an official National Historic District. Now any outwardly visible alterations to a structure—even the installation of air conditioners or a change in the color of paint—must conform to a rigid code.

The code's success shows in the fine harmony of the buildings, most covered in weathered-gray shingles, sometimes with a clapboard facade painted white or gray. Local wood being in short supply, clapboard facades were a sign of wealth. In town, which is more strictly regulated than the outskirts, virtually nothing jars. You'll find no neon, stoplights, billboards, or fast-food franchises. In spring and summer, when the many tidy gardens are in bloom and gray shingles are covered with cascades of roses, it all seems perfect.

The desire to protect Nantucket from change extends to the land as well. When the 1960s tourism boom began, it was clear that something had to be done to preserve the breezy, wide-openness of the island—its miles of clean, white-sand beaches and heath-covered moors—that is as beautiful as the historic town. A third of the 14- by 3-mi island's 30,000 acres are now protected from development, thanks to the ongoing efforts of several public and private organizations and the generosity of Nantucketers, who have donated thousands of acres to the cause. The Nantucket Conservation Foundation (NCF), established in 1963, has acquired through purchase or gift more than 8,200 acres, including working cranberry bogs and great tracts of moorland.

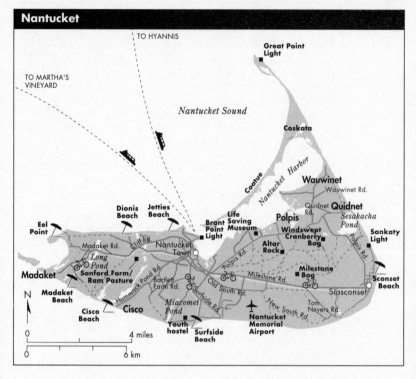

## Nantucket

TO HYANNIS

TO MARTHA'S VINEYARD

*Nantucket Sound*

Great Point Light

Coskata

*Nantucket Harbor*

Coatue

Wauwinet

Wauwinet Rd.

Dionis Beach

Jetties Beach

Quidnet Rd.

Quidnet

*Sesakacha Pond*

Eel Point

Brant Point Light

Life Saving Museum

Polpis

Windswept Cranberry Bog

Sankaty Light

Madaket Rd.

Cliff Rd.

Nantucket Town

Altar Rock

Long Pond

Polpis Rd.

Milestone Bog

Sconset Beach

Madaket

Sanford Farm/ Ram Pasture

Bartlett Farm Rd.

Milestone Rd.

Polpis Rd.

Siasconset

Madaket Beach

Hummock Pond Rd.

Old South Rd.

Tom Nevers Rd.

N

Cisco Beach

Cisco

*Miacomet Pond*

Surfside Rd.

New South Rd.

Youth hostel

Nantucket Memorial Airport

Surfside Beach

0        4 miles

0        6 km

A land bank funded by a 2% tax on real-estate transactions was instituted in 1984 and has since acquired more than 1,000 acres. Most of these are open to the public and marked with roadside signs.

The first Europeans came to the island to escape repressive religious authorities on the mainland—having themselves fled to the New World to escape persecution in England, the Puritans of the Massachusetts Bay Colony proceeded to persecute Quakers and those who were friendly with them. In 1659 Thomas Macy, who had obtained Nantucket through royal grant and a deal with the resident Wampanoag tribe, sold most of the island to nine shareholders for £30 and two beaverskin hats. These shareholders then sold half shares to people whose skills the new settlement would need. The names of these families— Macy, Coffin, Starbuck, Coleman, Swain, Gardner, Folger, and others— appear at every turn in Nantucket, where many descendants continue to reside three centuries later.

The first year, Thomas Macy and his family, along with Edward Starbuck and the 12-year-old Isaac Coleman, spent fall and winter at Madaket, getting by with the assistance of local Wampanoags. The following year, 1660, Tristram Coffin and others arrived, establishing a community—later named Sherburne—at Capaum Harbor on the north shore. When storms closed the harbor early in the 18th century, the center of activity was moved to the present Nantucket Town. Relations with the Wampanoags seem to have been cordial, and many of the tribe became expert whalers. Numbering about 3,000 when the settlers arrived, the native population was greatly reduced in 1763 by a plague. The last full-blooded Wampanoag on the island died in 1855.

The settlers tried their hand at farming, though their crops never did well in the sandy soil. In 1690 they sent for a Yarmouth whaler to teach them to catch right whales from small boats just offshore. In 1712 a boat was blown farther out to sea and managed to capture a sperm whale, whose oil was much more highly prized. Thus began the whaling era on Nantucket.

In the 18th century, whaling voyages never lasted much longer than a year. By the 19th century the usual whaling grounds had been so depleted that ships had to travel to the Pacific to find their quarry and could be gone for five years. Some Nantucket captains actually have South Sea islands named for them—Swain's Reef, Gardner Pinnacles, and so forth. The life of a whaler was very hard, and many never returned home. An account by Owen Chase, first mate of the Nantucket whaling ship *Essex*, of "the mysterious and mortal attack" of a sperm whale, which in 1820 ended in the loss of the ship and most of the crew, fascinated a young sailor named Herman Melville and formed the basis of his 1851 novel, *Moby-Dick*.

The fortunes of Nantucket's whaling industry rose and fell with the tides of three wars and ceased altogether in the 1860s. By the next decade tourism was being pursued, and hotels began springing up at Surfside, on the south shore. Developments at Siasconset to the east followed, and in the 1920s the fishing village became a fashionable resort for theater folk. The tourist trade waxed and waned until the 1960s. Since then it has been the island's bread and butter.

Like the original settlers, most people who visit Nantucket today come to escape—from cities, from stress, and in some ways from the 20th century. Nantucket has a bit of nightlife, including two raucous year-round dance clubs, but that's not what the island is about. It's about small gray-shingled cottages covered with pink roses in summer, about daffodil-lined roads in spring. It's about moors swept with brisk salt breezes and scented with bayberry, wild roses, and cranberries. Perhaps most of all, it's about rediscovering a quiet place within yourself and within the world, getting back in touch with the elemental and taking it home with you when you go.

## Pleasures and Pastimes

### Beaches
The shores of Nantucket are lined with beaches, most of which are open to the public. From mid-June into October the water is usually warm, especially on the Nantucket Sound side. The south and east shores have strong surf and undertow—those on the north and west sides are calmer and warmer; this variety allows for all types of activities, from wading with kids to surfing. Some beaches are accessible by bike path, others by shuttle bus, and others by foot or four-wheel-drive. Most have facilities and lifeguards.

### Bicycling
One of the best ways to tour Nantucket is by bicycle. Miles of paved bike paths wind through all types of terrain from one end of the island to the other; it is possible to bike around the entire island in a day. Several paths lead from town out to beaches, and have drinking fountains and benches placed in strategic spots along the way. The paths are also perfect for runners and rollerbladers.

### Bird-Watching
Nantucket is among the 100 best birding spots in the country—more than 354 species flock to the island's moors, meadows, and marshes in the course of a year. Birds that are rare in other parts of New England thrive here due to the lack of predators and the abundance of wide open, undeveloped space. Northern harriers, short-eared owls, and Savannah sparrows nest in the grasslands; oystercatcher, gulls, plovers, and tiny

least terns nest on sands and in beach grasses; snowy egrets, great blue herons, and osprey stalk the marshlands; and ring-necked pheasant, mockingbirds, and Carolina wrens inhabit the tangled bogs and thickets. Take a pair of binoculars and wander through the conservation areas or along the bike trails or beaches and you're sure to catch sight of interesting bird life.

## Conservation Areas

The island of Nantucket has a fascinating array of landscapes—wild, open tracts of bog land and moors, groves of scrub oak and pitch pine, and wide, sandy beaches that stretch seemingly forever. A diverse assortment of plant and animal species, many of which are endangered or threatened, thrive here as nowhere else. The windswept heathlands that comprise much of Nantucket's acreage are becoming globally rare, and are protected on the island by various conservation organizations who use carefully controlled fires to burn away encroaching shrubs and trees. Flowering plants are especially prolific after these burns, and include such rare species as bush rockrose, pink lady's slipper, and the Eastern silvery aster.

Other local species of flora and fauna to keep watch for: the white flowers of the Nantucket shadbush, brilliant red cardinal flowers, birdsfoot violets, false heather, pink Virginia roses, wild morning glories, and big, beautiful red-orange wood lilies; in the fall, asters, goldenrod, and the tall purple spikes and thistlelike flowers of the endangered New England blazing star. Although cranberries are the most abundant berry, there are also blueberries, raspberries, blackberries, and huckleberries. Beach plums, beach peas, and the fragrant non-native rosa rugosa grow well near the sea in the sandiest soil. In the meadows and groves are mice, moles, meadow voles, white-tailed deer, and cottontail rabbits—and a distinct lack of predatory mammals such as raccoons, skunks, and opossums due to Nantucket's distance from the mainland. On the beach and in the sea, mollusks such as oysters, mussels, steamers, quahogs, and blue crabs are plentiful, and can be harvested as long as you have the proper shellfishing permit.

Please remember that these conservation areas are set aside to preserve and protect Nantucket's fragile ecosystems—tread carefully.

## Dining

"You're looking for a great, cheap place to eat?" mused a Nantucket cabdriver. "Try Cape Cod."

There is sad truth to the joke. Nantucket has adopted Manhattan prices, a solid 20 percent higher than comparable fare on the Cape. For years, this was justified by saying that transportation costs to the remote island are high. But ferry charges for freight are not so exorbitant that a simple salad in virtually every restaurant must cost $7, and besides, no one thinks twice about making the logistical argument while simultaneously trumpeting their use of local fish and produce. Finally, there's the fact that Martha's Vineyard somehow manages to use the ferries, and keep prices within sight.

In the end, there's simply no getting around it: Nantucket is expensive, and basically takes the attitude that if you can't handle it, don't bother to come. What's more, it seems that "American" cuisine, complete with extensive linguistic explanations of every dish, is now obligatory. The vinaigrette is always balsamic, the chicken usually free-ranging, the mushrooms slow-roasted Portobellos, the black pepper invariably crushed and/or bruised. Even the local brewer, Cisco, can't resist the verbiage with their offering; Whales Tale Pale Ale.

But come dinnertime, it's best to grin, bear it, hand over the plastic, and don't let cynicism spoil the meal. Because the other truth is that on this competitive little island, there are some great chefs, sophisticated and well-schooled. Their venues are usually small, carefully appointed, and full of personality. Their wine lists are by and large superb. And there are a few cheaper, more family-style holdouts which can help hold the monthly service charges down.

Two schools of thought are generally offered for handling Nantucket on a fixed budget. One advocates going for the $12 breakfasts and $25 lunches, and then eating very lightly for dinner. The other basically suggests fasting though the day, and then thoroughly enjoying a beautiful supper at night. Who knows, perhaps this approach might be given a name one day: the Nantucket Diet.

Most restaurants open and close seasonally on Nantucket, although most also stay open or re-open for the Christmas stroll weekend in early December. Come off-season, and you'll find only a handful of the less expensive places still serving. Dining rooms associated with inns are the best bet year-round. Some restaurants outside of town offer taxi vouchers, the ultimate in responsible designated drivership.

For price ranges, *see* dining Chart B in On the Road with Fodor's.

## Fishing
Surf fishing is very popular on Nantucket, especially in the late spring when bluefish are running. Blues and bass are the main island catches—bluefishing is best at Great Point—but there are plentiful numbers of other flavorful fish as well. Freshwater fishing is also an option at many area ponds.

## Lodging
Nantucket has only 1,200 beds available at any given time, and in summer, as many as 40,000 people per day descend on the place. With such a captive audience swamping them year in and year out, it seems that some lodging places get lazy. At the same time, others are wonderful and have attentive innkeepers and staff. Apart from cottages and a few inns and hotels scattered across the island, Nantucket's lodging places are in the busy town. Those in the center are convenient, but houses are close together, right on the street. In season there may be street noise until midnight. Inns a 5- or 10-minute walk from the center are quieter. If appliances like coffeemakers, mini-refrigerators, or televisions are important to your stay, ask about them when you reserve a room.

For price ranges, *see* the lodging chart in On the Road with Fodor's.

## Shopping
From the early 1800s on, Nantucket has had an active shopping scene. Sea captains brought merchandise back from their sailing excursions from ports all over the world to trade and sell. When the whaling boom encouraged the men to take to the seas, the women of Nantucket ran the shops, earning the name "Petticoat Row" for a concentrated stretch of women-run shops on Centre Street.

Nantucket Town's commercial district—bounded approximately by the waterfront and Main, Broad, and Centre streets and continuing along South Beach Street—contains virtually all of the island's shops. Old South Wharf, built in 1770, hosts crafts, clothing, and antiques stores, a ship's chandlery, and art galleries in small, connected "shanties." Straight Wharf, where the Hy-Line ferry docks, is lined with T-shirt and other tourist-oriented shops, a gallery, a museum, and restaurants. (Phones and rest rooms are at the end of the wharf.) The majority of Nantucket's shops are seasonal, opening sometime after April and closing between Labor Day and November, though an active core stay open longer.

The island specialty is the Nantucket lightship basket, woven of oak or cane, with woven covers adorned with scrimshaw or rosewood. First made in the 19th century by crew members passing time between chores on the lightships that stood off Sankaty Head, the baskets are now used as purses by those who can afford them—prices range from $400 to well over $1,000. Miniature versions of the baskets are made by plaiting fine threads of gold or silver wire. Some have working hinges and latches, and some are decorated with plain or painted scrimshaw or small gems. Prices start at around $300 for gold versions. Another signature island product is a pair of all-cotton pants called Nantucket Reds, which fade to pink with washing. They're sold at Murray's Toggery Shop. There is something for nearly everyone here—apart from a few too many sweatshirts—from unique decorative items to hand-woven clothing, antiques, and island-made jams and jellies.

## Water Sports
Surfing and sailing are both popular on Nantucket; surfing is especially good on the south and east shores where the wild Atlantic crashes. Canoeing and kayaking on the inland ponds and in Nantucket Harbor are other wonderful ways to cruise the waters.

# Exploring Nantucket

The 14- by 3-mi island of Nantucket has one town, called Nantucket. The village of Siasconset, which is called 'Sconset, is in the southeast corner with a fair number of services and essential stores. North of 'Sconset, Wauwinet is an old residential enclave with a landmark inn. It is the gateway to the sprawling beach and reserves of Coatue, Coskata, and Great Point. The village of Madaket on the west end has a beach and harbor, great sunsets, a seasonal restaurant, and bluefishing off the point. Madaket also has a number of residential areas with no commercial or tourist facilities. Although major roads will take you to most of these areas, exploring them must often be done on dirt roads. Bike paths lead east to 'Sconset, south to Surfside Beach, and west to Madaket. The island is small enough to reasonably get around it all in four days, three if you really dash about.

*Numbers in the text correspond to numbers in the margin and on the Nantucket Town map.*

## Great Itineraries
IF YOU HAVE 2 DAYS

Spend the first day of your trip in Nantucket Town, absorbing the Island's history. Visit the **Museum of Nantucket History** ① and the **Whaling Museum** ⑤; to see them properly will take up much of the morning. Then stop off at one of the town's cafés for lunch—if you don't feel like dining inside, take lunch to **Lily Pond Park** for a picnic. Afterward, stroll along the streets and take note of the magnificent captains' homes built during the prosperous whaling era. Climb the tower at the **First Congregational Church** ⑨ and enjoy the panoramic view of moors, ponds, beaches, Sankaty Light, Great Point, and Muskeget and Tuckernuck islands. Visit the **Macy-Christian House** ㉔ to get an idea of how Nantucketers lived in past centuries. Wander among the cobblestone streets of the town center, or if you are eager for some swimming and sunning, Children's Beach is close by, and Surfside Beach is only a three-mi bike ride away.

Start day two with a visit to the **Hinchman House** ⑯, a natural history museum with examples of local flora and fauna. Then pack a lunch and head off (the best way to tour the island is by bicycle) to explore the walking trails at the **Sanford Farm and Ram Pasture.** Different trails cross diverse habitats, including wetlands, grasslands, forest, and former farmland. Continue on to spend the afternoon at beautiful **Eel Point** beach, walking, relaxing, and enjoying the views that surround you. There is good surf fishing here, and the area is great for birding as well.

You could otherwise head for **Siasconset,** with a stop at the Milestone Bog to get a good view of the famous Nantucket bogs. If you are here in the fall during the cranberry harvest, you'll want to spend more time at a cranberry bog. From 'Sconset, take the short trip to see Sankaty Light. As the day fades, walk around 'Sconset's little village. Be sure to stroll along Evelyn, Lily, and Pochick streets, which remain much as they were in the 1890s. Treat yourself to dinner at one of the village restaurants, and if possible, spend the night in a cottage by the beach.

IF YOU HAVE 4 DAYS
Spend the first day and night of your trip in **Nantucket Town,** following suggestions for the two-day itinerary above.

Cycle out toward **Madaket** the next day and take the 6½ mi (round-trip) trail that leads to the shore via **Hummock Pond**—a superb route for bird-watchers; surfers and swimmers will want to hang out at the beach here. Go on along a south shore road to **Madaket Beach** to catch the sunset, then head back to Nantucket Town for dinner.

In the morning on day three, walk out to **Brant Point Light** and take in the view of the harbor; then continue to **Jetties Beach** and rent a kayak or sailboat, or take a swim. Go back through town and take Polpis Road to the interesting and unique **Nantucket Life Saving Museum.** Go inland a bit to explore the trails that wind through the moor around **Altar Rock,** which has wonderful views. Cut across the moor to Sesachacha Pond and take the path that leads around the pond to the Audubon wildlife area, another great birding spot. Stop by the **Milestone Bog** to get a look at what was once the largest contiguous natural cranberry bog (until it was subdivided after 1959). During harvest season, you'll want to stay for awhile and watch the cranberry-collecting process. Then head for Siasconset for dinner.

Golfers might want to spend the next morning at the Siasconset Golf Club; others might enjoy tennis at the Siasconset Casino. 'Sconset Beach is a fine place to swim or surf, or simply relax in the sun. From 'Sconset, head to the gateway of **Coatue–Coskata–Great Point,** an unpopulated sand spit comprising three cooperatively managed wildlife refuges. Take a Jeep tour, or if you're feeling energetic, walk around Coskata Pond. The Trustees of Reservations sponsors a naturalist-led Great Point natural history tour, which is another option. Spend the afternoon exploring the area or on the beach. Coatue is open for shellfishing, surfcasting, and bird-watching, and though swimming is discouraged, the beach is a beautiful place to sit and read, or picnic. You may want to plan ahead for dinner at the Wauwinet Inn.

IF YOU HAVE 6 DAYS
If you're coming for nearly a week, you can spread out your activities and get to see all corners of the island. Spend the first day and night of your trip in **Nantucket Town.** You may want to take Roger Young's historic walking tour, chalk up a few museums, or head over to Jetties Beach. Then consider spending a couple of nights at the **Wauwinet Inn** to take advantage of the array of activities offered, or station yourself in **'Sconset** for a couple of days. Schedule a day or night for deep-sea fishing, or reserve a spot on the beautiful sloop *Endeavor* to sail around the harbor and out to **Coatue.** Put together a picnic and dig your feet

into the sand at **Eel Point.** With all the busy-ness you can find in town, be sure to compensate for it with some time in the secluded spots of the island.

## When to Tour Nantucket

Summer is Nantucket's most popular season, when everything is open and positively bustling. The Nantucket Harborfest in June celebrates Nantucket's ties to the sea with all sorts of festivities. There are house tours and sandcastle competitions, and there's always something going on.

Autumn is harvest time, and on Nantucket that means one thing—cranberries. The harvesting process is fascinating to watch, and the landscape is particularly lovely at this time of year.

Winter is quiet, with many shops and restaurants closed. It is a perfect time to visit if you enjoy roaming the streets and beaches in solitude. Around the holidays, the town comes to life and the throngs return for the Christmas Stroll, an old-time celebration including carolers, theatrical performances, art exhibitions, crafts fairs, and a tour of historical homes.

Spring is heralded by a profusion of blooms, most notably daffodils—they are everywhere. There is a four-day Daffodil Festival that culminates in a procession of narcissi-adorned antique cars that head to Siasconset for tailgate picnics. The island prepares for another summer, and it all begins again.

# NANTUCKET TOWN

*7 mi west of Siasconset, 3 mi north of Surfside, 6 mi east of Madaket.*

Nantucket Town is steeped in history, and it has one of the finest historical districts in the country. Beautiful 18th- and 19th-century architecture abounds, and museums have carefully preserved Nantucket's important whaling history. Located on a magnificent harbor, the town remains the center of island activity, as it has since the early 1700s. A small commercial area of a few square blocks leads up from the waterfront. Beyond it, quiet residential roads fan out. As you wander you may notice a small round plaque by some doorways. Issued by the Nantucket Historical Association, the plaques certify that the house dates from the 17th century (silver), 1700–1775 (red bronze), 1776–1812 (brass), 1813–1846 (green), or 1847–1900 (black). Unfortunately, they all seem to turn coppery green or black with age.

NOTE: Fourteen historic properties along Nantucket Town's streets are operated as museums by the **Nantucket Historical Association.** At any one of them you can purchase an NHA Visitor Pass ($10), which entitles you to free entry at all 14 museums, or you can pay single admission at each (prices vary). Most are open daily from Memorial Day to Columbus Day, except the one or two closed each year for maintenance. Two have longer seasonal hours: the ☞ **Museum of Nantucket History,** which reopens for Thanksgiving and Christmas Stroll weekends, and the ☞ **Whaling Museum,** which closes after the Christmas Stroll and reopens weekends after April 1. ☎ *508/228–1894 or 508/ 325–4015.* ☉ *NHA hours vary from year to year. The 1997 schedule was as follows: Memorial Day–Labor Day, daily 10–5; Labor Day– Columbus Day, 11–3. Most NHA properties close between Columbus Day and Memorial Day.*

★ ❶ The geological and historical overview given by the **Museum of Nantucket History** is a good introduction to Nantucket, and helps to put

# Nantucket Town

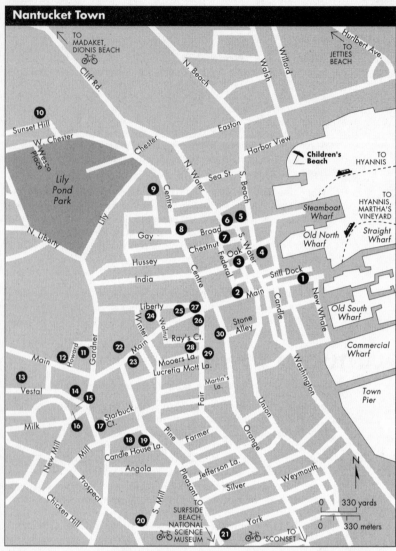

African Meeting
House, **21**

Atheneum, **3**

Coffin
Houses, **25**

Dreamland Theatre, **4**

1800 House, **18**

Fire Hose Cart
House, **11**

First Congregational
Church, **9**

Greater Light, **12**

Hadwen House, **23**

Hinchman House, **16**

Jared Coffin House, **8**

John Wendell Barrett
House, **26**

Macy-Christian
House, **24**

Maria Mitchell
Science Library, **15**

Mitchell House, **14**

Moors' End, **19**

Museum of
Nantucket History, **1**

Nantucket
Information Bureau, **7**

Old Gaol, **13**

Old Mill, **20**

Oldest House, **10**

Pacific Club/Chamber
of Commerce, **2**

Pacific National
Bank, **27**

Peter Foulger
Museum/Nantucket
Historical Association
Research Center, **6**

Quaker Meeting
House/Fair Street
Museum, **28**

St. Paul's Episcopal
Church, **29**

Starbuck Refinery and
Candle Works, **17**

"Three Bricks," **22**

Unitarian Universalist
Church, **30**

Whaling Museum, **5**

sights around the island in perspective. The brick building in which the museum is set was built by Thomas Macy after the Great Fire of 1846—which destroyed the wharves and 400 buildings, about a third of all those in the town—as a warehouse for the supplies needed to outfit whaling ships. It has been restored with historic accuracy down to period doors, hatchways, and hoists. Inside, audio and visual displays include an early fire-fighting vehicle, ship models, enlargements of old photographs, and a 13-ft diorama (with narration) showing the shops, ships, and activities of the bustling waterfront before the fire. The second floor features changing exhibits. Live demonstrations of such early island crafts as candle making are given daily from mid-June to Columbus Day. ⊠ *Straight Wharf.*

➋ The redbrick **Pacific Club** building still houses the elite club of Pacific whaling masters for which it is named. Understandably, since the last whaling ship was seen here in 1870, the club now admits whalers' *descendants,* who gather for the odd cribbage game or a swapping of tales. The building began in 1772 as the counting house of William Rotch, owner of the *Dartmouth* and *Beaver,* two of the three ships that hosted a famous tea party in Boston. According to the NHA, the plaque outside the Pacific Club identifying the third ship, the *Eleanor,* as Rotch's is incorrect. ⊠ *Main St.*

The most photographed view of **Main Street** is from the Pacific Club at the bottom end. The cobblestone square has a harmonious symmetry. The Pacific Club anchors the foot of it, and the Pacific National Bank, another redbrick building, squares off the head. The only broad thoroughfare in town, Main Street was widened after the Great Fire leveled all its buildings except those made of brick, to safeguard against flames hopping across the street in the event of another fire. The cobblestones were brought to the island as ballast in returning ships and laid to prevent the wheels of carts heavily laden with whale oil from sinking into the dirt on their passage from the waterfront to the factories.

At the center of Lower Main is an old horse trough, today overflowing with flowers. From here the street gently rises. At the bank it narrows to its pre-fire width and leaves the commercial district for an area of mansions that escaped the blaze. The simple shop buildings that replaced those lost are a pleasing hodgepodge of sizes, colors, and styles. Elm trees—thousands of which were planted in the 1850s by Henry and Charles Coffin—once formed a canopy over Main Street, but Dutch elm disease took most of them. In 1991 Hurricane Bob took two dozen more.

You might want to stop at the **Chamber of Commerce** to get your bearings. You'll find maps and island information available. ⊠ *Upstairs at 48 Main St.,* ☎ *508/228–1700.*

| NEED A BREAK? | You can breakfast or lunch inexpensively at several soup-and-sandwich places, including **David's Soda Fountain.** ⊠ *Congdon's Pharmacy, 47 Main St.,* ☎ *508/228–4549.* |
| --- | --- |

**Nantucket Pharmacy** has a lunch counter that offers a quick, inexpensive lunch of simple fare, mostly sandwiches. The local favorite is tuna salad—try one! ⊠ *45 Main St.,* ☎ *508/228–0180.*

➌ The great white Greek Revival building with the odd windowless facade and fluted Ionic columns is the **Atheneum,** Nantucket's town library. Completed in 1847 to replace a structure lost to the fire, this is one of the oldest libraries in continuous service in the United States. The famous astronomer, Maria Mitchell (☞ Mitchell House, *below*), was its first librarian. Opening ceremonies featured a dedication by Ralph Waldo Emerson, who—along with Daniel Webster, Henry David Thoreau, Frederick Douglass, Lucretia Mott, and John James Audubon—later delivered lectures in the library's second-floor Great Hall. In the 19th century the hall was the center of island culture, hosting public meetings, suffrage rallies, and county fairs.

Closed for more than a year for renovations, the Atheneum reopened in the spring of 1996; the inside is beautifully restored and the space arranged in a way that is far roomier and more functional. Check out the periodicals section on the second floor to get all the local news. A morning story hour is held in the new children's wing year-round, and a Saturday-morning children's film is offered off-season. The adjoining Atheneum Park is a wonderful spot to read. ⊠ *1 Lower India St.,* ☎ *508/228–1110.*

❹ Currently a summer cinema, the **Dreamland Theatre** was built as a Quaker meeting house, then became a straw factory, and later an entertainment hall. It was moved to Brant Point as part of the grand Nantucket Hotel in the late 19th century and then floated across the harbor by barge about 1905 and installed in its present location—a good illustration of early Nantucketers' penchant for the multiple use of dwellings as well as for moving houses around. Trees (and therefore lumber) were so scarce that Herman Melville joked in *Moby-Dick* that "pieces of wood in Nantucket are carried about like bits of the true cross in Rome." ⊠ *17 S. Water St.,* ☎ *508/228–5356.*

★ ❺ A wonderful way to learn about the whaling era is by visiting the **Whaling Museum,** set in an 1846 factory built for refining spermaceti and making candles. Spermaceti candles, incidentally, gave off a clean, steady light and only a slight fragrance, which is why they became such popular replacements for smelly tallow candles; due to the depletion of the sperm whale, authentic spermaceti candles are no longer made. This museum immerses you in Nantucket's whaling past with exhibits that include a fully rigged whaleboat, harpoons and other implements, portraits of sea captains, a large scrimshaw collection, a full-size tryworks once used to process whale oil aboard ship, the skeleton of a 43-ft finback whale, replicas of cooper, blacksmith, and other ship-fitting shops, and the original 16-ft-high glass prism from the Sankaty Light. The knowledgeable and enthusiastic staff gives a 20- to 30-minute introductory talk peppered with tales of a whaling man's life at sea (call for tour times). Don't miss the museum's gift shop next door. ⊠ *Broad St.,* ☎ *508/228–1894.* 🎫 *$5 or NHA pass. Museum closes after Christmas Stroll and reopens weekends April 1.*

❻ The **Peter Foulger Museum and Nantucket Historical Association Research Center** offer a peek into Nantucket's genealogical past. The museum displays changing exhibits from the permanent collection, including portraits, textiles, porcelains, silver, and furniture. The Research Center's extensive collection of manuscripts, photographs, ships' logs and charts, and genealogical records is open only to those doing research. ⊠ *Broad St.,* ☎ *508/228–1655.* 🎫 *$4 or NHA pass. Research permit: $10 (2 days).* 🕒 *Weekdays 10–4.*

┈┈┈┈┈┈┈┈

NEED A        The **Juice Bar** serves homemade ice cream with lots of toppings, waffle
BREAK?        cones, and baked goods. There are also fresh-squeezed juices, frozen
              yogurt, and specialty coffees. Long lines signal good things coming to
              those who wait. ⊠ *12 Broad St.,* ☎ *508/228–5799.* 🕒 *Apr.–mid-Oct.*

┈┈┈┈┈┈┈┈

❼ The **Nantucket Information Bureau** has public phones, rest rooms, and a bulletin board posting events. ⊠ *25 Federal St.,* ☎ *508/228–*

*0925.* ☉ *July–Labor Day, daily; Labor Day–June, weekdays. Call for hours.*

A stroll through town is the best way to absorb the history of Nantucket. Sea captains' houses lining the streets attest to the prosperity the Island once enjoyed. Most of the Greek Revival houses you will see were built to replace buildings lost in 1846's Great Fire.

**❽** The **Jared Coffin House** has operated as an inn since the mid-19th century. Coffin was a wealthy merchant, and he built this Georgian brick house with Ionic portico, parapet, hip roof, and cupola—the only three-story structure on the island at the time—for his wife, who wanted to live closer to town. They moved here in 1845 from their home on Pleasant Street, but (so the story goes) nothing would please Mrs. Coffin, and within two years they left the island altogether for Boston. ⊠ *29 Broad St.,* ☎ *800/248–2405.*

**★ ❾** The **First Congregational Church,** also known as the Old North Church, is Nantucket's largest and most elegant. Its tower—whose steeple is capped with a weather vane depicting a whale catch—rises 120 ft, providing the best view of Nantucket to be had. On a clear day the reward for climbing the 92 steps (many landings break the climb) is a panorama encompassing Great Point, Sankaty Light, Muskeget and Tuckernuck islands, moors, ponds, beaches, and the winding streets and rooftops of town. Peek in at the church's interior, with its old box pews, a chandelier 7 ft in diameter, and a trompe l'oeil ceiling done by an Italian painter in 1850 and since restored. The 1904 organ has 914 wood and metal pipes. (Organ aficionados may want to have a look at the 1831 Appleton organ—one of only four extant—at the United Methodist Church next to the Pacific National Bank on Main Street.) The Old North Vestry in the rear, the oldest house of worship on the island, was built in 1725 about a mile north of its present site. The main church was built in 1834. ⊠ *62 Centre St.,* ☎ *508/228–0950.* ⊠ *$3.* ☉ *Mid-June–mid-Oct., Mon.–Sat. 10–4.*

**❿** History and architecture buffs should be sure to get a look at the hilltop **Oldest House,** also called the Jethro Coffin House, built in 1686 as a wedding gift for Jethro and Mary Gardner Coffin. The most striking feature of the saltbox—the oldest house on the island—is the massive central brick chimney with brick horseshoe adornment. Other highlights are the enormous hearths and diamond-pane lead-glass windows. Cutaway panels show 17th-century construction techniques. The interior's sparse furnishings include an antique loom. ⊠ *Sunset Hill (a 10- to 15-minute walk out Centre St. from Main St.).* ⊠ *$3 or NHA pass.*

**Lily Pond Park,** a 5-acre Land Bank conservation area on the edge of the town center, is a prime bird-watching spot. Its lawn and wetlands—there is a trail, but it's muddy—foster abundant wildlife, including birds, ducks, and deer. You can pick blackberries, raspberries, and grapes in season wherever you find them. ⊠ *N. Liberty St.*

**⓫** Built in 1886 as one of several neighborhood fire stations—Nantucketers had learned their lesson—the **Fire Hose Cart House** displays a small collection of fire-fighting equipment used a century ago. ⊠ *8 Gardner St.* ⊠ *Free.*

**⓬** A whimsical blend of necessity and creativity, **Greater Light** is an example of the summer homes of the artists who flocked to Nantucket in its early resort days. In the 1930s two unusual Quaker sisters from Philadelphia—actress Hanna and artist Gertrude Monaghan—converted a barn into what looks like the lavish set for an old movie. The exotic decor includes Italian furniture, Native American artifacts and textiles, a wrought-iron balcony, bas-reliefs, and a coat of arms. The sisters also remodeled the private house next door, called Lesser Light, for their parents. Greater Light, closed for the past few years, is scheduled to be open for the 1998 season, but get the latest scoop at the Chamber of Commerce. ⊠ *8 Howard St.*

❸ It's tough to escape the law when you live on an island. Those who did not obey the rules and regulations ended up in the **Old Gaol,** an 1805 jailhouse in use until 1933. Shingles mask the building's construction of massive square timbers, plainly visible inside. Walls, ceilings, and floors are bolted with iron. The furnishings consist of rough plank bunks and open privies, but you needn't feel too much sympathy for the prisoners: Most of them were allowed out at night to sleep in their own beds. ⊠ *15R Vestal St.* ☞ *Free.*

The **Maria Mitchell Association,** established in 1902 by Vassar students and Maria's family, administers The Mitchell House, the Maria Mitchell Science Library, the Hinchman House, the Maria Mitchell Aquarium, and the Loines Observatory. A combination admission ticket to all sites is $5.

Women played a strong role in Nantucket's history. During the whaling days men would be gone for up to five years at a time—and it was up to the women to keep the town going. They became leaders in every arena, from religion to business. Mary Coffin Starbuck helped establish Quakerism on the island and was a celebrated preacher. Lucretia Coffin Mott was a powerful advocate of the antislavery and women's-rights movements. During the post–Civil War depression, Centre Street near Main Street became known as Petticoat Row, a reflection of the large number of women shopkeepers.

❹ Astronomy aficionados will appreciate the **Mitchell House,** birthplace of astronomer and Vassar professor Maria (pronounced mah-*rye*-ah) Mitchell, who in 1847, at age 29, discovered a comet while surveying the sky from the top of the Pacific National Bank. Her family had moved to quarters over the bank, where her father—also an astronomer—worked as a cashier. One of 10 children of Quaker parents, Mitchell was the first woman astronomy professor in the United States, and the first woman to discover a comet. The restored 1790 house contains family possessions and Maria Mitchell memorabilia, including the telescope with which she spotted the comet. The kitchen, of authentic wide-board construction, retains the antique utensils, iron pump, and sink of the time. Tours of the house and the roof walk are available. The adjacent observatory is used by researchers and is not open to the public. ⊠ *1 Vestal St.,* ☎ *508/228–2896.* ☞ *$3.* ☼ *Mid-June–Aug., Tues.–Sat. 10–4.*

❺ The **Maria Mitchell Science Library** contains an extensive collection of science books and periodicals, including field-identification guides and gardening books, as well as books on Nantucket history. ⊠ *Vestal St.,* ☎ *508/228–9219.* ☞ *Free.* ☼ *Mid-June–mid-Sept., Tues.–Sat. 10–4; mid-Sept.–mid-June, Wed.–Fri. 2–5, Sat. 9–12.*

❻ If you're interested in the natural history of Nantucket, the **Hinchman House Natural Sciences Museum**—displays specimens of local birds, shells, insects, and plants, exhibits habitats, and offers bird and wildflower walks and children's nature classes. ⊠ *7 Milk St.,* ☎ *508/228–0898.* ☞ *$3.* ☼ *Mid-June–Aug., Tues.–Sat. 10–4.*

☾ **Maria Mitchell Aquarium** presents local marine life in salt- and freshwater tanks. Family marine ecology trips are given four-times weekly in season. ⊠ *28 Washington St., near Commercial Wharf,* ☎ *508/228–5387.* 🎫 *$1.* ☉ *Mid-June–Aug., Mon., Tues., Wed. 10–4.*

**Loines Observatory** offers Wednesday-night stargazing mid-July to mid-August. ⊠ *Milk St. Ext.,* ☎ *508/228–9198 or 508/228–9273.*

Nantucket is packed with places that speak of its history. Off New Dollar Lane by Milk Street, and down a long driveway, the remains of the **17** **Starbuck refinery and candle works** are now used as apartments and garages.

**18** To see how the other half lived, stop by the **1800 House,** a typical Nantucket home—one not enriched by whaling money—of that time. Once the residence of the high sheriff, the house features locally made furniture and other household goods, a six-flue chimney with beehive oven, and a summer kitchen. ⊠ *10 Mill St.*

Built between 1829 and 1834, the handsome Federal brick house **19** known as **Moors' End** is where Jared Coffin lived before moving to what is now the **Jared Coffin House** (☞ *above*)—the proximity to the fumes from the Starbuck refinery was one of Mrs. Coffin's complaints. It is a private home, so you won't be able to see Stanley Rowland's vast murals of the whaling era on the walls or the scrawled notes about shipwreck sightings in the cupola (your best bet for a glimpse is a coffee-table book about Nantucket). Behind high brick walls is the largest walled garden on Nantucket. ⊠ *19 Pleasant St.*

**20** Several windmills sat on Nantucket hills in the 1700s, but only the **Old Mill,** a 1746 Dutch-style octagonal windmill built with lumber from shipwrecks, remains. The Douglas-fir pivot pole used to turn the cap and sails into the wind is a replacement of the original pole, a ship's foremast. The mill's wooden gears work on wind power, and when the wind is strong enough, corn is ground into meal that is sold here. ⊠ *S. Mill St.* 🎫 *$2 or NHA pass.*

As far back as the early 1700s, there was a small African-American population on Nantucket; the earliest blacks were slaves of the island's earliest settlers. When slavery was abolished on the island in 1770, Nantucket became a destination for free blacks and escaping slaves. Today nine sights associated with the island's African-American heritage are on the self-guided **Black Heritage Trail** tour (☞ Island Activities in Nantucket A to Z, *below*). Of these, the **African Meeting House** is the only **21** extant public building constructed and occupied by the island's African-American populace in the 19th century. Built in the 1820s, the meeting house also served first as a schoolhouse and then, after 1848 when the schools were integrated, as a church, up until 1911. Restoration began in 1997 under the direction of its owner, Boston's Museum of Afro-American History; at press time the property is slated to open to the public in 1998. Plans are to incorporate a museum and space for lectures, concerts, and readings that will help to portray the experiences of African-Americans on Nantucket. 🏛 *York and Pleasant Sts.,* ☎ *508/228–4058.*

Many of the mansions of the golden age of whaling were built on Upper ★ **22** Main Street. The well-known **"Three Bricks,"** identical redbrick mansions with columned, Greek Revival porches at their front entrances, were built between 1836 and 1838 by whaling merchant Joseph Starbuck for his three sons—one house still belongs to a Starbuck descendant. The Three Bricks are similar in design to the Jared Coffin House but have only two stories. ⊠ *93–97 Upper Main St.*

While touring Main St., the white, porticoed Greek Revival mansions you'll see—referred to as the Two Greeks—were built in 1845 and 1846 by wealthy factory owner William Hadwen, a Newport native. Number 94, built as a wedding gift for his adopted niece, was modeled on the Athenian Tower of the Winds, with Corinthian capitals on the entry columns, a domed-stair hall with statuary niches, and a round oculus. The Hadwens' own domicile, at No. 96, is now a museum, called the

**㉓ Hadwen House.** A two-year program to restore the house to its mid-19th-century origins—including the addition of classic Victorian gas chandeliers and furnishings, as well as reproduction wallpapers and window treatments—was completed in 1993. A guided tour points out such architectural details as the grand circular staircase, fine plasterwork, and carved Italian-marble fireplace mantels. Behind the house are period gardens. ⊠ *96 Main St.,* ✉ *$3 or NHA pass.*

**㉔** The **Macy-Christian House,** a two-story lean-to, built circa 1740 for another Thomas Macy, features a great open hearth flanked by brick beehive ovens and old paneling. Part of the house reflects the late-19th-century Colonial Revival style of the 1934 renovation by the Christian family. The formal parlor and upstairs bedroom are in authentic Colonial style. ⊠ *12 Liberty St.,* ✉ *$2 or NHA pass.*

Built for brothers, and facing each other, are the two attractive brick
**㉕ Coffin houses:** the **Henry Coffin House** and the **Charles G. Coffin House.** Wealthy shipping agents and whale-oil merchants, the Coffins used the same mason for these 1830s houses and the later Three Bricks. ⊠ *75 and 78 Main St.*

**㉖** Another grand Main Street home is the **John Wendell Barrett House.** Legend has it that Lydia Mitchell Barrett stood on the steps and refused to budge when, during the Great Fire, men tried to evacuate her so they could blow up the house to stop the spread of the fire. Luckily, a shift in the wind settled the showdown. ⊠ *72 Main St.*

**㉗** The 1818 **Pacific National Bank,** like the Pacific Club it faces, is a monument to the far-flung voyages of the Nantucket whaling ships it financed. Inside, above old-style teller cages, are murals of street and port scenes from the whaling days. ⊠ *Main St.*

**Murray's Toggery,** near the top of the hill in town, marks the site of R. H. Macy's first retail store (☞ Shopping, *below*). ⊠ *62 Main St.*

**㉘** Built around 1838 as a Friends school, the **Quaker Meeting House** is now a Quaker place of worship in summer. A small room of quiet simplicity, with white-and-gray walls, 12-over-12 windows with antique glass, and unadorned wood benches, it is in keeping with these peaceful people who believe that the divine spirit is within each person and that one does not require an intermediary to worship God. Attached to the meeting house is an unattractive 1904 concrete building that houses the **Fair Street Museum.** The museum has rotating exhibits, such as the NHA's collection of antique lightship baskets and portraits of historical Nantucket figures. ⊠ *1 Fair St.* ✉ *$3 or NHA pass.*

**㉙** On a hot day, peek into the 1901 **St. Paul's Episcopal Church,** a massive granite structure adorned at the front and back by beautiful Tiffany windows. The interior is cool and white, with dark exposed beams, and offers a quiet sanctuary from the crowds and the heat. ⊠ *Fair St. at Mooers La.*

**㉚** The 1809 **Unitarian Universalist Church,** also known as South Church, has a gold-domed spire that soars above town, just as the First Congregational Church's slender white steeple does. Also like First Congregational, South Church has trompe-l'oeil ceiling painting, this one simulating an intricately detailed dome, executed in 1840 by a European painter. Here, however, illusion is taken to greater lengths: The curved chancel and paneled walls you see are also creations in paint. The 1831 mahogany-cased Goodrich organ in the loft is played at services and concerts (☞ Music *in* Nightlife and the Arts, *below*). In the octagonal belfry of the tower, which houses the town clock, there is a bell cast in Portugal that has been ringing out the noon hour since it was hung in 1815. ⊠ *11 Orange St., at Stone Alley.*

Looking beyond the Island horizons, a bit of whimsy is found on the wall of the last building on the left on **Washington Street**—notice the sign listing distances from Nantucket to various points of the globe. It's 14,650 mi to Tahiti.

☾ **Murray Camp of Nantucket** (⊠ Box 3437, Nantucket 02584, ☎ 508/325–4600, ℻ 508/325–4646) offers weekly summer day camps for children 5–14. Activities include French instruction, sailing and other water sports, arts and crafts, drama and music, and environmental-awareness classes.

## Dining and Lodging

$$$$ ✕ **Company of the Cauldron.** There's only one menu each night, so it's important to pick your night with care—then again, just about anything coming out of this kitchen is excellent. A typical night would begin with roasted sweet red and white pepper soup, arugula salad with rosemary red-wine vinaigrette, mesquite grilled swordfish with warm tomatillo scallion salsa, red beans and rice with summer squash, and for dessert, homemade rhubarb and strawberry ice cream in a waffle cup with chocolate sauce. The interior has a dark, tavern feel, as befits the name, with tables lined up in rows that are less intimate than comradely. Gentle harp music sometimes accompanies the meal. ⊠ 7 *India St.,* ☎ *508/228–4016. Reservations essential. MC, V. No smoking. No lunch.*

$$$–$$$$ ✕ **21 Federal.** The epitome of sophisticated, gentrified island dining,
★ this is a place to see and be seen, as well as to enjoy some of the island's best new and traditional American cuisine. An informal dining room extends into the dark-paneled bar. Beyond are two other rooms, with traditional gray wainscoting, black-suede banquettes, and damask-covered tables; a curving staircase leads to a similar second floor, although the first floor has better ambience (and more people exposure). Entrées include muscovy duck breast, pistachio-crusted halibut, and often items off a country grill, like aged sirloin steak and veal chop. It's a virtual guarantee for a eye-pleasing, memorable night out. ⊠ *21 Federal St.,* ☎ *508/228–2121. AE, MC, V. Closed Jan.–Mar.; may close Mon. in shoulder seasons.*

$$$ ✕ **American Seasons.** The culinary context here is geographic: From
★ the four corners of the continental U.S., chef and owner Michael Getter gathers specialties. You can mix and match, taking Nantucket lobster for an appetizer and then a loin of pork for the main course. Anywhere you travel on the menu, you'll find excellent preparations. Razor clams with a green curry broth sweetened with coconut is a daring and tasty starter. A brioche apple charlotte with praline and vanilla ice cream could come from anywhere south of the Mason-Dixon line, and bursts with flavor. Pastoral murals add a rustic charm, although eating on the patio is a fine alternative. As is fitting, expect only American wines. ⊠ *80 Centre St.,* ☎ *508/228–7111. AE, MC, V. No smoking. Closed mid-Dec.–early Apr. No lunch.*

$$$ ✕ **DeMarco.** Winner of the 1997 award from *Cape Cod Life* as "Best Ethnic Restaurant on Nantucket," it perhaps says more about Nantucket than DeMarco that basic, good, northern Italian food is considered "eth-

nic" here. The antipasto menu includes *vitello tonnato* (tender poached veal with a creamy tuna-caper-anchovy sauce). The selection of pasta dishes is strong, especially the *maltagliata alla boschaiola* ("badly cut" fresh triangles of pasta in a tangy sauce of pomodoro and wild mushrooms). One favorite is the *filleto di rombo* (an olive-crusted halibut steak served over a delightful roasted tomato salsa). The dining room is a bit generic, especially upstairs. Dining downstairs near the curving bar is definitely a better bet. ⊠ *9 India St.,* ☎ *508/228–1836. AE, MC, V. Closed Jan.–Apr. No lunch.*

$$$  ✕ **India House.** A short stroll from downtown brings you to a handsome
★   old inn with small dining rooms and an outdoor garden that makes for lovely dining in good weather. Chefs Michael Caracciolo and Dereck Brewley take their cooking very seriously, but also manage to have some fun: One recent menu featured "northern Italian hedgehog with white truffle and Gorgonzola Asiago strozapretti." It was a joke, of course, but the island's health inspector stopped by just to make sure. The real food is splendid, particularly the calamari appetizer with penne and olives, and the swordfish with a topping of glazed roasted pecans and cashews. An excellent brunch, usually served outdoors, starts Fourth of July and includes such favorites as three-berry French toast and poached eggs with smoked salmon and caviar. Reservations not accepted for brunch. ⊠ *37 India St.,* ☎ *508/228–9043. AE, D, MC, V. Closed Jan.–Mar. No lunch.*

$$$  ✕ **Kendrick's at the Quaker House.** Kendrick's two small, unassuming rooms are truly Quakerish in their sparse but handsome simplicity. The revamped menu has taken on the island patina, with appetizers like clams with bacon, wilted spinach, and sundried tomatoes. Entrées run the usual gamut. Although the Quakers might not have approved, there is a small bar with an intriguing menu served until 11 PM, including a soft-shell crab sandwich, grilled eggplant, and a warm steak salad. ⊠ *5 Chestnut St.,* ☎ *508/228–9156. AE, MC, V. Closed Columbus Day–Memorial Day. No lunch.*

$$$  ✕ **Le Languedoc.** This busy little place has a café downstairs, a somewhat more formal dining room upstairs, and a garden terrace that's pleasant in good weather. The cuisine blends Continental and American styles, presenting plenty of seasonal fish, soft-shell crabs, and local produce. The Cornish game hen is roasted under brick. The backroom is particularly handsome, with dark red walls over dove grey paneling and white tablecloths. Reservations are for one of the two seatings per night, at 7 or 8:45 PM. ⊠ *24 Broad St.,* ☎ *508/228–2552. AE, MC, V. BYOB. Closed Jan.–mid-Apr. and Sun. off-season. No lunch July–Aug.*

$$  ✕ **Black-Eyed Susan's.** Don't let the old-style counter with swiveling stools fool you. Black-Eyed Susan's is not cheap: Pea soup with avocado and mint goes for nine bucks, the chicken comes with Black Mission fig and cabernet foie gras sauce and costs $20. They're hard prices to swallow—even if the food isn't—for what is supposed to be the island's homey alternative. ⊠ *11 India St.,* ☎ *508/325-0308. Closed mid-Oct.–Mar. No dinner Sun.*

$$  ✕ **Brotherhood of Thieves.** Plenty of brotherhood, some sisterhood, and not much thievery here. The sign out front harkens back to the days of slavery, with images of money and the devil lurking around a Nantucket trader. Inside, the trading in burgers and brews is brisk. The line rarely disappears, and it's more about fun than the "English-style" food. But the chowder is good, the sandwiches big, the noise level commendable. Once the place fills up, strangers sit together at long tables, and the live folk music adds to the friendly ambience. ⊠ *23 Broad St., no phone. Reservations not accepted. No credit cards.*

**$$** ✕ **Cioppino's.** Although the house specialty is the eponymous rich Italian seafood soup, the menu is much more expansive than that, with plenty of meat and new seafood specials every night. The menu changes every two weeks, and an early-bird program, "Twilight at Cioppino's," features a three-course menu for $20, briefly administered between 5:30 and 6:15. Owner Tracy Root worked for years at Chanticleer before opening his own place seven years ago, and remains a passionate, engaged restaurateur, proud of his unusual American wine selections and his food. Although not as inventive as elsewhere, the food is often more filling. Locals pack the small, eight-stool bar up front. ⊠ *20 Broad St.,* ☎ *508/228–4622. D, DC, MC, V. Closed Oct.–Apr.*

**$$** ✕ **Obadiah's Native Seafood.** Bluefish, yellowtail sole, scallops, lobster, swordfish, scrod—fish has been Obadiah's stock in trade for more than 20 years, and remains so despite the stock's decline. Andrew and Lynda Willauer transformed the back of an old sea captain's house into a restaurant that has not forgotten its roots (or in this case, given the maritime menu, its ballast). The patio is especially nice, and many of the main dishes have children's portions at half price plus two dollars. This is easily the best straight-ahead seafood value on the island. ⊠ *2 India St., at Independence La.,* ☎ *508/228–4430. AE, MC, V.*

**$$** ✕ **Straight Wharf.** The wide-open dining room looks right out on the water from Harbor Square. But as you might expect in such a setting, this is a better stop for a drink and appetizer than a full-blown meal. The good-looking bar occupies one half of the restaurant and serves a lighter, less expensive "grill" menu. ⊠ *Harbor Sq.,* ☎ *508/228–4499. AE, MC, V. Closed Mon. and Labor Day–Memorial Day.*

**$** ✕ **Atlantic Cafe.** With its booths in the back and serious beer-drinking
★ up front, the Atlantic has been the island's belly-up-to-the-bar place since 1978. Two big, salty codfish cakes with plenty of slaw and fries helps the beer go down, and you can still actually order food like nachos and potato skins without endangering your reputation. "Mary and the Boys" is a Bloody Mary with six shrimp, and it's a good deal. The painting over the big bar reveals a few old, bearded, Nantucket salts in a card game; any rainy afternoon, you'll find some young salts in the corner mimicking the scene. ⊠ *15 S. Water St.,* ☎ *508/228–0570. AE, DC, MC, V.*

**$** ✕ **Espresso Café.** People-watching is one of the pastimes at this fast-paced, fun little sandwich shop. Breakfast amounts to the traditional muffin and a cappuccino or latté, especially if you're eager to hit the beach early. The day menu is full of interesting sandwiches, crisp thin-crust pizzas (the feta and black olive is the best), curried lamb stew, veggie quesadillas, and jalapeño corn bread. ⊠ *40 Main St.,* ☎ *508/ 228–6930. Reservations not accepted. No credit cards. Closes at 8 PM.*

**$** ✕ **Off Centre Café.** The walk through a little minimall is kind of strange, but Chris and Liz Holland's café makes up for that with a cheery, friendly aspect, especially for breakfast or a quick lunch. There's only seating for 20, and the tables are jammed together. Outdoor seating is basically on the sidewalk, making for great people watching if not for much privacy. Blackboard specials carry the day, especially the fish chowder or the ratatouille. It's maddening that even here, in the most casual and down-home of places, it is hard to justify the prices. Off Centre Café also runs a bread and dessert retail and wholesale business. ⊠ *29 Centre St.,* ☎ *508/228–8470. No credit cards. No lunch in season.*

**$** ✕ **Picnic Lunches.** Your best bet at lunch, especially if you're on a schedule, is to get a bunch of goodies from **Provisions** (⊠ Straight Wharf, ☎ 508/228–3258) and find your way to Eel Point Beach or some other remote spot.

$ ✕ **Something Natural.** Of all of the few really good take-out places on the island—for sandwiches, desserts, loaves of bread—this one deserves special mention, if only for the picnic area that comes in so handy for an occasional quick, inexpensive lunch. ✉ *50 Cliff Rd.,* ☎ *508/228–0504. No credit cards.*

$ ✕ **Vincent's.** Pizza under a warmer, calzones, plenty of pasta, free delivery—these are hardly island staples but Vincent's does them all, and has since 1954. Don't get into anything fancy—in this case, "fancy" means fettuccine Alfredo—just order up some pizzas, pitas, and focaccia and fill up. Whether you eat in or out, enjoy it, because this place is an endangered species on Nantucket. ✉ *21 S. Water St.,* ☎ *508/228–0189. MC, V.*

$$$$ ✕▦ **White Elephant.** Long a hallmark of service and style on the island, the White Elephant offers, above all, a choice location—right on Nantucket Harbor, separated only by a wide lawn. The main hotel, wrapped by a deck with a fine view of the bobbing boats, has a formal restaurant with a waterside outdoor café, a lounge with entertainment, and a large harbor-front pool. Rooms have an English country look, with stenciled pine armoires, sponge-painted walls, and floral fabrics. A similar decor characterizes the one- to three-bedroom cottages (some with full kitchens). Waterfront cottages are especially handsome, rimmed in roses. ✉ *Easton St., Box 1139, Nantucket 02554,* ☎ *508/228–2500; for reservations,* ☎ *800/475–2637,* ℻ *508/325–1195. 48 rooms, 32 cottages. Restaurant, lounge, room service, pool, putting green, croquet, dock, concierge, meeting rooms. AE, D, DC, MC, V. Closed late Oct.–early May.*

$$–$$$$ ✕▦ **Harbor House.** This family-oriented complex, like its more upscale
★ sibling, the White Elephant, prides itself on service. The restaurant, **Hearth at the Harbor House,** is a traditional island favorite with a lovely outdoor patio. The simple New England fare includes filet mignon, lamb chops, baked stuffed shrimp, and crabmeat-stuffed baked flounder with parsley butter. A lavish Sunday brunch buffet includes a raw bar and a dessert table (brunch reservations are essential). The three-course "sunset special" is Nantucket's bow to the early-bird special (served 5–6:30). The 1886 main inn and several "town houses" are set on a flower-filled quadrangle very near the town center. Standard rooms are done in English-country style, with bright floral fabrics and queen-size beds. Some have French doors that open onto decks. The generally larger town house rooms, in buildings grouped around the pool, have a more traditional look, with upscale pine and pastels. Some have cathedral ceilings, sofa beds, and decks. All rooms have phones and TVs. The Garden Cottage has its own garden and a private-house feel, but its rooms (some with pressed-tin ceilings) are smaller. ✉ *S. Beach St., Box 1139, Nantucket 02554,* ☎ *508/228–1500; for reservations,* ☎ *800/ 475–2637,* ℻ *508/228–7197. 113 rooms. Restaurant, lounge, room service, putting green, concierge, business services. AE, D, DC, MC, V. Closed mid-Dec.–mid-Apr.*

$$–$$$ ✕▦ **Jared Coffin House.** This collection of four buildings is a longtime favorite of many visitors to Nantucket (☞ Exploring, *above*). **Jared's** restaurant ($$$)—famous, formal, and safe—is one of the island's elegant dining rooms. Dinner is straight ahead and predictable, never veering far from perfectly grilled swordfish and herb-encrusted tenderloin. Service is impressive, and the wine list is expansive. (Dinner is not served January–April.) Downstairs, the Tap Room ($$) is dark and cozy, with lots of exposed wood and nautical art. The fare is mostly prime rib, fried clams, and pasta—things you'll never find upstairs at Jared's. Enjoy lunch on the outdoor patio, weather permitting. The main inn building, a three-story brick mansion built in 1845 by a wealthy shipowner and

topped by a cupola, has a historic tone that the others don't. The public and guest rooms are furnished with period antiques (the other buildings, with reproductions), Oriental carpets, and lace curtains. The Harrison Gray House, an 1842 Greek Revival mansion across the street, offers larger guest rooms with large baths and queen-size canopy beds. All rooms have phones and cable TV. Small, inexpensive single rooms are available. ✉ *29 Broad St., Box 1580, Nantucket 02554,* ☎ *508/228–2400 or 800/248–2405,* 🖷 *508/228–8549. 60 rooms. Restaurant, bar, café, concierge. AE, D, DC, MC, V. BP.*

$$$$ 🏨 **Cliffside Beach Club.** Although the cedar-shingle exterior, landscaped
★ with climbing roses and hydrangeas, and the pavilion on the private sandy beach 1 mi from town reflect the club's 1920s origins, the interiors are done in contemporary summer style, with white walls, fine woodwork, white or natural wood furniture, cathedral ceilings, and local art. Some have kitchenettes, fireplaces, wet bars, or private decks. Two large "town house suites" have full kitchens and decks overlooking dunes, moors, and Nantucket Sound. ✉ *Jefferson Ave., Box 449, Nantucket 02554,* ☎ *508/228–0618,* 🖷 *508/325–4735. 19 rooms, 8 apartments, 1 cottage. Restaurant, piano bar, exercise room, beach, playground. AE. Closed mid-Oct.–May. CP.*

$$$$ 🏨 **Wharf Cottages.** These weathered shingle cottages sit on a wharf in Nantucket Harbor, with yachts tied up just steps away. Each unit has a little garden and sitting area, a fully equipped kitchen, and attractive modern decor with a nautical flavor: white walls, navy-blue rugs, light-wood floors and furniture. Studios have a sofa bed for sleeping. Other cottages have one to three bedrooms. Some have large water-view windows, and all have water views. There's a three-night minimum in high season. ✉ *New Whale St., Box 1139, Nantucket 02554,* ☎ *508/228–1500, ext. 4928; for reservations, 800/475–2637,* 🖷 *508/228–7197. 23 cottages. AE, D, DC, MC, V. Closed mid-Oct.–Memorial Day.*

$$$–$$$$ 🏨 **Beachside at Nantucket.** Those who prefer rooms-around-a-pool motels with all the creature comforts will find a very nice one here, a bit of a walk from the town center. Each unit in the one- and two-story buildings is furnished in wicker and florals and has a queen-size or two double beds, a tiled bath, and appliances. Some rooms have French doors opening onto pool-view decks. ✉ *30 N. Beach St., Nantucket 02554,* ☎ *508/228–2241 or 800/322–4433,* 🖷 *508/228–8901. 90 rooms, 3 2-bedroom suites. Pool, meeting rooms. AE, D, DC, MC, V. Closed mid-Oct.–mid-Apr. CP.*

$$$–$$$$ 🏨 **Manor House.** From the screened and open porches or the little front yard, you can watch the town go by from this spot in the heart of the historic district. Rooms in this 1846 house are spacious, with reproduction rice-carved beds (king or queen) and Waverly or Schumacher wallpapers. Seven rooms have king canopy beds and working fireplaces. All have TV, air-conditioning, and phones. In winter, a guest room is converted into a common room with a fireplace. A small cottage next door done in wicker and chintz has two bedrooms and a fully equipped kitchen. Packages are available. ✉ *31 Centre St., Box 1436, Nantucket 02554–1436,* ☎ *508/228–0600 or 800/673–4559,* 🖷 *508/325–4046. 15 rooms, 1 cottage. No smoking. AE, D, MC, V. CP.*

$$$–$$$$ 🏨 **Roberts House and Meeting House Inns.** In the midst of streets lined with shops and restaurants, these attractive inns in historic buildings are some of the most solid accommodations in town. Many of the guest rooms have a mix of antique and reproduction furniture, and high ceilings and a spacious, Victorian feeling. All have queen-size or larger beds, TV, phones, and air-conditioning; some have working fireplaces and small refrigerators. Bathrooms are generally not as attractive as the

rooms. Common rooms downstairs reflect the 19th-century atmosphere of the houses that the inns occupy. ⊠ *11 India St., at Centre St., Box 1436, Nantucket 20554–1436,* ☎ *508/228–9009 or 800/872–6817,* FAX *508/325–4046. 24 rooms. AE, D, MC, V.*

**$$–$$$$**    🛏 **Westmoor Inn.** Built in 1917 as a Vanderbilt summer house, this yellow Federal-style mansion with widow's walk and portico is a mile from town, a short walk to a quiet ocean beach, just off the Madaket bike path. The many common areas include a wide lawn set with Adirondack chairs and a garden patio secluded behind 11-ft hedges. Beyond the entry hall and grand staircase, the large, gracious living room with piano and game table is the setting for an early evening wine and cheese reception. A wicker-filled sunroom has a common TV. Guest rooms are beautifully decorated in French-country style, most with soft florals and stenciled walls. One first-floor suite has a giant bath with extra-large Jacuzzi and French doors opening onto the lawn. A large third-floor room has a fantastic view of Nantucket Sound. A high point of any stay is breakfast in the dining room, with glass walls and ceiling. ⊠ *Cliff Rd., Nantucket 02554,* ☎ *508/228–0877,* FAX *508/228–5763. 14 rooms. Bicycles. No smoking. AE, MC, V. Closed early Dec.–mid-Apr. CP.*

**$$$**    🛏 **Centerboard Guest House.** The look and polish of this inn a few blocks
★    from the center of town is different from any other. White walls, some with murals of moors and sky in soft pastels, blond-wood floors, and natural woodwork create a cool, dreamy atmosphere. There is yet more white in the lacy linens and puffy comforters on the feather beds. Touches of color are added by stained-glass lamps, antique quilts, and fresh flowers. The first-floor suite (right off the entry hall) is stunning, with 11-ft ceilings, a Victorian living room with fireplace and bar, inlaid parquet floors, superb furnishings and decor, and a green-marble bath with Jacuzzi. ⊠ *8 Chester St., Box 456, Nantucket 02554,* ☎ *508/228–9696. 5 rooms, 1 suite. No smoking. AE, MC, V. CP.*

**$$$**    🛏 **Nantucket Settlements.** These attractively decorated properties consist of three apartment houses and a complex of seven cottages, all with kitchens and access to free laundry facilities. Right in town are the **Nantucket Whaler** (⊠ 8 N. Water St.), an 1846 Greek Revival with pilastered white clapboard facade and a large deck, and the 1822 **Grey Goose** (⊠ 24 Hussey St.), with high ceilings, old moldings, and antique-looking furnishings. Within ½ mi of town are the **Orange Suites** (⊠ 95 Orange St.), in a renovated late-18th-century house with new kitchens and baths and a brick patio with barbecue grill. Also ½ mi out are the gray-shingled **Brush Lane Cottages,** in a quiet compound with lots of flowers and greenery. The newest cottages have cathedral ceilings, lots of white and light wood, oak cabinets in large kitchens, and French doors that open onto decks. There's daily maid service. ⊠ *Office: 8 N. Water St., Box 1337, Nantucket 02554,* ☎ *508/228–6597 or 800/462–6882,* FAX *508/228–6291. 25 units (studios to 3-bedroom cottages). Baby-sitting. MC, V.*

**$$$**    🛏 **Seven Sea Street.** This inn on a quiet side street in the center of town was built in 1987 by Mary and Matthew Parker, former publishers of *Nantucket Journal* magazine. Though the furnishings and colors are in the Colonial style, the place has a somewhat Scandinavian look, with tongue-in-groove light pine and red oak, exposed-beam ceilings, stenciled white walls with pine trim, and highly polished wide-board floors. Most rooms have braided rugs, queen-size beds with fishnet canopies and quilts, modern baths, and a desk area. All have TV, phones, air-conditioning, and mini-refrigerators. The deluxe Honeymoon Suite, good for longer stays, has a cathedral post-and-beam ceiling, a full kitchen, and a view of the harbor. Relax in the garden patio or on the harbor-view widow's walk. ⊠ *7 Sea St., Nantucket 02554,* ☎ *508/228–*

*3577, ⟨FAX⟩ 508/228–3578. 11 rooms, 2 suites. Hot tub, library. No smoking. AE, D, MC, V. CP.*

$$–$$$ 🏠 **The Century House.** This 1833 late Federal-style sea captain's home became a rooming house in the 1870s and has operated as a guest house ever since. Present owners Jean Heron and Gerry Connick have lovingly refurbished the house to create a comfortable haven for their guests. The bright, homey rooms each have a distinctive flair. Some have light floral Laura Ashley wallcoverings, wide-plank floors, Oriental rugs, and canopy four-poster beds; others have spool or sleigh beds. Common areas include an elegant sitting room with a fireplace and plenty of cozy couches. A wraparound veranda, set with rocking chairs, is a lovely place to sit and admire the gardens. A lavish buffet breakfast is served in the pine-paneled country kitchen. The innkeepers also manage two cottages, one in 'Sconset across from the beach, another on Nantucket harbor. 🏠 *10 Cliff Rd., Nantucket 02554,* ☎ *508/228–0530. 14 rooms. MC, V. CP.*

$$–$$$ 🏠 **Cliff Lodge.** Owners Debby and John Bennett have made the big guest
★ rooms at their B&B bright and airy, decorating them with pastel hooked rugs on spatter-painted floors, country curtains and furnishings, and down comforters. Built in 1771, the lodge preserves lots of old-house feeling, in part with moldings, wainscoting, and wide-board floors. Some baths are quite small. The very pleasant apartment has a living room with a fireplace, a private deck and entrance, and a large eat-in kitchen. In addition to the attractive common rooms, guests may enjoy the wicker sunporch, the garden patio, or the roof walk with a great view of the harbor. ⊠ *9 Cliff Rd., Nantucket 02554,* ☎ *508/228–9480,* ⟨FAX⟩ *508/228–6308. 11 rooms, 1 apartment. Grill. No smoking. MC, V. Closed Jan.–Feb. CP.*

$$–$$$ 🏠 **18 Gardner Street.** Set in two antique buildings, including an 1835 main house with 9-ft ceilings, this B&B has good-size rooms—10 of which have working fireplaces. Rooms are furnished in an elegant yet comfortable style, with satin wall coverings, wide-board floors, mostly queen-size beds (some canopy or four-poster) with eyelet sheets and handmade quilts, some nice antique pieces, and brass lamps. All have TV and air-conditioning. The spacious sitting room with a fireplace is a good place to mingle with other guests. ⊠ *18 Gardner St., Nantucket 02554,* ☎ *508/228–1155 or 800/435–1450. 18 rooms (2 share bath), 3 suites. Refrigerators, bicycles. No smoking. AE, MC, V. CP.*

$$–$$$ 🏠 **Ten Lyon Street Inn.** A five-minute walk from the town center, Ann Marie and Barry Foster's mostly new house has been rebuilt with historical architectural touches such as variable-width plank floors, salvaged Colonial mantels on the nonworking fireplaces, and hefty ceiling beams of antiqued red oak. The white walls and blond woodwork provide a clean stage for exquisite antique Oriental rugs in deep, rich colors; choice antiques, such as Room No. 1's French tester bed draped in white mosquito netting; and English floral fabrics, down comforters, and big pillows. Bathrooms are white and bright. Several have separate showers and antique tubs, and all have antique porcelain pedestal sinks and brass fixtures. The garden is a pleasant place to sit and relax. ⊠ *10 Lyon St., Nantucket 02554,* ☎ *508/228–5040. 7 rooms. MC, V. Closed mid-Dec.–mid-Apr. CP.*

$–$$$ 🏠 **Martin House Inn.** Ceci and Channing Moore's friendly and very
★ nicely refurbished B&B in an 1803 house offers a great variety of mostly spacious rooms with four-poster beds or canopies, pretty linens, and fresh flowers; several have queen-size beds, couches, and/or writing tables. A few rooms have fireplaces, including No. 21 on the second floor, which also has a queen-size canopy bed and a private porch overlooking the backyard. Third-floor shared-bath rooms are sunny and bright, with a quirky under-eaves feel. The large living room with a fireplace and

the wide porch with hammock invite lingering, and the breakfast-dining room is very pleasant. Single rooms are available. ✉ *61 Centre St., Box 743, Nantucket 02554,* ☎ *508/228–0678,* FAX *508/325–4798. 13 rooms, 9 with private bath (4 rooms share 1 bath). Piano. AE, MC, V. CP.*

$$ 🏠 **Chestnut House.** At this centrally located guest house, innkeepers Jeannette and Jerry Carl's hand-hooked rugs and paintings, along with their son's Tiffany-style lamps, are everywhere, creating homey guest rooms. Suites have a sitting room with sofa. The guest parlor reflects the Arts and Crafts style, and some rooms have William Morris–theme wallpapers. A spacious, cheery cottage sleeps four (queen-size bed and sofa bed) and has a full kitchen and bath and a small deck—a convenient option for a family here in the center of town. Rates include a voucher for breakfast at Arno's restaurant in season or the Jared Coffin House year-round. ✉ *3 Chestnut St., Nantucket 02554,* ☎ *508/228–0049,* FAX *508/228–9521. 1 room, 4 suites, 1 cottage. No smoking. AE, MC, V. BP.*

$$ 🏠 **Hawthorn House.** Innkeeper Mitch Carl and his wife, Diane, have filled their 1850 house with art, hooked rugs, and stained glass. The smallish rooms are decorated with antiques, William Morris–style wallpapers, and homey, personal touches, like Mitch's stained-glass lamps, and Diane's handmade quilts. A dark but conveniently located cottage sleeps two and has air-conditioning. Breakfast vouchers are included for Arno's restaurant in season or the Jared Coffin House year-round. ✉ *2 Chestnut St., Nantucket 02554,* ☎ FAX *508/228–1468. 9 rooms, 7 with private bath, 1 cottage. MC, V. BP.*

$$ 🏠 **76 Main Street.** Built in 1883 by a sea captain, just beyond the bustle of the shops, this B&B carefully blends antiques and reproductions, Oriental rugs, handmade quilts, and lots of fine woods. The cherry-wood Victorian entrance hall is dominated by a long, elaborately carved staircase. Room No. 3, originally the dining room, also has wonderful woodwork, a carved-wood armoire, and twin four-posters. Spacious No. 1, once the front parlor, has three large windows, massive redwood pocket doors, and a bed with eyelet spread and canopy. Every room has its own charm and a simple, homespun comfort. The motel-like rooms in the 1955 annex out back, set around a flagstone patio and gardens, have low ceilings but are quite spacious, and are perfect for families. Owner Shirley Peters usually officiates over breakfast, which she serves in the bright kitchen; it provides a chance to chat with other guests and to sample her wonderful homemade scones. ✉ *76 Main St., Nantucket 02554,* ☎ *508/228–2533. 18 rooms. Refrigerator. No smoking. AE, D, MC, V. Closed Jan.–Mar. CP.*

$ 🏠 **Hostelling International–Nantucket.** This 49-bed facility occupies a former lifesaving station and is one of the most picturesque hostels in the country. It is in Surfside Beach, a 3-mi ride from town on the bike path. There are common areas and a kitchen, and the hostel offers a variety of educational, cultural, and recreational programs. Reservations are essential in July and August, strongly recommended always. ✉ *31 Western Ave., Nantucket 02554,* ☎ *508/228–0433 in off-season;* ✉ *Box 158, W. Tisbury 02575,* ☎ *508/693–2665). Grill, picnic area, volleyball, piano. MC, V. Closed mid-Oct.–mid-Apr.*

$ 🏠 **Nesbitt Inn.** This family-run guest house in the center of town offers comfortable, shared-bath rooms (including inexpensive singles) sweetly done in authentically Victorian style, with lace curtains, some marble-top and brass antiques, and a sink in each room. Some beds are not as firm as they should be, and the location (next door to a popular bar-restaurant) means it gets runoff noise (ask for a room on the quieter side), but the Nesbitt is a very good buy in this town. The backyard's perfect for children. ✉ *21 Broad St., Box 1019, Nantucket 02554,* ☎

*508/228–0156 or 508/228–2446. 9 doubles and 3 singles share 4 baths.*
*Grill, no-smoking rooms, refrigerator. No pets. MC, V. CP.*

## Nightlife and the Arts

### BARS AND CLUBS

**Brotherhood of Thieves** (⊠ 23 Broad St., no phone) has live folk music
year-round. The well-stocked bar offers an interesting selection of
beers and ales, plus dozens of cordials and liqueurs.

**The Tap Room** (⊠ Jared Coffin House, 29 Broad St., ☎ 508/228–2400)
has live guitar in season.

**Hearth at the Harbor House** (⊠ S. Beach St., ☎ 508/228–1500) has
dancing to live music (country to folk) on weekends year-round in its
attractive lounge with a fireplace; on most nights in season, you can
also hear Top-40 tunes by a piano-and-vocal duo.

**Brandt Point Grill at the White Elephant** (⊠ Easton St., ☎ 508/228–
2500) has a formal, harbor-view lounge with a pianist playing show
tunes most nights from Memorial Day to mid-September. Proper dress
is suggested. It's also a good place to sit and watch the boats over afternoon drinks and hors d'oeuvres.

**The Box** (⊠ a.k.a. Chicken Box, 6 Dave St., off Lower Orange St., ☎
508/228–9717) is open daily year-round and has live music every
night in season, weekends off-season.

**The Muse** (⊠ 44 Surfside Rd., ☎ 508/228–6873 or 508/228–8801)
is where all ages dance to rock, reggae (especially popular on the island), and other music, live or recorded. It also has a take-out pizza
shop (☎ 508/228–1471).

**The Rose & Crown** (⊠ 23 S. Water St., ☎ 508/228–2595) is a friendly,
noisy seasonal restaurant with a big bar, a small dance floor, live
bands, DJs, and karaoke nights.

### FILM AND PHOTOGRAPHY

**Dreamland Theatre** (⊠ 19 S. Water St., ☎ 508/228–5356) is one of
two first-run theaters in Nantucket Town; call for films and times.

**Gaslight Theatre** (⊠ 1 N. Union St., ☎ 508/228–4435) is a first-run
theater; call for film dates and times.

**Nantucket Filmworks** presents a different slide show on Nantucket every
other year, created by one of the island's best photographers, Cary Hazle-
grove. Shows are given at the Methodist church (⊠ Centre and Main
Sts., ☎ 508/228–3783) mid-June–mid-September.

**The Nantucket Film Festival** runs for five days, with film screenings,
staged readings, panel discussions, and hobnobbing with actors and
directors. Day- and weeklong passes should be purchased early for pri-
ority seating.(⊠ Box 688, Prince St. Station, New York, NY 10012,
☎ 212/642–6339, E-mail: ackfest£aol.com).

### THEATER

**Actors Theatre of Nantucket** (⊠ Methodist Church, 2 Centre St., ☎
508/228–6325) presents several Broadway-style plays Memorial Day–
Columbus Day, plus children's post-beach matinees in July and August,
comedy nights, late-night productions, and readings.

**Theatre Workshop of Nantucket** (⊠ Bennett Hall, 62 Centre St., ☎
508/228–4305), a community theater since 1956, offers plays, musicals,
and staged readings year-round.

### MUSIC

**Band concerts** (☎ 508/228–7213) are held at 6:30 on various summer Sundays at Children's Beach.

**Nantucket Arts Council** (⊠ Coffin School, Winter St., ☎ 508/228–1216)
sponsors a music series (jazz, country, classical) September–June.

**Nantucket Chamber Music Center** (⊠ Coffin School, Winter St., ☎ 508/228–3352) offers year-round choral and instrumental concerts as well as instruction.

**Nantucket Musical Arts Society** (⊠ Box 897, Nantucket 02554, ☎ 508/228–1287) holds Tuesday-evening concerts in July and August featuring internationally acclaimed musicians (past participants include Virgil Thomson and Ned Rorem) at the First Congregational Church (⊠ 62 Centre St.) and free, informal "Meet the Artists" gatherings the previous evening elsewhere. 1998 marks the society's 40th season.

**Noonday Concerts** on an 1831 Goodrich organ are performed Thursdays at noon in July and August at the Unitarian Universalist Church (⊠ 11 Orange St., ☎ 508/228–5466).

## Outdoor Activities and Sports

### BEACHES

**Dionis Beach** is, at its entrance, a narrow strip of sand that turns into a wider, more private strand with high dunes—it is the only Nantucket beach with dunes—and fewer children. The beach has a rocky bottom and calm, rolling waters, lifeguards, and rest rooms. Take the Madaket bike path to Eel Point Road and look for a white rock pointing to the beach, about 3 mi west of town. ⊠ *Eel Point Rd.*

A calm area by the harbor, **Children's Beach** is an easy walk from town, and it is perfect for small children. There is a grassy park with benches, a playground, lifeguards, food service, picnic tables, showers, and rest rooms. Tie-dyeing "lessons" are offered at noon mid-July–August. ⊠ *S. Beach St.*

**Cisco** is a long, sandy south-shore beach with heavy surf, lifeguards and rest rooms. It's a popular spot for body and board surfers. ⊠ *Hummock Pond Rd.*

**Jetties,** a short bike or shuttle-bus ride from town, is the most popular beach for families because of its calm surf, lifeguards, bathhouse, rest rooms, and snack bar. It's a lively scene, especially with ferries passing, water-sports rentals (Windsurfer, sailboat, kayak), a playground and volleyball nets on the beach, and tennis courts adjacent. Swim lessons for children 6 and up are offered 9:30–noon, July 4–Labor Day. The concessions and rest rooms are wheelchair-accessible, and there is a boardwalk to the beach. ⊠ *Hulbert Ave.*

**Surfside** is the premier surf beach, with lifeguards, rest rooms, a snack bar, and a wide strand of sand. It attracts college students as well as families, and is great for kite-flying and surf casting. ⊠ *Surfside Rd.*

### BIKING

The easy 3-mi **Surfside Bike Path,** which begins on the Surfside Road (from Main Street take Pleasant Street; then turn right onto Atlantic Avenue), leads to the island's premier ocean beach. Benches and drinking fountains are placed at strategic locations along the path.

**Nantucket Bike Shop** (⊠ Steamboat Wharf, ☎ 508/228–1999; Apr.–Oct.) rents bicycles and mopeds and provides an excellent touring map. Daily rentals typically cost $15–$30 for a bicycle, $30–$60 for a moped, though half-, full-, and multiple-day rates are available.

**Nantucket Cycling Club** (☎ 508/228–1164) holds open races in summer.

**Young's Bicycle Shop** (⊠ Steamboat Wharf, ☎ 508/228–1151) rents several different types of bicycle, plus mopeds, cars and Jeep Wranglers in season. The shop provides an excellent touring map. Daily rentals typically cost $15–$30 for a bicycle, $30–$60 for a moped, though half-, full-, and multiple-day rates are available.

### BIRD-WATCHING

**Birding Adventures** (☎ 508/228–2703) offers tours led by a naturalist who knows his raptors from his oystercatchers.

**The Maria Mitchell Association** (✉ Vestal St., ☎ 508/228–9198) organizes wildflower and bird walks from June to Labor Day.

### BOATING

**Force 5 Watersports** (✉ Jetties Beach, ☎ 508/228–5358; 37 Main St., ☎ 508/228–0700) rents Sunfish, Windsurfers, kayaks, surfboards, boogie boards, and other water gear; Sunfish and Windsurfer lessons are also available.

**Nantucket Boat Rentals** (✉ Slip 1, ☎ 508/325–1001) rents powerboats.

**Nantucket Harbor Sail** (✉ Swain's Wharf, ☎ 508/228–0424) rents sailboats and outboards.

### FISHING

**Barry Thurston's Fishing Tackle** (✉ Harbor Sq., ☎ 508/228–9595) will give you fishing tips, and rents all the gear you'll need.

**Beach Excursions Ltd.** (☎ 508/228–3728) provides guided surf-casting trips and rents the gear you'll need.

**Bill Fisher Tackle** (✉ 14 New La., ☎ 508/228–2261) rents equipment and can point you in the direction of the best fishing spots.

*Herbert T.* (✉ Slip 14, ☎ 508/228–6655) and other boats are available for seasonal charter from Straight Wharf.

**Whitney Mitchell** (☎ 508/228–2331) leads guided surf-casting trips with tackle by four-wheel-drive.

### GOLF AND MINIATURE GOLF

**J. J. Clammp's** is an 18-hole miniature golf course set in gardens, reached by a path connecting with the 'Sconset bike path. There's also a restaurant and a free shuttle from downtown. ✉ *Nobadeer Farm and Sun Island Rds., off Milestone Rd. on the way to 'Sconset,* ☎ *508/228– 8977.* 🎫 *$6.* 🕐 *July–Aug., daily 8 AM–10 PM; June and Sept., daily (hrs vary widely).*

**Miacomet Golf Club** (✉ Off Somerset Rd., ☎ 508/325–0335), a public course owned by the Land Bank and abutting Miacomet Pond and coastal heathland, has nine holes (expansion to 18 is in the works) on very flat terrain.

### HEALTH AND FITNESS CLUB

**Club N. E. W.** (✉ 10 Young's Way, ☎ 508/228–4750) offers StairMasters, Lifecycles, treadmills, rowers, Airdyne bikes, New Generation Nautilus, and free weights; aerobics, dance, and yoga classes; plus a nutritionist, personal trainers, and baby-sitting.

### ROLLERBLADING

**Nantucket Sports Locker** (✉ 14 Cambridge St., ☎ 508/228–6610) offers rentals.

### SCUBA DIVING

**The Sunken Ship** (✉ Broad and S. Water Sts., ☎ 508/228–9226) offers complete dive-shop services, including lessons, equipment rentals, and charters; it also rents water skis, boogie boards, tennis rackets, and fishing poles.

### SWIMMING

**The Olympic-size, indoor Nantucket Community Pool** (✉ Nantucket High School, Atlantic Ave., ☎ 508/228–7262) is open daily year-round for lap swimming and lessons.

TENNIS

There are six asphalt **town courts** (☎ 508/325–5334) at Jetties Beach. Sign up at the Park and Recreation Commission building for one hour of court time (usually the limit), or for lessons or tennis clinics.

**Brant Point Racquet Club** (✉ 48 N. Beach St., ☎ 508/228–3700), a short walk from town, has nine fast-dry clay courts and a pro shop and offers lessons, rentals, playing programs, and round-robins.

WHALE WATCHING

**Nantucket Whalewatch** offers naturalist-led full-day excursions every Tuesday mid-July–August. ✉ Hy-Line dock, Straight Wharf, ☎ 508/283–0313 or 800/942–5464. ⌫ $65. Reservations essential.

## Shopping

ANTIQUES AND AUCTIONS

**Forager House Collection** (✉ 20 Centre St., ☎ 508/228–5977) specializes in folk art and Americana, including whirligigs, wood engravings, vintage postcards, Nantucket lightship baskets, and antique maps, charts, and prints.

**Janis Aldridge** (✉ 50 Main St., ☎ 508/228–6673) has beautifully framed antique engravings, including architectural and botanical prints, and home furnishings.

**Nina Hellman Antiques** (✉ 48 Centre St., ☎ 508/228–4677) carries scrimshaw, ship models, nautical instruments, and other marine antiques, plus folk art and Nantucket memorabilia.

**Paul La Paglia** (✉ 38 Centre St., ☎ 508/228–8760) has moderately priced antique prints, including Nantucket and whaling scenes, botanicals, and game fish.

**Rafael Osona** (✉ American Legion Hall, 21 Washington St., ☎ 508/228–3942; for a schedule, write to Box 2607, Nantucket 02584) holds auctions of fine antiques from Memorial Day to early December. Items range from the 18th to the 20th century and include everything from furniture and art to Nantucket baskets and memorabilia.

**Tonkin of Nantucket** (✉ 33 Main St., Box 996, Nantucket 02584, ☎ 508/228–9697) has two floors of fine English antiques—including furniture, china, art, silver, marine and scientific instruments, and Staffordshire miniatures—as well as new sailors' valentines and lightship baskets.

BOOKS

**Mitchell's Book Corner** (✉ 54 Main St., ☎ 508/228–1080) has an excellent room full of books on Nantucket and whaling, many ocean-related children's books, plus the usual bookstore fare. Authors appear weekly in summer.

**Nantucket Bookworks** (✉ 25 Broad St., ☎ 508/228–4000) carries an extensive assortment of hardcover and paperback books, with an emphasis on literary works over Nantucket-specific titles, as well as a children's-book room and unusual gift and stationery items.

CLOTHING

**Cordillera Imports** (✉ 18 Broad St., ☎ 508/228–6140) sells jewelry, affordable clothing in natural fibers, and crafts from Latin America, Asia, and elsewhere.

**Handblock** (✉ 42 Main St., ☎ 508/228–4500) carries women's and children's apparel with lots of lovely floral patterns, linens, giftware, jewelry, and candles.

**Michelle's Romantic Clothing** (✉ 7 Centre St., ☎ 508/228–4409) offers lacy vintage dresses, antique jewelry, flower-bedecked hats, and shoes to match.

**Murray's Toggery Shop** (✉ 62 Main St., ☎ 508/228–0437) sells traditional footwear and clothing—including the famous Nantucket Reds—for men, women, and children. An **outlet store** (✉ 7 New St., ☎ 508/228–3584) has discounts of up to 50%.

**Nantucket Looms** (✉ 16 Main St., ☎ 508/228–1908) allows customers to watch weavers hand-fashioning sweaters, scarves, throws, et cetera, at two large wooden looms. The shop is open year-round and also carries distinctive furnishings for home and garden.

**The Peanut Gallery** (✉ 8 India St., ☎ 508/228–2010) has a discriminating collection of children's clothing, including Cary, Flapdoodles, and island-made items.

**Vis-a-Vis** (✉ 34 Main St., ☎ 508/228–5527) has unique, funky, and classic women's and children's clothing, accessories, and decorative objects, including hooked rugs, quilts, and collectibles.

CRAFTS

**Claire Murray** (✉ 11 S. Water St., ☎ 508/228–1913 or 800/252–4733) carries the designer's Nantucket-theme and other hand-hooked rugs and kits, quilts, and knitting and needlework kits.

**Erica Wilson Needle Works** (✉ 25 Main St., ☎ 508/228–9881) sells the famed designer's kits, as well as home decorative items, clothing, hats and accessories, and handmade jewelry by Heidi Weddendorf.

**Four Winds Craft Guild** (✉ 6 Ray's Ct., ☎ 508/228–9623) carries a large selection of antique and new scrimshaw and lightship baskets, as well as ship models, duck decoys, and a kit for making your own lightship basket.

**Golden Basket** (✉ 44 Main St., ☎ 508/228–4344) and its affiliated shop, Golden Nugget (✉ Straight Wharf, ☎ 508/228–1019), sell miniature gold and silver lightship baskets, pieces with starfish and shell motifs, and other fine jewelry.

**Nantucket lightship basket makers** include **Michael Kane** (✉ 18½ Sparks Ave., ☎ 508/228–1548) and **Bill and July Sayle** (✉ 112 Washington St., ☎ 508/228–9876).

**Rosa Rugosa** (✉ 10 Straight Wharf, ☎ 508/228–5597) has home decorative items, including furniture painted with roses, and antiques.

**Scrimshander Gallery** (✉ 19 Old South Wharf, ☎ 508/228–1004) deals in new and antique scrimshaw.

**The Spectrum** (✉ 26 Main St., ☎ 508/228–4606) sells distinctive art glass, wood boxes, jewelry, kaleidoscopes, and more.

FARM STANDS

Monday through Saturday in season, colorful farm stands are set up on Main Street to sell local produce and flowers.

At **Bartlett's Ocean View Farm & Greenhouses** (✉ Bartlett Farm Rd. off Hummock Pond Rd., ☎ 508/228–9403), a 100-acre farm run by eighth-generation Bartletts, a farm stand is open in season. In June you can pick your own strawberries.

FOOD

**Chanticleer to Go** (✉ 15 S. Beach St., ☎ 508/325–5625) offers prepared gourmet foods from the 'Sconset French restaurant **Chanticleer** (☞ *below*), with instructions for home use, as well as pastries, salads, sandwiches, imported cheeses and pâtés, wine, and espresso and cappuccino.

GALLERIES

**Robert Wilson Galleries** (✉ 34 Main St., ☎ 508/228–6246 or 508/228–2096) carries outstanding contemporary American marine, impressionist, and other art.

**Sailors' Valentine Gallery** (✉ Macy's Warehouse, Lower Main St., ☎ 508/228–2011) has contemporary fine and folk art (including international "outsider art"), South American furnishings, sculpture, and exquisite sailors' valentines.

**William Welch Gallery** (✉ 14 Easy St., ☎ 508/228–0687) exhibits Welch's signature watercolors, pastels, and oils of Nantucket scenes, as well as the Nantucket oil paintings of Jack Brown.

GIFT SHOPS

**Museum Shop** (✉ Broad St., next to the Whaling Museum, ☎ 508/228–5785) has island-related books, antique whaling tools, reproduction furniture, and toys, including reproduction 18th- and 19th-century whirligigs.

**Seven Seas Gifts** (✉ 46 Centre St., ☎ 508/228–0958) stocks all kinds of inexpensive gift and souvenir items, including shells, baskets, toys, and Nantucket jigsaw puzzles. The building itself is unique, with addition after addition tacked on as the contents of the store required more space. Check out the pressed tin ceilings and the fireplaces tucked in amongst the merchandise.

# SIASCONSET, THE SOUTH SHORE, AND MADAKET

Beyond the hustle and bustle of Nantucket Town, the island also has a serene side. Several bike paths meander through the low shrubs and soft-colored heather of the moors and bogs, where all you'll hear is the song of birds. At the northern tip of the island, the vast, windswept beaches of the Coatue–Coskata–Great Point reserve stretch for miles—and offer limitless sun, sea, and solitude. A sleepy little town with rose-covered cottages lining narrow streets, Siasconset also has unparalleled bluffs overlooking the wild Atlantic. To the west, the dune-backed beaches of Madaket and Eel Point are the ideal place from which to watch the sun slide into the sea in a burst of brilliant color. The many conservation areas on this part of the island showcase a variety of landscapes, from open heathlands to pine woods and sandy beaches. Even in the height of season, you'll be amazed to discover just what quiet solitude awaits.

## Siasconset, Polpis, and Wauwinet

*7 mi east of Nantucket Town, 9 mi northeast of Surfside.*

★ Originally a community of cod and halibut fishermen and shore whalers from the 17th century, **Siasconset** was already becoming a summer resort during whaling days, when people from Nantucket Town would come here to get away from the smell of burning whale oil in the refineries. In 1884 the narrow-gauge railway—built three years earlier to take spiffily clad folk from the New Bedford steamers to the beach at Surfside—came to 'Sconset, bringing ever more off-islanders. These included writers and artists from Boston in the 1890s, followed soon by Broadway actors on holiday during the theaters' summer hiatus. Attracted by the village's beauty, remoteness, sandy ocean beach, and cheap lodgings—converted one-room fishing shacks, and cottages built to look like them—they spread the word, and before long 'Sconset became a thriving actors' colony.

Today a charming village of pretty streets with tiny rose-covered cottages and driveways of crushed white shells, 'Sconset is almost entirely a summer community—the local postmaster claims that about 150 families live here through the winter, but you'd never know it. At the central square are the post office, a liquor store, a bookstore, a market, and two restaurants.

Siasconset makes a lovely day trip from town. Off-season there's not a lot to do in the village, but it still has its attractions, taking a stroll through some of its narrow streets being among them, and there are plenty of beautiful beaches and conservation areas close by to enjoy.

NEED A **The Nantucket Bake Shop** (⊠ 79 Orange St., ☎ 508/228–2797) is a
BREAK? great place to stop on your way out of Nantucket Town to fill a knapsack with Portuguese bread and pastries. It's closed January–March.

★ The **Milestone Bog,** more than 200 acres of working cranberry bogs surrounded by conservation land, is always a beautiful sight to behold, especially during the fall harvest. Cultivated since 1857, the bog was the world's largest contiguous natural cranberry bog until it was subdivided after 1959. The land was donated to the NCF in 1968. The bogs are leased to a grower, who harvests and sells the crops.

The harvest begins in late September and continues for six weeks, during which time harvesters work daily from sunup to sundown in flooded bogs. The sight of floating bright red berries and the moors' rich autumn colors is not to be missed. At other times the color of the bog may be green, rust-red, or, in June and early July, the pale pink of cranberry blossoms. Any time of year, the bog and the moors have a remarkable quiet beauty that's well worth seeing.

In 'Sconset, a stroll down three of the side streets—**Evelyn, Lily, and Pochick,** south of Main Street—will give you an idea of what Siasconset was like a century ago. The streets remain much as they were in the 1890s, when a development of tiny rental cottages in the fishing-shack style was built here. The summer blossoms of roses climbing all over many of the cottages are fantastic carpets of color.

An interesting reminder of the past is the 'Sconset Pump, a nicely preserved well marked with a plaque proclaiming it "dug in 1776." ⊠ *Off Broadway St.*

Despite its name, the 1899 **Siasconset Casino** is in fact a tennis club and bowling alley that was never used for gambling. During the actors'-colony heyday, it became a venue for theater productions. Some theater is still seen here, but the casino is mostly a summer tennis club and cinema. Opposite the casino is the much-photographed entryway of the Chanticleer restaurant: A trellis arch topped by a sculpted hedge frames a rose garden with a flower-bedecked carousel horse at its center. ⊠ *New St.,* ☎ *508/257–6661.*

The **'Sconset Union Chapel,** the village's only church, holds Roman Catholic mass at 8:45 AM and Protestant services at 10:30 on summer Sundays. ⊠ *Chapel St.*

The red-and-white-striped **Sankaty Lighthouse,** overlooking the sea on one side and the Scottish-looking greens of the private Sankaty Head Golf Club on the other, is one of the Cape and Islands' many endangered lighthouses. Standing on a 90-ft-high bluff that has lost as much as 200 ft of shoreline in the past 75 years, the 1849 Sankaty Light could be lost, as Great Point Light was in 1984, to further erosion. A fragile piece of land, Nantucket loses more of its shoreline every year, especially at Sankaty and on the south shore, where no shoals break the ocean waves as they do on the north shore. Some of that sand is simply moved along shore to the other end of the island, but that's no great consolation to Sankaty and 'Sconset dwellers.

Alongside **Polpis Road,** which heads north out of 'Sconset on its loop back toward town, hundreds of thousands of daffodils bloom in spring. A million Dutch bulbs donated by an island resident were planted along Nantucket's main roads in 1974, and more have been planted every year since.

A good spot for bird-watching, **Sesachacha Pond** (pronounced seh-*sah*-kah-cha or, more often here, where long words seem to be too much trouble, just *sah*-kah-cha) is circled by a walking path that leads to an Audubon wildlife area. The pond is separated from the ocean by a narrow strand on its east side. A couple of years ago, the water level of the pond was lowered, and several arrowheads were found. From the pond, there's a good view of Sankaty Light high above.

Throughout the year, the 205-acre **Windswept Cranberry Bog,** part working bog, part conservation land, is a beautiful tapestry of greens, reds, and golds—and a popular hangout for many different bird species. The bog is especially vibrant in mid-October, when you can watch the cranberry harvest. A map, as well as the 30-page "Handbook for Visitors to the Windswept Cranberry Bog," is available for $4 ($5 by mail) from the NCF (☞ Island Activities *in* Nantucket A to Z, *below*). The bog is off Polpis Road northwest of 'Sconset.

**Wauwinet** is a hamlet of beach houses on the northeastern end of Nantucket. European settlers found the neck of sand above it to be the easiest way to get to the ocean for fishing. Instead of rowing around Great Point, fishermen would go to the head of the harbor and haul their dories over the narrow strip of sand and beach grass separating Nantucket Harbor from the ocean. Hence the name for that strip: the haulover. Various storms have washed it out, allowing fishermen to sail clean through the area, but the moving sands have continued to fill it back in.

In the 1870s, the first Wauwinet House started luring townspeople to its 75¢ shore dinners. Local historian Jane Lamb, resident of "Chaos Corner" on Polpis Road, relates the story of July 4th, 1877, revelers who, while sailing back to town after dancing until 11 PM, were grounded on shoals, rolling back and forth until the tide rose with the sun. Happily, Wauwinet dinner cruises didn't die with that night, and you can still sail out to Wauwinet for dinner at the current inn.

**Coatue–Coskata–Great Point,** an unpopulated spit of sand comprising three cooperatively managed wildlife refuges, is a great place to spend a day exploring, relaxing, or pursuing a favorite activity, like bird-watching or fishing. **Coatue,** the strip of sand enclosing Nantucket Harbor, is open for many kinds of recreation—shellfishing for bay scallops, soft-shell clams, quahogs, and mussels (license required), surf casting for bluefish and striped bass (spring through fall), picnicking, or just enjoying the crowdless expanse. **Coskata**'s beaches, dunes, salt marshes, and stands of oak and cedar attract marsh hawks, egrets, oystercatchers, terns, herring gulls, plovers, and many other birds, particularly in spring and fall migration. A successful program has brought osprey here to nest on posts set up in a field by Coskata Pond. Don't come without field glasses.

Because of dangerous currents and riptides and the lack of lifeguards, swimming is strongly discouraged, especially within 200 yards of the 70-ft stone tower of **Great Point Light.** Those currents, at the same time, are fascinating to watch at the Great Point tide rip. Seals and fishermen alike benefit from the unique feeding ground that it creates. The lighthouse is a 1986 re-creation of the light destroyed by a storm in 1984, which in turn replaced one burned down by a drunk. The new light was built to withstand 20-ft waves and winds of up to 240 mi an hour, and it was fitted with eight solar panels to power it.

The area may be entered only on foot or by four-wheel-drive vehicle, for which a permit ($85 for a year; $20 a day for a rental vehicle; ☎ 508/228–2884 for information) is required. Issued only for a vehicle that is properly registered and equipped—confirm this with the rental agent if you plan to enter the area—the permits are available at the gatehouse at Wauwinet (☎ 508/228–0006) June–September or off-season from a ranger patrolling the property. If you enter on foot, be aware that Great Point is a 5-mi walk from the entrance on soft, deep sand. Jeepless people often hitchhike here. Another alternative is a Jeep tour.

★ An unmarked dirt track off Polpis Road between Wauwinet and town leads to **Altar Rock,** from which the view is spectacular. The rock sits on a high spot in the midst of open moor and bog land—technically called lowland heath—which is very rare in the United States. The entire area, of which the Milestone Bog is a part, is laced with paths leading in every direction. Don't forget to keep track of the trails you travel in order to find your way back. ⊠ *Polpis Rd.*

Housed in an interesting re-creation of an 1874 Life Saving Service station, the **Nantucket Life Saving Museum** displays items including original rescue equipment and boats, as well as photos and accounts of daring rescues. Late in 1995, the museum acquired a surfboat, one of four still extant, and a horse-drawn carriage from the Henry Ford Museum, and a beach cart (the only known original), all in mint condition. ⊠ *Fulling Hill Rd., off Polpis Rd.,* ☎ *508/228–1885.* ☞ *$3.* ☉ *Mid-June–mid-Sept., Tues.–Sun. 9:30–5; mid-Sept.–mid-June, to tour groups with reservations.*

## Dining and Lodging

**$$$$** ✕ **Chanticleer.** First, Chanticleer deserves its due. For more than 20 years, Anne and Jean-Charles Berruet have been serving superb French food in a lovingly-maintained, formal country setting, complete with gorgeous clematis cascading down the weathered shingles. The wine cellar is among the best in the country, reputed to hold more than 40,000 bottles. All that said, some have come to feel that the food and the ambience have both become heavy and overbearing. Like all things, this is subjective, and for many a Nantucket trip would not be complete without getting dressed up and making the trip here. As for freshness and presentation, the food is unimpeachable and the desserts are profoundly rich. Dining here is a commitment to a classic form of elegance—think of if as entering the set of an elaborate play. For the full effect, try to stay on center stage—the main dining room in front. ⊠ *9 New St.,* ☎ *508/ 257–6231. Jacket required. AE, MC, V. Closed Mon.*

**$$–$$$** ✕ **'Sconset Café.** A great location, and BYOB to boot. If you stop in
★ and there is no table (as is often the case), they will give you a beeper, send you down to the beach or over to the Summer House for a drink, and beep you when your table is ready. You can relax, wear jeans, and enjoy what is becoming both a better and a more expensive menu, featuring local fish and a few other items such as duck. If you've carted a couple of bottles of your favorite wine to the island, this is the place to do them in. Lunchtime menu items are reasonable, including lots of sandwiches and salads. ⊠ *Post Office Sq.,* ☎ *508/257–4008. BYOB (liquor store next door). No credit cards. Closed 1st week of Oct.–mid-May.*

**$** ✕ **Picnic Lunches.** For a great picnic in 'Sconset, stop in at **Claudette's** ('Sconset center, ☎ 508/257–6622) and pick up something for a nearby beach.

**$$$$** ✕🏠 **Summer House.** Here, across from 'Sconset Beach and clustered around a flower-filled lawn, are the rose-covered cottages associated with Nantucket summers. For the restaurant, compliments and complaints abound—divided between those who love the place for its 'Sconset bluff ocean views and live piano music, and those who are put off by the pompous, over-fussy service (reservations are essential for dinner no matter which camp you're in). The pre–World War II whitewashed beach cottage look lends the dining room a sort of *Great Gatsby* vibe. The daily seafood specials are often the highlights on the menu. Lunch at the poolside café is very relaxing. Each one- or two-bedroom cottage at Summer House is furnished in romantic English country style: trompe-l'oeil-bordered white walls, Laura Ashley floral accents, white eyelet spreads, and stripped English-pine antique furnishings. Some cottages have fireplaces or kitchens, and most have marble baths with whirlpools. Adirondack chairs set about the lawn offer relaxation with a view. ⊠ *Ocean Ave., Box 880, Siasconset 02564, ☎ 508/257–4577, ₣ₐₓ 508/257–4590. 8 cottages. 2 restaurants, bar, piano bar, pool. AE, MC, V. Closed Nov.–late Apr. CP.*

**$$$$** ✕🏠 **Wauwinet.** Some people would say that this historic 19th-century
★ hotel is the only place to stay on Nantucket, and its friendliness, exquisite location, impeccable furnishings, first-rate restaurant, and extensive services make it easy to see why. **Topper's** restaurant is lovely, tasteful, and elegant, a perfect match for the inn and its setting (dinner reservations are essential). Lobster-crabcakes with corn, jalapeño olives, and mustard sauce are a deserving favorite appetizer, and others, such as warm Nantucket oysters with caviar, are equally alluring. Pan-seared sea scallops with lobster risotto, or roasted rack of *cervena* venison (a lean, farm-raised New Zealand deer) with creamy polenta and forest mushroom glaze are superb entrées. The sommelier is a master at matching bottles from his cellar with any of Topper's dishes, and you can ask him to serve you a different glass with each of your dinner courses. Brunch comes highly recommended, and a wonderful breakfast is included in room rates, as is afternoon port or sherry and cheese. Around the inn itself, a sweeping lawn with white chaise longues leads to a pebbly private harbor beach, where you can play with a life-size wooden chess set. Miles of harbor and ocean beaches are steps away. Guest rooms and cottages are individually decorated in country-beach style, with pine antiques; some have views of the sunset over the water. Activities here abound: tennis, mountain biking, croquet, and sailing in Nantucket catboats. There are also boat shuttles to Coatue beach across the harbor and, perhaps best of all, the innkeeper runs a Land Rover tour of the Great Point reserve—all of which are included in the room rate. Jitney service to and from town 8 mi away, plus Steamship pickup, make it convenient if you don't have a car. In season, you can cruise to the inn in the gorgeous *Wauwinet Lady* skiff. ⊠ *Wauwinet Rd., Box 2580, Nantucket 02584, ☎ 508/228–0145 or 800/426–8718, ₣ₐₓ 508/228–7135. 25 rooms, 5 cottages. Restaurant, bar, room service, 2 tennis courts, croquet, boating, mountain bikes, library, concierge, business services. AE, DC, MC, V. Closed Nov.–Apr. BP.*

**$$** 🏠 **Wade Cottages.** On a bluff overlooking the ocean, this complex of guest rooms, apartments, and cottages in 'Sconset couldn't be better located for beach lovers. The buildings, in the same family since the 1920s, are arranged around a central lawn with a great ocean view. Most inn rooms and cottages have sea views, and all have phones. Furnishings are generally in somewhat worn beach style, with some antique pieces. ⊠ *Shell St., Box 211, Siasconset 02564,* ☎ *508/257–6308,* FAX *508/257–4602; off-season, 212/989–6423. 8 rooms (3-night minimum), 4 with private bath; 6 apartments (1-wk minimum); 3 cottages (2-wk minimum). Refrigerator, badminton, Ping-Pong, beach, coin laundry. MC, V. Closed mid-Oct.–late May. CP.*

## Nightlife and the Arts

**Nantucket Island School of Design and the Arts** (⊠ 23 Wauwinet Rd., ☎ 508/228–9248; for schedule, ⊠ Box 958, Nantucket 02554) offers a year-round program of multicultural, environmental, nature, and arts-and-crafts classes, lectures, and slide shows for adults and children.

The **Siasconset Casino** (⊠ New St., Siasconset, ☎ 508/257–6661) shows first-run films in season—and it has finally replaced its old metal seats with upholstered ones.

## Outdoor Activities and Sports

### BEACHES

**'Sconset Beach** (also known as Codfish Park) has a golden-sand beach with moderate to heavy surf, a lifeguard, showers, rest rooms, and a playground. Restaurants are a short walk away.

### BIKING

The 6-mi **'Sconset Bike Path** starts at the rotary east of town and parallels Milestone Road, ending in the village. It is mostly level, with some gentle hills, and benches and drinking fountains at strategic locations along the way.

The 8-mi **Polpis Bike Path,** a long trail with gentle hills, winds alongside Polpis Road almost all the way into Siasconset. It goes right by the Life Saving Museum (☞ *above*), and has great views of the moors, the cranberry bogs, and Sesachacha Pond.

### GOLF

**Sankaty Head Golf Club** (⊠ Sankaty Rd., Siasconset, ☎ 508/257–6391), a private 18-hole course, is open to the public from late September to mid-June. This challenging Scottish-style links course cuts through the moors and has spectacular views of the lighthouse and ocean from practically every hole.

**Siasconset Golf Club** (⊠ Milestone Rd., ☎ 508/257–6596), begun in 1894, is an easy-walking nine-hole public course surrounded by conservation land.

### NATURE TOURS

The Trustees of Reservations sponsor naturalist-led **Great Point Natural History Tours** (☎ 508/228–6799).

### SWIMMING

Besides the island's many ponds and beaches, the **Summer House** (☞ Dining and Lodging, *above*) in 'Sconset offers its pool, on the bluff above the ocean beach, to diners at its poolside café in season.

### TENNIS

**Siasconset Casino** (⊠ New St., ☎ 508/257–6661) is a private club with seven clay courts and one poor hard court. Infrequently the club has openings at 1 or 2 PM. Call ahead to check.

# The South Shore

**Miacomet Pond** is a freshwater pond surrounded by grass and heath, separated from the ocean by a narrow strip of sandy Land Bank beach. The pond—in whose reedy fringes swans and snapping turtles are sometimes seen, along with the resident ducks—is a peaceful setting for a picnic or quiet time. A right turn off Surfside Road onto Miacomet Road (which begins paved and turns to dirt) will take you there.

**Cisco** is a beautiful area not far from Nantucket Town, with a stretch of beach that suffers the battering of wild wind-tossed seas. Several ocean-front houses at Cisco have been lost or moved since 1990 because of massive bluff erosion due to many unusually fierce winter storms. To get to the area, take Main Street west to Milk Street and continue at its end on Hummock Pond Road to its end.

**Nantucket Vineyard.** Five varieties of vinifera grapes grow at this vineyard and winery 2½ mi south of town. Tastings of red, white, and rosé wines are available year-round, as are bottles for purchase. ⊠ *3 Bartlett Farm Rd., off Hummock Pond Rd., ☎ 508/228–9235. ☞ Free. ☉ Mon.–Sat. 10–6 (call first in winter).*

A 5- to-10 minute bike ride from town on the way to Madaket, the **Sanford Farm, Ram Pasture, and the Woods** offer a wonderful combination of habitats to explore. The area comprises more than 900 contiguous acres of wetlands, grasslands, forest, and former farmland off Madaket Road. Interpretive markers border the 6½-mi (round-trip) walking trail that leads to the shore via Hummock Pond and offers great ocean and heath views. It begins off the Madaket Road parking area near the intersection of Cliff Road, as do a 1.7-mi loop and a 3.1-mi round-trip to a barn on high ground with views of the south shore. Stop at **Something Natural** (☞ Dining and Lodging *in* Nantucket, *above*) on Cliff Road on your way out of town to pick up a picnic. Maps are available through the **NCF** (☞ Island Activities *in* Nantucket A to Z, *below*), which owns 767 acres, or the **Land Bank** (⊠ 22 Broad St., Nantucket 02554, ☎ 508/228–7240), which owns 165.

# Madaket

*6 mi west of Nantucket Town, 15 mi west of Siasconset, 4 mi southwest of Dionis.*

This rural, residential village at the westernmost point of Nantucket doesn't have much to offer by way of a town, but beautiful beaches and conservation areas well worth exploring are close by, including some of the most scenic places to observe the island's wildlife. A bike ride out from town over gentle hills takes 30 to 45 minutes.

A ¾-mi **Land Bank walking trail** begins at a set of picnic tables on the north side Madaket Road, ½ mi west of the Eel Point Road intersection. It wanders upland to an overview of a swamp, through a hawthorn grove and blueberry patches, and across meadows. A short jaunt, it is a perfect introduction to the many different habitats and environments found on Nantucket.

**Long Pond** is a 64-acre Land Bank property with a diversity of habitats and terrain that makes birding especially good. A 1-mi walking path along the pond, past meadows and a natural cranberry bog, makes for a pleasant, relaxing stroll.

An unspoiled conservation area perfect for a leisurely walk, especially appealing to bird-watchers, **Eel Point** is a spit of sand with harbor on one side and shoal-protected ocean on the other. Covered here and there with goldenrod, wild grapes, roses, bayberries, and other coastal plants, the area is a nesting place for gulls and also attracts great numbers of other birds, including herons and egrets, which perch on small islands formed by a sandbar that extends out 100 yards or more. The water is shallow, and the surf fishing is good. Nature guides on Eel Point are available from the **Maria Mitchell Association** (✉ 2 Vestal St., Nantucket 02554, ☎ 508/228–9198; $4.50) or the **NCF** (☞ *Island Activities in Nantucket A to Z, below*).

## Dining

$$$  ✕ **Westender.** Location really is everything in a resort locale, as the Westender proves. The 100-yard walk from Madaket Beach to the bar is one of this place's prime attractions. Thanks to a new building that blocks them, the ocean views aren't what they used to be. The upstairs is airy and pleasant, but the downstairs is more fun, with a bar that plays reggae music or Bonnie Raitt songs. Bar food and lighter fare are the way to go, although big dinners are served (and priced accordingly). Even if it's not perfect, sometimes it just plain beats battling into town. ✉ *Madaket Rd.,* ☎ *508/228–5100. AE, MC, V. Closed Columbus Day–Memorial Day.*

## Outdoor Activities and Sports

### BEACHES

Six miles from town and accessible only by foot, **Eel Point** has one of the island's most beautiful and interesting beaches for those who don't necessarily need to swim—a sandbar extends out 100 yards, keeping the water shallow, clear, and calm. There are no services, just lots of birds, wild berries and bushes, and solitude.

Known for great sunsets and surf, **Madaket Beach** is reached by shuttle bus or the Madaket bike path (6 mi) and has lifeguards and rest rooms. The **Westender** restaurant (☞ *Dining, above*), a short walk away, has a take-out window, a bar, and a deck for drinks.

### BICYCLING

★  The **Madaket Bike Path,** reached via Cliff Road, is a hilly but beautiful 6-mi route to the western tip of the island. There are picnic tables by Long Pond, and benches and drinking fountains along the way.

# NANTUCKET A TO Z

## Arriving and Departing

### By Ferry

Year-round service is available from Hyannis only. Hy-Line has two boats, one of which runs between Nantucket and the Vineyard in summer only. The only way to get a car to Nantucket is on the Steamship Authority. To get a car from the Vineyard to Nantucket, you would have to return to Woods Hole, drive to Hyannis, and ferry it out from there.

#### FROM HYANNIS

The **Steamship Authority** runs car-and-passenger ferries to the island from Hyannis year-round. The trip takes 2¼ hours. For policies and restrictions, ☞ Martha's Vineyard A to Z *in* Chapter 3. ✉ *South St. dock,* ☎ *508/477–8600; on Nantucket,* ☎ *508/228–3274 for reservations or 508/228–0262; TTD 508/540–1394.* 🎫 *One-way: $10, $5 bicycles. Cars, one-way: mid-May–mid-Oct., $90; mid-Mar.–mid-May and mid-Oct.–Nov., $70; Dec.–mid-Mar., $50.*

**Hy-Line**'s high-end, high-speed boat, *The Grey Lady,* ferries passengers from Hyannis and back year-round. The trip takes just over an hour. That speed has its down-side in rough seas—lots of bouncing and lurching that some find nauseating. Seating ranges from benches on the upper deck to airlinelike seats in side rows of the cabin to café-style tables and chairs in the cabin front. There is a snack bar on board. ✉ *Ocean*

*St. dock,* ☎ *508/778–0404 or 800/492–8082.* 🎫 *One-way: $29, $4.50 bicycles.*

**Hy-Line**'s slower ferry makes the 1¾- to 2-hour trip from Hyannis early May–October 28. The MV *Great Point* offers a first-class section ($21 one-way) with a private lounge, rest rooms, upholstered seats, carpeting, complimentary Continental breakfast or afternoon cheese and crackers, a bar, and a snack bar. ✉ *Ocean St. dock,* ☎ *508/778–2602 for reservations or 508/778–2600 for information; on Nantucket,* ☎ *508/228–3949.* 🎫 *One-way: $11, $4.50 bicycles.*

FROM HARWICH PORT

**Freedom Cruise Line** runs passenger ferry service from June through October. The trip takes just under two hours and offers an alternative to the crowds in Hyannis during the busy summer months. ✉ *Saquatucket Harbor,* ☎ *508/432–8999 for reservations.* 🎫 *One-way: $16, $5 bicycles.*

FROM MARTHA'S VINEYARD

**Hy-Line** makes 2¼-hour runs to and from Nantucket from early June to mid-September—the only inter-island passenger service. ☎ *508/778–2600 in Hyannis, 508/228–3949 on Nantucket, 508/693–0112 in Oak Bluffs.* 🎫 *One-way: $11, $4.50 bicycles.*

## By Plane

**Nantucket Memorial Airport** (☎ 508/325–5300) is about 3½ mi southeast of town via Old South Road. A taxi from the airport to town costs about $6.

**Business Express/Delta Connection** (☎ 800/345–3400) flies from Boston year-round and from New York (La Guardia) in season.

**Cape Air/Nantucket Air** (☎ 508/771–6944, 508/228–6252, 508/228–6234, 800/352–0714 or 800/635–8787 in MA) flies from Hyannis year-round and offers charters.

**Coastal Air** (☎ 508/228–3350, 203/448–1001 or 800/262–7858) is a year-round charter company with an island base.

**Colgan Air** (☎ 508/325–5100 or 800/272–5488) flies from Newark and Hyannis year-round.

**Continental Express** (☎ 800/525–0280) has nonstops from Newark in season.

**Island Airlines** (☎ 508/228–7575, 800/698–1109 in MA, or 800/248–7779) flies from Hyannis year-round and offer charters.

**Northwest Airlink** (☎ 800/225–2525) flies from Boston year-round, from Newark in season.

**Ocean Wings** (☎ 508/325–5548 or 800/253–5039) offers year-round charters from its island base.

**Westchester Air** (☎ 914/761–3000 or 800/759–2929) flies charters out of White Plains, New York.

# Getting Around

One of the attractions of a Nantucket vacation is escape from the fast lane. You might find yourself walking a lot more than usual and taking advantage of the island's miles of scenic bike paths. Even so, in high season the main streets are clogged with traffic (and parking spaces are filled), and residents beg you to leave your car at home. If your visit will be short and spent mostly in town and on the beaches, taxis and beach shuttles can supplement foot power adequately.

Some of the island's most beautiful and least touristed beaches are accessible only by foot or four-wheel-drive vehicles. Yearlong permits are available for $20 for private vehicles, $100 for rental vehicles, at the police department (☎ 508/228–1212) on South Water Street. Coatue–

Coskata–Great Point is open to Jeeps but requires a separate NCF permit.

### By Bicycle and Moped

Mountain bikes are best if you plan to explore the dirt roads. To drive a moped you must have a driver's license and a helmet. You may not use the vehicle within the town historic district between 10 PM and 7 AM, and you may never drive it on the bike paths. Moped accidents happen often on the narrow or dirt roads—watch out for loose gravel.

### By Bus

From mid-June to Labor Day, **Barrett's Tours** (⊠ 20 Federal St., ☎ 508/228–0174 or 800/773–0174), across from the Information Bureau, runs beach shuttles to 'Sconset and Madaket ($5 round-trip, $3 one-way), Surfside ($3 round-trip, $2 one-way), and Jetties ($1 one-way) several times daily. Children pay half fare to 'Sconset and Surfside.

### By Car

The proposition of having a car on the island is invariably expensive, whether you bring one over on the ferry or rent one once you arrive (☞ Car Rentals, *below*). Yet you might find the experience of not having a car to bother with quite pleasant. With the seasonal island shuttle, having a car may even be unnecessary if you're staying in a hotel. If you're staying in a house and need to shop for groceries, a car could be of use, but you should ask yourself whether it's worth the price of admission. Nantucket *is* small, so you won't be using much gas, but as a warning, parking in town and traffic getting in and out is just ghastly in season.

### By Island Shuttle

The Nantucket Regional Transit Authority (☎ 508/228–7025, TDD 508/325–0788) runs shuttle buses around the island between June 1 and September 30. There are five routes with different timing on each; most service begins at 7 AM and ends at 11 PM. Call for schedules. Fares are 50¢ in town, $1 to 'Sconset and Madaket, $10 for a 3-day pass, $15 for a 7-day pass, and $30 for a 1-month pass. Seasonal passes are also available.

### By Taxi

Taxis usually wait outside the airport or at the foot of Main Street by the ferry. Rates are flat fees, based on one person with two bags before 1 AM: $4 within town (1½-mi radius), $8 to the airport, $12 to 'Sconset, $13 to Wauwinet.

**A-1 Taxi** (☎ 508/228–3330 or 508/228–4084).
**Aardvark Cab** (☎ 508/228–2223).
**All Point Taxi** (☎ 508/228–5779).
**B. G.'s Taxi** (☎ 508/228–4146). A seven-passenger van is available upon request.
**Peterson's Taxi** (☎ 508/228–9227).

## Contacts and Resources

### Baby-sitting Service

**Nantucket Babysitters' Service** (☎ 508/228–4970) is offered by the South Suburban Nurses Registry.

### B&B Reservation Agencies

**DestINNations** (☎ 800/333–4667) handles a limited number of Nantucket hotels and B&Bs but will arrange any and all details of a visit.

**Heaven Can Wait Accommodations** (⊠ Box 622, Siasconset 02564, ☎ 508/257–4000) books inns and B&Bs and also plans island honeymoons.

**House Guests Cape Cod and the Islands** (⊠ Box 1881, Orleans 02653, ☎ 508/896–7053 or 800/666–4678, ℻ 508/896–7054) books B&Bs, cottages, and efficiencies.

**Martha's Vineyard and Nantucket Reservations** (⊠ Box 1322, Lagoon Pond Rd., Vineyard Haven 02568, ☎ 508/693–7200 or 800/649–5671 in MA) books inns, hotels, B&Bs, and cottages.

**Nantucket Information Bureau** (⊠ 25 Federal St., ☎ 508/228–0925) maintains a list of room availability in season and at holidays for last-minute bookings. At night, check the lighted board outside for available rooms.

## Car Rentals
If you decide to rent a car, reserve one well in advance in season. **Budget** (508/228–5666) rents cars, vans, and Ford Explorers. **Hertz** (508/228–9421) has cars and Ford Explorers available. The local company **Nantucket Windmill** (508/228–1227 or 800/228–1227) rents cars, vans, Ford Explorers, and Jeeps at low rates and with free mileage.

## Cash Machines
ATMs are at the **airport, Steamboat Wharf, Nantucket Bank** (⊠ 2 Orange St. or 104 Pleasant St.), **Pacific National Bank** (⊠ 61 Main St.), **A&P** (⊠ Straight Wharf), and **Finast** (⊠ Lower Pleasant St.).

## Emergencies
Dial 911 to reach the police or fire department.

The **Nantucket Cottage Hospital** (⊠ 57 Prospect St., ☎ 508/228–1200) has a 24-hour emergency room.

## Guided Tours
### CRUISES
Boats of all kinds leave from Straight Wharf on **harbor sails** throughout the summer; many are available for charter as well. **Anna W. II** (⊠ Slip 12, ☎ 508/228–1444) is a renovated lobster boat with shoreline, sunset, and moonlight cruises, as well as lobstering demonstrations and winter seal cruises. The **Endeavor** (⊠ Slip 15, ☎ 508/228–5585), a beautiful 31-ft Friendship sloop, offers harbor tours, sunset cruises, and sails to Coatue, where you are rowed ashore to spend a private morning beachcombing. The 40-ft sailing yacht **Sparrow** (⊠ Slip 18, ☎ 508/228–6029), with a teak, brass, and stained-glass interior, offers 1½-hour sails for six guests, plus charters.

The **"Around the Sound"** cruise, a one-day round-trip from Hyannis with stops at Nantucket and Martha's Vineyard and six hours at sea, is available June–mid-September. ⊠ Hy-Line, Ocean St. Dock, Hyannis, ☎ 508/778–2600 or 508/778–2602; on Nantucket, ☎ 508/228–3949. ⊠ $33, $13.50 bicycles.

### GENERAL TOURS
**Barrett's Tours** (⊠ 20 Federal St., ☎ 508/228–0174 or 800/773–0174) gives 75- to 90-minute narrated bus tours of the island from spring through fall; the Barrett family has lived on Nantucket for generations. Buses meet the ferries.

**Carried Away** (☎ 508/228–0218) offers narrated carriage rides through the town historic district in season.

**Gail's Tours** (☎ 508/257–6557) are lively 1½-hour van tours narrated by sixth-generation Nantucketer Gail Johnson, who knows all the inside stories.

**Great Point Natural History Tours** (☎ 508/228–6799), led by naturalists, are sponsored by the Trustees of Reservations.

**"Historic Nantucket Walking Tours,"** a free self-guided tour pamphlet published by the Nantucket Historical Association, is available at the Information Bureau.

**Nantucket Island Tours** (Straight Wharf, ☎ 508/228–0334) gives 75- to 90-minute narrated bus tours of the island from spring through fall; buses meet the ferries.

**Roger Young's historic walking tours** (☎ 508/228–1062; in season) of the town center are entertaining and leisurely.

## Harbor Facilities

Harbor facilities are available in town year-round at the **Town Pier** (☎ 508/228–7260). The **Boat Basin** (☎ 508/228–1333 or 800/626–2628) offers harbor facilities year-round, with shower and laundry fa- cilities, electric power, cable TV, phone hookups, a fuel dock, and summer concierge service. **Madaket Marine** (☎ 508/228–9086 or 800/564–9086) has moorings May–October and fuel and slips year-round for boats up to 29 ft in Hither Creek.

## House Rentals

A number of realtors (complete lists are provided by the chamber and the Information Bureau (☞ Visitor Information, *below*) offer rentals ranging from in-town apartments in antique houses to new waterfront houses. **Congdon & Coleman** (✉ 57 Main St., Nantucket 02554, ☎ 508/325–5000, FAX 508/325–5025) has properties island-wide. **'Scon- set Real Estate** (✉ Box 122, Siasconset 02564, ☎ 508/257–6335, or 508/228–1815 in winter) lists properties in and around 'Sconset.

## Island Activities

### HISTORICAL TOURS

The self-guided **Black Heritage Trail** tour covers nine sites in and around Nantucket Town with associations to Nantucket's African-American population, including the African Meeting House, the Whaling Museum, and the Atheneum. The trail guide is free from the Friends of the African Meeting House on Nantucket. ✉ *Box 1802, Nantucket 02544,* ☎ *508/228–4058.*

### NATURE TOURS

For a map of the **Nantucket Conservation Foundation**'s properties, visit or write to NCF headquarters. The map costs $3, or $4 by mail. ✉ *118 Cliff Rd., Box 13, Nantucket 02554,* ☎ *508/228–2884.* ☉ *Week- days 8–5.*

### NIGHTLIFE AND THE ARTS

For listings of events, see the free seasonal weekly *Nantucket Map & Legend,* the ferry companion paper *Yesterday's Island,* and both island newspapers.

### SHELLFISHING

The **shellfish warden** (✉ 38 Washington St., ☎ 508/228–7260) issues digging permits for littleneck and cherrystone clams, quahogs, and mussels.

## Pharmacies

**Congdon's** (✉ 47 Main St., ☎ 508/228–0020) is open nightly until 10 from mid-June to mid-September. **Island Pharmacy** (✉ Finast Plaza, Sparks Ave., ☎ 508/228–6400) is open daily until 8 or 9 during the summer. **Nantucket Pharmacy** (✉ 45 Main St., ☎ 508/228–0180) stays open until 10 nightly from Memorial Day to Labor Day.

## Visitor Information

The **Chamber of Commerce** (✉ 48 Main St., ☎ 508/228–1700) is open Labor Day–Memorial Day, weekdays 9–5; Memorial Day–Labor Day, weekdays 8:30–5:30, Saturday 10–2. The **Nantucket Visitors Ser-**

**vice and Information Bureau** (✉ 25 Federal St., ☎ 508/228–0925) is open daily July–Labor Day; weekdays Labor Day–June. Hours vary. Also look for information kiosks at Steamboat and Straight wharves. The **Helpline** (☎ 508/228–7227) has information on island health and human services.

# 5 Portraits of the Cape and Islands

*A Brief History of the Cape*

*Cape and Islands Books*

# A BRIEF HISTORY OF THE CAPE

**T**HE FORTUNES of Cape Cod have always been linked to the sea. For centuries, fishermen in search of a livelihood, explorers in search of new worlds, and pilgrims of one sort or another in search of a new life—down to the beach-bound tourists of today—all have turned to the waters around this narrow peninsula arcing into the Atlantic to fulfill their needs and ambitions.

Although some maintain that the Viking Thorvald from Iceland broke his keel on the shoals here in 1004, European exploration of Cape Cod most likely dates from 1602, when Bartholomew Gosnold sailed from Falmouth, England, to investigate the American coast for trade opportunities. He first anchored off what is now Provincetown and named the cape for the great quantities of cod his crew managed to catch. He then moved on to Cuttyhunk in the Elizabeth Islands (which he named for the queen); on leaving after a few weeks, he noted that the crew were "much fatter and in better health than when we went out of England." Samuel de Champlain, explorer and geographer for the king of France, visited in 1605 and 1606; his encounter with the resident Wampanoag tribe in the Chatham area resulted in deaths on both sides.

None of these visits, however, led to settlement; that began only with the chance landing of the Pilgrims, some of whom were Separatists rebelling from enforced membership in the Church of England, others merchants looking for economic opportunity. On September 16, 1620, the *Mayflower*, with 101 passengers, set out from Plymouth, England, for an area of land granted them by the Virginia Company (Jamestown had been settled in 1607). After more than two months at sea in the crowded boat they saw land; it was far north of their intended destination, but after the stormy passage and considering the approach of winter, they put in at Provincetown Harbor on November 21. Before going ashore they drew up the Mayflower Compact, America's first document establishing self-governance, because they were in an area under no official jurisdiction and dissension had already begun to surface.

Setting off in a small boat, a party led by Captain Myles Standish made a number of expeditions over several weeks, seeking a suitable site for a settlement in the wilderness of woods and scrub. Finally they chose Plymouth, and there they established the colony, governed by William Bradford, that is today re-created at Plimoth Plantation.

Over the next 20 years, settlers spread north and south from Plymouth. The first parts of Cape Cod to be settled were the bay side sections of Sandwich, Barnstable, and Yarmouth (all incorporated in 1639), along an old Indian trail that is now Route 6A. (Martha's Vineyard was first settled in 1642, and Nantucket in 1659.) Most of the newcomers hunted, farmed, and fished; salt hay from the marshes was used to feed cattle and roof houses.

The first homes built by the English settlers on Cape Cod were wigwams built of twigs, bark, hides, cornstalks, and grasses, which they copied from those of the local Wampanoag Indians who had lived here for thousands of years before the Europeans arrived. Eventually, the settlers stripped the land of its forests to make farmland, graze sheep, and build more European-style homes, though with a New World look all their own. The steep-roofed saltbox and the Cape Cod cottage—still the most popular style of house on the Cape, and copied all over the country—were designed to accommodate growing families.

A newly married couple might begin by building a one- or two-room half-Cape, a rather lopsided 1½-story building with a door on one side of the facade and two windows on the other; a single chimney rose up on the wall behind the door. As the family grew, an addition might be built on the other side of the door large enough for a single window, turning the half-Cape into a three-quarter-Cape; a two-window addition would make it a symmetrical full Cape. Additions built onto the sides and back were called warts. An interesting feature of some Cape houses is the graceful bow roof, slightly curved like the bottom of a boat (not surprising, since ships' carpenters did much of the house building as well). More noticeable, often in the older houses, is a profusion of small, irregularly shaped and located windows in the gable ends; Thoreau wrote of one such house that it looked as if each of the various occupants "had punched a hole where his necessities required it." Many ancient houses have been turned into historical museums. In some, docents take you on a tour of the times as you pass from the keeping room—the heart of the house, where meals were cooked at a great hearth before which the family gathered for warmth—to the nearby borning room, in whose warmth babies were born and the sick were tended, to the "showy" front parlors where company was entertained. Summer and winter kitchens, backyard pumps, beehive ovens, elaborate raised wall paneling, wide-board pine flooring, wainscoting, a doll made of corn husks, a spinning wheel, a stereopticon, a hand-stitched sampler or glove—each of these historic remnants gives a glimpse into the daily life of another age.

The Wampanoags taught the settlers what they knew of the land and how to live off it. Early on they showed them how to strip and process blubber from whales that became stranded on the beaches. To coax more whales onto the beach, men would sometimes surround them in small boats and make a commotion in the water with their oars until the whales swam to their doom in the only direction left open to them. By the mid-18th century, as the supply of near-shore whales thinned out, the hunt for the far-flung sperm whale began, growing into a major New England industry and making many a sea captain's fortune. Wellfleet, Truro, and Provincetown were the only ports on the Cape that could support deep-water distance whaling (and these were overtaken by Nantucket and New Bedford), but ports along the bay conducted active trade with packet ships carrying goods and passengers to and from Boston. Cape seamen were in great demand for ships sailing from Boston, New York, and other deep-water ports. In the mid-19th century, the Cape saw its most prosperous days, thanks largely to the whaling industry.

The decline in whaling hit the economies of Martha's Vineyard and Nantucket first and hardest, and both islands began cultivating tourism in the 19th century. Martha's Vineyard had played host to annual Methodist Camp Meetings since 1835, and Siasconset, on Nantucket, became a summer haven for New York theater folk when train service reached that remote end of the island in 1884. Whereas previously people traveled from Boston to and along the Cape only by stagecoach or packet boat, in 1848 the first train service from Boston began, reaching to Sandwich; by 1873 it had been extended little by little to Provincetown. In the 1890s President Grover Cleveland made his Bourne residence (now gone) the Cape's first "summer White House" (to be followed in the 1960s by JFK). Grand seaside resorts grew up for summering families, and Cape Cod began to actively court visitors.

ARTISTS WERE DRAWN to Provincetown starting in the early part of the century with the establishment of several art schools. Writers (including John Reed and the young Eugene O'Neill) started the Provincetown Players, which would be the germ of Cape community theater and professional summer-stock companies. The Barnstable Comedy Club, founded in 1922 and still going strong, is the most notable of the area's many amateur groups; novelist Kurt Vonnegut acted in its productions in the 1950s and 1960s and had some of his early plays produced by the group. Professional summer stock began with the still-healthy Cape Playhouse in Dennis in 1927, and its early years featured the likes of Bette Davis (who was first an usher there), Henry Fonda, Ruth Gordon, Humphrey Bogart, and Gertrude Lawrence. In 1928, the University Players Guild (today called Falmouth Playhouse) opened in Falmouth, attracting the likes of James Cagney, Orson Welles, Josh Logan, Tallulah Bankhead, and Jimmy Stewart (who, while on summer vacation from Princeton, had his first bit part during Falmouth's first season).

The idea of a Cape Cod canal, linking the bay to the sound, was studied as early as the 17th century, but not until 1914 did the privately built canal merge the waters of the two bays. It was not, however, a thunderous success; too narrow and winding, the canal allowed only one-way traffic and created dangerous currents. The federal government bought it in 1928 and had the U.S. Army Corps of Engineers rebuild it. In the 1930s three bridges—two traffic and one railroad—went up, and the rest is the latter-day history of tourism on Cape Cod.

The building of the Mid Cape Highway (Route 6) in the 1950s marked the great boom in the Cape's growth, and the presidency of John F. Kennedy, who summered in Hyannis Port, certainly added to the area's allure. Today the Cape's summer population is over 500,000, 2½ times the year-round population.

Though tourism, construction, and light industry are the mainstays of the Cape's economy these days, the earliest inhabitants' occupations have not disappeared. There are still more than 100 farms on the Cape, and the fishing industry—including lobstering, scalloping, and oyster aquaculture, as well as the fruits of fishing fleets such as those in Provincetown and Chatham—brings in $2 million a month.

Visitors are still drawn here by the sea: Scientists come to delve into the mysteries of the deep, artists come for the light, and everyone comes for the charm of the beach towns, the beauty of the white sand, the soft breezes, and the roaring surf.

# CAPE AND ISLANDS BOOKS

## General

CAPE COD➤ The classic works on Cape Cod are Henry David Thoreau's readable and often entertaining *Cape Cod,* an account of his walking tours in the mid-1800s, and Henry Beston's 1928 *The Outermost House,* which chronicles the seasons during a solitary year in a cabin at ocean's edge. Both reveal the character of Cape Codders and are rich in tales and local lore, as well as observations of nature and its processes. *Cape Cod: Henry David Thoreau's Complete Text with the Journey Recreated in Pictures,* by William F. Robinson, is a handsome New York Graphics Society edition, illustrated with prints from the period and current photographs. *Cape Cod Pilot,* by Josef Berger (alias Jeremiah Digges), is a WPA guidebook from 1937 that is filled with "whacking good yarns" about everything from religion to fishing, as well as a lot of still-useful information. *A Place Apart: A Cape Cod Reader,* edited by Robert Finch, includes writings about the Cape from dozens of writers, from Melville to Adam Gopnik.

NANTUCKET➤ *Nantucket Style,* by Leslie Linsley and Jon Aron (published by Rizzoli), is a look at 25 houses, from 18th-century mansions to rustic seaside cottages, with 300 illustrations. Recently reprinted is Henry Chandler Forman's architectural classic *Early Nantucket and Its Whale Houses.*

## History

CAPE COD➤ *Cape Cod, Its People & Their History,* by Henry C. Kittredge (first published in 1930), is the standard history of the area, told with anecdotes and style as well as scholarship. *Sand in Their Shoes,* compiled by Edith and Frank Shay, is a compendium of writings on Cape Cod life throughout history. *Of Plimoth Plantation* is Governor William Bradford's description of the Pilgrims' voyage to and early years in the New World. *Art in Narrow Streets,* by Ross Moffett, and *Time and the Town: A Provincetown Chronicle,* by Provincetown Playhouse founder Mary Heaton Vorse, paint the social landscape of Provincetown in the first half of this century. A new illustrated history for children,

*The Story of Cape Cod,* by Kevin Shortsleeve (available through *Cape Cod Life,* ☎ 800/645-4482), is written in rhyming verse and sure to entertain. *In the Footsteps of Thoreau: 25 Historic and Nature Walks on Cape Cod,* by Adam Gamble, is a useful guide for naturalists and Thoreau admirers.

MARTHA'S VINEYARD➤ The many books written by Henry Beetle Hough, the Pulitzer Prize–winning editor of the *Vineyard Gazette* for 60 years, include his 1970 *Martha's Vineyard* and his 1936 *Martha's Vineyard, Summer Resort.*

NANTUCKET➤ Alexander Starbuck's 1924 *History of Nantucket,* now out of print, is the most comprehensive work on early Nantucket. *Nantucket: The Life of an Island,* by Edwin P. Hoyt, is a lively and fascinating popular history.

## Fiction

Herman Melville's *Moby-Dick,* set on a 19th-century Nantucket whaling ship, captures the spirit of the whaling era. *Cape Cod,* by William Martin, is a historical novel and mystery following two families from the *Mayflower* voyage to the present, with lots of Cape history and flavor along the way. *Murder on Martha's Vineyard,* by David Osborn, and Alice Hoffman's lovely *Illumination Night* are set on the island. *Dark Nantucket Noon,* by Jane Langton, is a mystery full of island atmosphere. *Nantucket Daybreak* is set in off-season Nantucket and portrays the life of scallopers in a story of love and betrayal. Francine Mathews's *Death in Rough Water,* stars detective Merry Folger and involves a fisherman accidentally killed during a gale. Other mystery writers who set their books on the Cape and islands are Margot Arnold, Rick Boyer, Philip A. Craig, Virginia Rich, and Phoebe Atwood Taylor.

## Nonfiction

Robert Finch, editor of *A Place Apart* (☞ *above*) has also written three memoirs: *Outlands: Journeys to the Outer Edges of Cape Cod, Common Ground: A Naturalist's Cape Cod,* and *The Primal Place,* a meditation about life on the Cape, especially its natural rhythms and history. In *Heaven's Coast,* Mark Doty, a Provincetown writer, recounts the death of his lover, Wally Roberts, from complications caused by AIDS. In a similar vein, David

Gessner's *A Wild, Rank Place: One Year on Cape Cod* combines insights about the Cape with reminiscence about battling cancer and confronting his father's death.

## Photography

*A Summer's Day* (winner of the 1985 Ansel Adams Award for Best Photography Book) and *Cape Light* present color landscapes, still lifes, and portraits by Provincetown-associated photographer Joel Meyerowitz. *Martha's Vineyard* and *Eisenstaedt: Martha's Vineyard* are explorations of the island by *Life* magazine photographer Alfred Eisenstaedt, who summered on the island until his death in 1995.

## Periodicals

Glossy magazines on the area include *Cape Cod Life* (✉ Box 1385, Pocasset 02559–1385, ☎ 508/564–4466 or 800/645–4482), *Provincetown Arts* (✉ Box 35, Provincetown 02657, ☎ 508/487–3167), *Martha's Vineyard Magazine* (✉ Box 66, Edgartown 02539, ☎ 508/627–4311), and *Nantucket Journal* (☎ 508/228–8700).

# INDEX

X = *restaurant*, ⌂ = *hotel*

# NOTES

# Fodor's Travel Publications

*Available at bookstores everywhere, or call 1–800–533–6478, 24 hours a day.*

## Gold Guides

### U.S.

Alaska

Arizona

Boston

California

Cape Cod, Martha's Vineyard, Nantucket

The Carolinas & Georgia

Chicago

Colorado

Florida

Hawai'i

Las Vegas, Reno, Tahoe

Los Angeles

Maine, Vermont, New Hampshire

Maui & Lāna'i

Miami & the Keys

New England

New Orleans

New York City

Pacific North Coast

Philadelphia & the Pennsylvania Dutch Country

The Rockies

San Diego

San Francisco

Santa Fe, Taos, Albuquerque

Seattle & Vancouver

The South

U.S. & British Virgin Islands

USA

Virginia & Maryland

Walt Disney World, Universal Studios and Orlando

Washington, D.C.

### Foreign

Australia

Austria

The Bahamas

Belize & Guatemala

Bermuda

Canada

Cancún, Cozumel, Yucatán Peninsula

Caribbean

China

Costa Rica

Cuba

The Czech Republic & Slovakia

Eastern & Central Europe

Europe

Florence, Tuscany & Umbria

France

Germany

Great Britain

Greece

Hong Kong

India

Ireland

Israel

Italy

Japan

London

Madrid & Barcelona

Mexico

Montréal & Québec City

Moscow, St. Petersburg, Kiev

The Netherlands, Belgium & Luxembourg

New Zealand

Norway

Nova Scotia, New Brunswick, Prince Edward Island

Paris

Portugal

Provence & the Riviera

Scandinavia

Scotland

Singapore

South Africa

South America

Southeast Asia

Spain

Sweden

Switzerland

Thailand

Toronto

Turkey

Vienna & the Danube Valley

## Special-Interest Guides

Adventures to Imagine

Alaska Ports of Call

Ballpark Vacations

Caribbean Ports of Call

The Complete Guide to America's National Parks

Disney Like a Pro

Europe Ports of Call

Family Adventures

Fodor's Gay Guide to the USA

Fodor's How to Pack

Great American Learning Vacations

Great American Sports & Adventure Vacations

Great American Vacations

Great American Vacations for Travelers with Disabilities

Halliday's New Orleans Food Explorer

Healthy Escapes

Kodak Guide to Shooting Great Travel Pictures

National Parks and Seashores of the East

National Parks of the West

Nights to Imagine

Rock & Roll Traveler Great Britain and Ireland

Rock & Roll Traveler USA

Sunday in San Francisco

Walt Disney World for Adults

Weekends in New York

Wendy Perrin's Secrets Every Smart Traveler Should Know

Worldwide Cruises and Ports of Call

# Fodor's Special Series

## Fodor's Best Bed & Breakfasts

America

California

The Mid-Atlantic

New England

The Pacific Northwest

The South

The Southwest

The Upper Great Lakes

## Compass American Guides

Alaska

Arizona

Boston

Chicago

Colorado

Hawaii

Idaho

Hollywood

Las Vegas

Maine

Manhattan

Minnesota

Montana

New Mexico

New Orleans

Oregon

Pacific Northwest

San Francisco

Santa Fe

South Carolina

South Dakota

Southwest

Texas

Utah

Virginia

Washington

Wine Country

Wisconsin

Wyoming

## Citypacks

Amsterdam

Atlanta

Berlin

Chicago

Florence

Hong Kong

London

Los Angeles

Montréal

New York City

Paris

Prague

Rome

San Francisco

Tokyo

Venice

Washington, D.C.

## Exploring Guides

Australia

Boston & New England

Britain

California

Canada

Caribbean

China

Costa Rica

Egypt

Florence & Tuscany

Florida

France

Germany

Greek Islands

Hawaii

Ireland

Israel

Italy

Japan

London

Mexico

Moscow & St. Petersburg

New York City

Paris

Prague

Provence

Rome

San Francisco

Scotland

Singapore & Malaysia

South Africa

Spain

Thailand

Turkey

Venice

## Flashmaps

Boston

New York

San Francisco

Washington, D.C.

## Fodor's Gay Guides

Los Angeles & Southern California

New York City

Pacific Northwest

San Francisco and the Bay Area

South Florida

USA

## Pocket Guides

Acapulco

Aruba

Atlanta

Barbados

Budapest

Jamaica

London

New York City

Paris

Prague

Puerto Rico

Rome

San Francisco

Washington, D.C.

## Languages for Travelers (Cassette & Phrasebook)

French

German

Italian

Spanish

## Mobil Travel Guides

America's Best Hotels & Restaurants

California and the West

Major Cities

Great Lakes

Mid-Atlantic

Northeast

Northwest and Great Plains

Southeast

Southwest and South Central

## Rivages Guides

Bed and Breakfasts of Character and Charm in France

Hotels and Country Inns of Character and Charm in France

Hotels and Country Inns of Character and Charm in Italy

Hotels and Country Inns of Character and Charm in Paris

Hotels and Country Inns of Character and Charm in Portugal

Hotels and Country Inns of Character and Charm in Spain

## Short Escapes

Britain

France

New England

Near New York City

## Fodor's Sports

Golf Digest's Places to Play

Skiing USA

USA Today The Complete Four Sport Stadium Guide

# WHEREVER YOU TRAVEL, *H*ELP IS NEVER FAR AWAY.

From planning your trip to

providing travel assistance along

the way, American Express®

Travel Service Offices are

always there to help

you do more.

American Express Travel Service
Offices are found in central locations
throughout the United States.
For the office nearest you, please
call 1-800-AXP-3429.

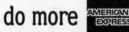